MANAGING MOBILITY IN AFRICAN RANGELANDS

MANAGING MOBILITY IN AFRICAN RANGELANDS

The Legitimization of Transhumance

Edited by
MARYAM NIAMIR-FULLER

Food and Agriculture Organization of the United Nations
Beijer International Institute of Ecological Economics
IT Publications
1999

Intermediate Technology Publications Ltd
103–105 Southampton Row, London WC1B 4HH, UK

© FAO and the Beijer International Institute of
Ecological Economics 1999

A CIP record for this book is available from the British Library

IT Publications ISBN 1 85339 473 4
FAO ISBN 92 5 104297 7

Typeset by J&L Composition Ltd, Filey, North Yorkshire
Printed in the UK by SRP, Exeter

Contents

vi

List of Tables

List of Maps

List of Figures

Foreword

The Food and Agriculture Organization of the United Nations (FAO) has long been concerned with pastoral development issues, both through a variety of field-based activities and through its Regular Programme work.

The primary focus of FAO activities is to improve food security; one way to achieve this is through the development of sustainable agricultural and pastoral systems. Pastoral development, particularly in arid and semi-arid areas, and the study of transhumant systems are focus areas of the Grassland and Pasture Crops Group of FAO's Crop and Grassland Service.

The collection of papers in this publication focuses on pastoral mobility and transhumant systems in Africa, building on the work of Behnke, Scoones and Kerven in *range ecology at disequilibrium: new models of natural variability and pastoral adaptation in African savannas*. Published in 1993 by the United Kingdom's Overseas Development Institute, this book supported the view that many dry-land pastoral areas are characterized by non-equilibrium dynamics and that many pastoral livestock and land management strategies may be a direct response to this. The present collection also elaborates on the UN Sudano-Sahelian Office Technical Consultations on Pastoral Development (1993–1997) and on the book entitled *Living with uncertainty: new directions in pastoral development in Africa* (Scoones, ed., 1995). Published by Intermediate Technology Publications (ITP), this last work further illustrated the new paradigms of thinking in range ecology and examined both the practical and policy implications for pastoral development in dryland areas. Together, these publications brought these important issues in pastoral development to the attention of the international community.

Within the framework of collaboration between FAO and the secretariat of the United Nations Convention to Combat Desertification, this book contributes to our understanding both of the obstacles to and benefits from improved management of pastoral mobility in dryland Africa.

The contributions of authors and the considerable input made by the editor, Maryam Niamir-Fuller, are much appreciated by FAO in its efforts to disseminate information on pastoral systems and transhumance. The collaboration of ITP Managing Editor, Neal Burton, is particularly noted, especially in early discussions with the Grassland and Pasture Crops Group. The cooperation of Shalini Dewan, Chief, Publishing and Multimedia Service of FAO is also appreciated. Funds for publication were made available by the FAO Regional Office for Africa. Thanks are particularly due to Stephen Reynolds and Caterina Batello of the Grassland and Pasture Crops Group for ensuring that the book was brought to publication.

MARCIO C.M. PORTO
Chief
Crop and Grassland Service
Plant Production and Protection Division
FAO, Rome, Italy

x

Acknowledgements

The research programme *Property rights and the performance of natural resource systems* was conducted at the Beijer International Institute of Ecological Economics, Royal Swedish Academy of Sciences, from 1993 to 1996. The programme received support from the World Environment and Resources Program of the John D. and Catherine T. MacArthur Foundation and the Environmental Economics Division of the World Bank. The goal of the programme was to understand how ecological and human systems interact through the mechanism of property rights, and how to design and implement effective property-rights systems to promote sustainability of natural resources. The programme was divided into three major areas, and this book is one of the results of the efforts within the area of 'Social and Ecological Linkages'. In addition to assistance with the publication, resources were made available from the Institute to fund new fieldwork, without which this book would have lost its originality.

Several colleagues helped to review each chapter and brought their own analytical strengths and insights to this book: Neal Artz, Ben Cousins, Carol Kerven, Robin Mearns, the late Richard Moorehead (1951–1998), Elinor Ostrom, Gregory Perrier, Alain Seznec, Ann Stroud, Brent Swallow, Jeremy Swift, Djeidi Sylla, Camilla Toulmin and Trond Vedeld. Special thanks go to Susan Hanna, Coordinator of the Beijer Institute sub-project, who provided the backbone to the entire process and helped to see it from the beginning through to its end. Stephen Reynolds and Caterina Batello, of the Grassland and Pasture Crops Group of FAO, gave timely and much appreciated support in bringing the book to publication. Many thanks also go to Elizabeth Daniel de Almeida and Elizabeth Speller for their professional editing efforts, and to Georg Turner for her artistic advice. The University of Wisconsin's Cartography Laboratory has upgraded the excellent maps and figures. Entirely indispensable were the critical comments, and crucial support, from Richard Watts Fuller.

The late Mansoureh Maria Marguerita Niamir (1950–1995), anthropologist and archaeologist, first introduced me to pastoralism when I visited her on an excavation site in northern Kurdistan, Iran. She has supported me in all the ways a loving sister would and, in particular, encouraged me to put my thoughts on pastoral development down on paper.

MARYAM NIAMIR-FULLER

Contributors

ROY BEHNKE holds a Ph.D. in Social–Cultural Anthropology from the University of California, Los Angeles. He has conducted anthropological fieldwork on pastoralists in Libya, Sudan and Somalia, as well as numerous research activities on policy and project-design implications of the spontaneous commercialization of pastoral economies, and on problems of state–community co-management of rangeland resources. He has worked on several projects as a development socio-logist in Namibia, Zambia, Iraq, Somalia, Botswana and several Central Asian countries. From 1989 to 1996 he was a Research Fellow at the Overseas Develop-ment Institute (London), responsible for research on livestock development and coordinating ODI's Pastoral Development Network. He is currently the Senior Rural Sociologist on the Northern Namibia Livestock Development Project, and can be reached through the Ministry of Agriculture and Water Development, Windhoek, Namibia.

ALAIN BOURBOUZE is a professor at the INA Paris-Grignon in the Department of Animal Production. He currently works at the International Center for Advanced Mediterranean Agronomic Studies (CIHEAM), attached to the Mediterranean Agronomic Institute of Montpellier. His activity in rangeland management and development of dryland and mountain land-use systems emerged from work carried out in marginal areas of North Africa, notably in Morocco. He has a particular interest in land tenure, local organizations, and all socio-economic factors affecting farming systems, especially animal production systems. He has been active in coordinating the network 'Reseau Parcours' and is currently working on transformations of agropastoral systems in the south of Tunisia. He can be contacted at: IAMM, BP 5056, 34033 Montpellier cedex 1, France.

WAL DUANY was born in Akobo, Sudan. He received his M.A. degree in Inter-national Politics with a concentration in the Middle East from Syracuse University and his Ph.D. in Public Policy from Indiana University. He has held a number of positions in the Government of Sudan, and was formerly Regional Minister for Cabinet Affairs in the High Executive Council, Regional Minister of Finance and Development, Chairman and Managing Director of Southern Sudan Regional Development Corporation, and a member of the National and Regional Assemblies. Currently he is a mediator and consultant on negotiation and conflict resolution, where he helps resolve disputes and assists parties to be more effective negotiators. He is a Research Associate in the Workshop in Political Theory and Policy Analysis of Indiana University, where he is working on a manuscript on indigenous peace processes among the Nuer of Sudan. He can be contacted at: 513 North Park, Indiana University, Bloomington, Indiana 47408, USA.

PASCAL LEGROSSE is an anthropologist at the Centre d'Études Africaines, of L'Ecole des Hautes Études en Sciences Sociales, in Paris. His paper is based on data he collected in Mali in 1996 for his dissertation. He can be contacted at: 71–73 Avenue d'Italie, 75013 Paris, France.

MARYAM NIAMIR-FULLER received her Bachelors degree in Sociology from Queen's University, Canada, a Masters degree in Regional Planning from Harvard University, and a Ph.D. in Range Management with a minor in Soils from the University of Arizona. She has been working on pastoral development since 1980 both as a full-time researcher/development expert and as a consultant for various international and non-governmental organizations. She has worked in all of Africa's Sahelian countries, as well as in Mongolia, Patagonia (Chile and Argentina), Brazil, Haiti and Morocco. She is a member of the Advisory Committee on Ecological Economics for SCOPE (Scientific Committee on the Problems of the Environment, ICSU). Since 1993 she has been an Adviser to UNSO (Office to Combat Drought and Desertification), for whom she has developed a global pastoral programme, and organized the biannual International Technical Consultations on Pastoral Development. She is currently an adviser on the cross-cutting theme of land degradation to the Global Environment Facility (UNDP). She can be contacted care of FAO, Lusaka, Zambia.

MOHAMED OULD ZEIDANE was born in 1959 in Nouakchott, Mauritania. After higher studies in applied mathematics and rural economy, he began his career in 1988 as an agro-economist responsible for pastoral studies for the Government of Mauritania. Between 1992 and 1996 he supervised agricultural economic activities within the Prime Minister's Office. He has carried out studies in pastoral land tenure, and has worked on pastoral production systems in many parts of the Western Sahel for such organizations as FAO, World Bank, and the International Institute for Economic Development (IIED). He is currently pursuing studies at the French National Administration School (ENA) in Paris. He can be contacted at: ENA, 13 rue de l'Université, F-75343 Paris cedex 07, France.

IAN SCOONES is a Fellow of the Institute of Development Studies at the University of Sussex, UK and a member of the Environment Group. He is an agricultural ecologist by training, with a wide interest in social, economic, and ecological dimensions of environmental change in Africa. He is a contributor to and editor of the books *Range ecology at disequilibrium* (ODI, 1993) and *Living with uncertainty* (IT Publications, 1995). His work has concentrated in southern Africa, where he has carried out extensive fieldwork on dryland agricultural and pastoral development issues. He can be contacted at: IDS, University of Sussex, Brighton, BN1 9RE, UK.

MATTHEW TURNER is an Assistant Professor of Geography at the University of Wisconsin. His research interests concern the relationship between socio-economic and environmental change in dryland areas of Africa. His dissertation work was on a politico-ecological analysis of transhumance in the Maasina region of Mali with a particular emphasis on political economic factors affecting the nature of grazing pressure, resulting in vegetative changes along floodplain's edge. His post-doctoral research was funded by the Rockefeller Foundation and was conducted with the International Livestock Research Institute team in Niger. Research there focused on the changing relationship between herd management, livestock-mediated nutrient transfer, and cropland productivity in the Sahelian zone. This GIS (geographic information system)-based research is on-going. Another on-going research project concerns the effect of intra-household social dynamics on regional shifts in livestock herd composition in the Sahel. He can be contacted at: Department of Geography, University of Wisconsin-Madison, 384 Science Hall, 550 North Park Street, Madison, WI 53706–1491, USA.

1

Introduction

MARYAM NIAMIR-FULLER

Is PASTORAL MOBILITY an archaic remnant of the past, or is it the foundation for future sustainability? After decades of trial and error, a clearer definition of the complexities of pastoral development is arising based on work done by African and international researchers. They are refuting the causality of notions normally attributed to pastoralism, such as land mismanagement and degradation, famines and wars. Current research suggests that we have less to fear from pastoral land-stewardship than was previously thought. On the one hand, the natural environment exploited by pastoralists is more resilient than expected, and on the other, pastoral techniques for land management are not as dysfunctional as was once widely assumed (Behnke, 1994). Moreover, we have badly underestimated the contribution of pastoralists to the local, regional and national economy.

The term 'pastoralist' is defined as a mode of production where livestock make up 50 per cent or more of the economic portfolio of a small holder. This book focuses on mobile or transhumant pastoralists. The term 'transhumance' refers to regular seasonal movements of livestock between well-defined pasture areas (dry to wet season, or low to highland). Pastoralism in Africa can encompass a very wide range of production systems, ranging from sedentary populations who use extensive communal land (such as in Scoones's case study on southern Zimbabwe in this book), to semi-transhumants who are sedentary for only parts of the year (e.g. the Dinka of the Sudan in Duany's case study), to full transhumants (the northern Mauritanian pastoralists in Zeidane's chapter, and Behnke's contribution from Namibia)[1]. When referring to 'mobility', livestock movements can be considered separately from movements by humans. The family may be sedentary for most of the year, but its livestock will be moved, sometimes for vast distances, by members of the family, village, clan or other social unit.

Transhumant pastoralists are a rather large and significant minority in the world (see Map 1.1). They constitute an estimated 16 per cent of the population of the Sahelian Zone (Bonfiglioli and Watson, 1992). In a few countries, such as Somalia and Mauritania, they are the majority of people. Mobile pastoralists almost invariably have established symbiotic socio-economic relationships with sedentary or less mobile populations.

---

[1] It would be impossible to depict the many variations that exist among pastoral production systems and livelihoods in Africa. A good overall review does not exist, but the following references combine to give a good picture: Galaty *et al.* 1981, Sandford 1983, Niamir 1990, Bonfiglioli & Watson 1992, and articles and debates published within the Pastoral Development Network (ODI), and *Nomadic Peoples.*

1

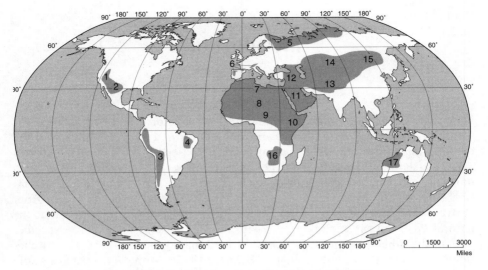

1. North American Indians (e.g. Navajo)
2. North and Central Mexican herders
3. Andean pastoralists of Equador, Peru, Bolivia, and Chile
4. Northeast Brazil Vaqueiros
5. Laplanders, Chookchi, Koryak, Yakout, Toungounz, Samoyed, Khanty
6. French, Italian, Spanish, Hungarian, and Portuguese herders
7. North African and Egyptian pastoralists
8. Sahara desert nomads
9. Sahelian pastoralists
10. Pastoralists of the Horn of Africa
11. Arabian peninsula and Middle East pastoralists
12. Turkic, Kurdish, and Iranian pastoralists
13. Afghani, Pakistani, Rajasthani, Nepali and Tibetan pastoralists
14. Turkmen, Kazakh, Uzbek and Tajik pastoralists
15. Mongolian and Chinese pastoralists
16. Zimbabwe and Botswanan pastoralists and hunter-gatherers
17. West Australian Aborigine pastoralists and hunter-gatherers

Map 1.1 *Schematic global distribution of transhumant pastoralists*

Development efforts in pastoral areas have evolved from privatization (ranching) to a greater acceptance of the viability and applicability of common-property management systems, particularly those based on traditional techniques and indigenous pastoral knowledge. However, in practice, ecological and political issues are rarely taken into account in designing common-property management regimes (CPR) in development projects, thereby aggravating the problems of sustainable use of common pool resources. Appropriate local institutions are rightly seen as indispensable for successful local-level natural-resource management, but there is more concern about establishing the structure of the institution, rather than considering its role and functions. As a result, most efforts at local-level capacity building have been static and unable to adapt to emerging needs. Land-tenure security is seen as one of the major incentives for promoting sustainable resource management, however current laws usually cannot cope with the needs of livestock mobility and flexibility in resource use. Biases inherent in the 'modern' judicial and administrative systems against pastoral mobility actually may increase resource competition and conflict.

2

At the heart of the problem has been the inability of 'normal' science[2] to deal adequately with the high degree of complexity, uncertainty and unpredictability inherent in arid and semi-arid lands. The book accepts the premise of the 'new paradigm' that some form of livestock mobility is an ecological and economic necessity in arid and semi-arid lands, and suggests ways in which the social, political and economic environment can be moulded to fit with this necessity.

Even though pastoral production systems vary tremendously from one ecosystem and economic system to another, the trends and perturbations faced by pastoralists in the latter half of this century appear strikingly similar across the world. Among these we can cite:

o desiccation of the environment due to recurrent droughts
o regional population increase and agricultural encroachment
o preferential subsidies and policies supporting crop cultivation
o state and private expropriation of so-called 'vacant' pastoral land
o declining ownership of livestock among livestock managers
o breakdown of the traditional, local institutions and systems for managing natural resources
o separation of the legal ownership of natural resources from their users
o negative stereotypes of pastoralists that lead to misguided national policies and inappropriate development approaches
o sedentarization and concentration of livestock leading to overgrazing around settlements and undergrazing in remote areas
o inadequate or non-existent delivery of social services and technical inputs to remote and mobile populations
o granting of greater political authority by colonial and post-colonial states to sedentary populations.

Although a few of these constraints are related to environmental trends, most are influenced directly or indirectly by policies, administrative procedures, and legal mechanisms.

The focus of this book is on the arid and semi-arid lands of Africa, most of which are under some form of common property regime. The world's drylands include hyper-arid, arid, semi-arid and dry sub-humid ecosystems, and are estimated to constitute 32 per cent of global land area (UNEP, 1991), but in Africa, drylands are 66 per cent of the total land area (Botterweg, 1994). Map 1.2 provides a general description of the vegetation of dryland Africa. Arid lands are often considered to be marginal in terms of their economic productivity, ecological diversity, political strength, and social importance (biases, minority group, low social services). Because of these perceptions, little attention and development aid has been directed to these areas.

The main objective of this book is to bring to the attention of scientists, students, development experts, government officials, and bilateral and multi-

---

[2] As opposed to 'post-normal science'; Funtowitcz, S.O. & J.R. Ravetz, 1993. 'Science for the post-normal age', *Futures*, 25(7): 739–755.

3

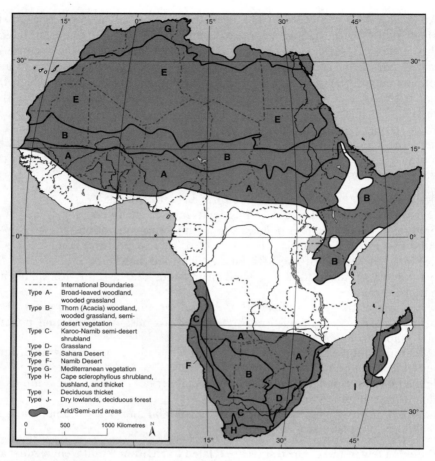

Map 1.2 *Vegetation map of arid and semi-arid Africa*
(Source: *Adapted from White, 1983*)

lateral development organizations, the state of the art in data and analyses of pastoral development in Africa. The book takes the main elements of the 'new paradigm' in pastoral development – whereby livestock mobility is encouraged rather than discouraged, and local communities are empowered to manage their own natural resources – and transforms it into practical guidelines. The book carries forward the theoretical debate initiated through the efforts of the Overseas Development Institute, Commonwealth Secretariat, IIED[3], IDS[4], and the so-called 'Woburn Workshops', as well as the works of CIHEAM[5] and CIRAD, which were debated within the biannual Technical Consultations on Pastoral

---

[3] International Institute for Environment and Development, London.
[4] Institute for Development Studies, University of Sussex, UK.
[5] Centre International de Hautes Etudes Agronomiques Mediterraneennes, Montpellier.

Development organized by UNSO[6] and hosted by various bilateral donors (1993–1998).

A central purpose of this book is to provide a synthesis of the different scientific fields and paradigms engaged in pastoral development that have emerged in recent years, and to provide practical tools and guidelines for adjusting the approaches and concepts to the context. In certain respects, words are practical tools. The politician uses them in his rhetoric, the researcher gains fame by inventing new ones, and the development worker obtains funding for her projects by using the latest buzz-words. In this book, we have attempted to sift through the many 'key words' in the literature, to find those that are value-added, i.e. truly connote new ideas and concepts, and have immediate practical applications. In this way, it is hoped that the book will contribute effectively to showing both the obstacles to and the benefits from improved management of pastoral mobility in dryland Africa.

In Chapter 2, we briefly present a review of the literature on the major historical changes in pastoral systems and trace how these changes have in most cases disrupted customary systems of transhumance, effectively reducing livestock mobility. The major livestock and pastoral development paradigms are then presented, and it is shown that the evolution of these paradigms has largely been both the cause and the effect of changes occurring in pastoral systems.

Chapters 3 to 10 are case studies representing a cross-section of Africa. The case studies range from arid to semi-arid systems, from semi-mobile to fully mobile production systems, and run the gamut from viable common property regimes to those that have broken down.

Chapter 3 by Mohamed Ould Zeidane draws insights from four carefully selected case studies in Mauritania. He shows that there is considerable variety not only in pastoral production systems, but also in the kinds of problems they are faced with. In some cases customary institutions and symbiotic relationships continue to persist, while in areas of higher grazing value the systems are breaking down. Modern institutions, such as pastoral cooperatives and associations, have only partially filled the void. The author points out the bias in past and present laws that favour farming relative to pastoral activities, thus aggravating conflicts between the two. He highlights the contradiction between the need to reduce the geographical scale for ease of administration, and the need to increase the scale, especially in arid lands, in order to respond to ecological variability and requirements of mobility. He suggests that a model where small administrative units have strong agreements and alliances with other units for reciprocal land use, could provide a solution. The author asserts that the more permanent the institution, the more general its activities and responsibilities will be, while the more an institution is temporary and *ad hoc*, the more specific its functions. He concludes by saying that land-tenure security is at the moment the most pressing issue confronting pastoralists.

Chapter 4, by Pascal Legrosse, focuses on a widely publicized and intensely

---

[6] United Nations Sudano-Sahelian Office, now called the Office to Combat Drought and Desertification, New York.

5

studied case where a traditional system of political and institutional control has been applied to a variable resource. The author has a keen insight into the complexities of the Maasina Fulani of the Niger Delta in Mali, and using new field data redefines and, in some places, debunks many terms and concepts attached to this pastoral 'empire'. He asserts that the Diina Code of the mid-1800s (the land-tenure code imposed by Sheikh Sekou Amadou), rather than being an example of successful customary pastoral common property management, was actually an attempt to sedentarize the pastoral Fulani so as to improve political stability in this volatile region. The Code and its institutions did not survive the French colonial rule, but vestiges of the old system still persist. The post-colonial government tried to control what was left, but instead their rules and regulations created more resource conflicts and overgrazing. Recently, some of the local élite are treating the natural resources as their own private property, imposing fees for use both on other Fulani and on outsiders. This chapter flows naturally into the next chapter, the first stating the problem, and the second providing guidelines toward solutions.

Chapter 5, by Matt Turner, compares the CPR of two Fulani groups in different parts of the western Sahel, both transhumants, but one more formalized (Maasina) than the other. He shows how the impact of the marriage of the 'tragedy of the commons' theory (Hardin, 1968) with the conventional range-management focus on stocking rates has resulted in 'tragic real-world consequences of ivory-tower epistemology'. He shows how, in a rush to comply with the needs of the tragedy of the commons paradigm, researchers have erroneously characterized the Maasina system of the Niger Delta, romantically portraying it as exclusive control over natural resources, when it is not much different from the unbounded and point-centred CPR systems of other agropastoral groups. According to the author, land tenure should not be used as an incentive for prohibiting the accumulation of livestock, but as a mechanism to govern flexible access to variable resources. Restrictions on livestock mobility have in effect created an 'artificial shortage' of grazing resources in this region. He cautions that while the non-legalistic nature of land tenure among transhumants is more consistent with the variable and uncertain environment, it also makes resource management and control more vulnerable to socio-economic change and political manipulation. He suggests that in those contexts where the transhumant mobility system is no longer part of a *social* organization, then a 'host–transhumant' model of CPR that strengthens symbiotic relationships between settled agropastoralists and transhumants might be more viable.

Chapter 6, by Wal Duany, is set in southern Sudan, among the Nuer agropastoralists of this semi-arid ecosystem. He describes the Nuer as a 'stateless society' within the larger nation. He shows how the British colonial administration found no *formal* code of law among the Nuer, and therefore assumed that their traditional legal system lacked accountability. By imposing a British system of law, they effectively eliminated the customary mechanisms of contestation and argumentation in the process of conflict resolution. He points out that having property rights depends on the recognition of those rights by others. Conflicts, like the civil war between North and South Sudan, arise when this right is not recognized. The Nuer see the State as a centralized and 'predatory' institution that claims ownership of Nuer land but has no institutions to manage them sustainably. The author succeeds in the difficult task of interpreting the traditional symbols, language and customs that form his own cultural background, in a way that can be understood by outsiders with the modern language of covenants and legal equity. Negotiation is only possible if lines of communication between adversaries are kept open through institutional arrangements and, more importantly, through mutual interdependence. He provides interesting examples of recent conflict resolution internally among the Nuer to show that self-governance is possible despite (or perhaps because of) Northern Sudanese hegemony and civil war.

Chapter 7, by Maryam Niamir-Fuller, on the case of the Karimajong in Uganda focuses on the past history of conflict between the tribes and sub-tribes in this part of eastern Uganda, and the recent attempts by the Karimajong and the government to resolve and prevent these conflicts. The author refers to the Karimajong world view which is based on a tertiary classification of all non-Karimajong peoples, where one is either a friend, an enemy, or 'the others' the latter being a derogatory concept used to describe those who do not have live-stock (therefore 'wealth'). Development workers, having been classified as the latter, have little chance of being effective change agents, unless they find ways to approach the Karimajong as 'friends'. The author describes the traditional and modern mechanisms available for both conflict prevention and conflict resolution. She reports on recent attempts by the Karimajong intelligentsia (parliamentarians, researchers) to divert the conflictual process towards peaceful negotiations and control of banditry. She points out that the process of conflict-management is just as important as the goal: agreeing to go to a negotiation table is in itself an important concession and commitment. She provides several guidelines borrowed from other contexts that could be adapted to this pastoral situation.

Chapter 8, by Roy Behnke, is set in very arid Namibia, where pastoralists raise goats and cattle adapted to the harsh environment. Households are almost fully mobile. He shows how the production system is primarily constrained by water, and only secondarily by vegetation, and how 'guests' or outsiders are often welcome because of labour shortages for building and maintaining water points. He describes the pattern of mobility as one akin to an irregular form of rotation that exaggerates the natural heterogeneity of the vegetation. He contends that it is the absence of a central authority (which would try to impose rigid rules) and reliance on *ad hoc* decisions at the household level based on reciprocity, that allow a fine-tuning and adjustment of stocking rate to the ecosystem's variability. However, the more critical the resource, the more tightly it is controlled and managed at 'higher' institutional levels. The traditional system is, he stresses, exceedingly professional and technically competent. Reforming or improving it would not be easy, nor would the changes be dramatic. It is not even clear whether the cost of change would be compensated by any benefits from a classical ranching model. For the moment, development is a 'screaming contest'; that is, those who lobby the most effectively, get the most benefits.

Chapter 9, by Ian Scoones, brings in a case study from a sub-humid zone of southern Africa inhabited by agropastoralists who raise livestock on both extensive (communal lands) and intensive resources (crop residues). The author reflects on the 'patchiness' of natural resources in semi-arid lands, and draws on the central tenets of the 'opportunistic management' approach. The paper shows how the relative value of different resources (value in terms of economic and political criteria) determines the type of property regime attached to that resource, and how this can vary with time and the socio-political context. Low-value resources are managed by *ad hoc* institutions, while high-value resources are managed and controlled by formal institutions. The same resource may be of high value and under tight control when it is scarce (e.g. due to drought or population increase), and of low value and under less formal control when it is abundant (high rainfall, low pressure). Such a dichotomy is useful analytically, but the author points out that in reality there are many degrees of 'value', and many overlapping and nested institutions managing common pool resources. The author suggests six focal areas to assist in the design of better institutions and common property regimes.

Chapter 10, by Alain Bourbouze, is based on several case studies in the 'Maghreb' of North Africa. The area is inhabited by a mixture of semi-nomadic and transhumant pastoralists and is one of the driest and more variable ecosystems in Africa. The author traces the changes in the production, social and ecological systems in the last century, concluding that at present the management of natural resources in most cases is no longer a communal effort to increase the value of the resources, but a means to control neighbours in order to define one's individual niche in the commons. He points out the distinction between 'communal exploitation' of resources and 'communal management of competition for resources by individuals', stressing the importance of the latter in sustainable use of natural resources. He briefly describes the 'pick-up' technology where feed and water are brought to livestock rather than livestock brought to the feed which has resulted in land degradation, but points out that at the moment it is only a viable option for rich pastoralists. He presents one case in the High Atlas region, where a 'Transhumant Charter' designed by government and based on modern legal principles, has been applied successfully. It enshrines the rights of all resource users, but has not yet been able to deal with decreasing rangelands, and increasing land conflicts, because of its generality and inflexibility. The pastoralists have used the charter as a tool for recognizing property rights, while day-to-day management of resources has been based on more traditional structures and techniques. He shows how most conflicts are prevented through the process of consensus-building in decision-making, as well as tolerance of minor infringements, so that only serious conflicts are referred to the formal judiciary (whether customary or modern). The author feels that the formality of the modern judiciary is a weakness because it can be used by individuals for their own advantage. Fines imposed on expropriators actually strengthen their future claim to the common resource because they already 'paid' for it. He cautions that the concepts of flexibility and participation can be used for both good and bad.

Most of the case studies focus on the micro-level, i.e. local-level attempts at managing resources or conflicts. But they all acknowledge that there are important links, potentials and constraints to the macro-level, including appropriate national policies, administrative procedures, legal mechanisms, and regional/ international markets and wars.

In the last chapter, the book argues that research lends growing support for the economic and ecological rationale for creating and strengthening institutions that facilitate livestock mobility in dryland environments. This represents a new perspective not only on pastoral development but also on the nature of social institutions governing natural resource-management in dryland areas. It incorporates into resource-management new research on non-equilibrium ecological theory, managed commons, conflict-resolution, participatory planning, and both formal and informal institutions. This framework provides practical tools, solutions, key words, and underlying principles to allow proponents to create the social, legal and institutional conditions that would legitimize transhumance. The chapter concludes by putting the African case in the wider perspective of the recent resurgence of both common property-management and transhumance in other continents.

1

Introduction

MARYAM NIAMIR-FULLER

LA MOBILITÉ PASTORALE est-elle un héritage archaïque du passé, ou l'assise d'un développement à venir? Après des décades d'essais et d'erreurs, émerge une définition plus claire des complexités du développement pastoral basé sur les travaux des chercheurs africains et internationaux. Ces chercheurs refusent d'attribuer au pastoralisme la responsabilité des phénomènes tels que la dégradation et la mauvaise gestion des terres, les famines et les guerres.

La recherche actuelle suggère que nous avons beaucoup moins à craindre de la gestion des terres effectuée par les pasteurs qu'on ne le pensait auparavant. D'une part, l'environnement naturel exploité par les pasteurs s'avère plus résilient que prévu, c'est à dire susceptible de revenir à un nouvel équilibre, d'autre part, les techniques pastorales de gestion des terres ne sont pas si inadaptées qu'on l'avait communément imaginé (Behnke, 1994). De plus, nous avons gravement sous-estimé la contribution des pasteurs aux économies locales, régionales et nationales.

Le pastoralisme se définit comme un mode de production où le bétail représente 50 pour cent ou plus de l'économie du petit éleveur. Ce livre est consacré aux pasteurs mobiles ou transhumants. Le terme de 'transhumance' se rapporte à des mouvements saisonniers, réguliers, du bétail entre des zones de pâturages bien définies (de la saison sèche vers la saison humide, ou des basses terres vers les hautes terres). En Afrique, le pastoralisme recouvre une grande variété de systèmes de production, allant des populations sédentaires qui utilisent des espaces communaux (comme dans l'étude de cas de Scoones sur le Zimbabwe du sud dans cet ouvrage) aux semi-transhumants, sédentaires seulement une partie de l'année (les Dinka du Soudan dans l'étude de cas du Duany), et aux grands transhumants (les pasteurs Mauritaniens du nord dans le chapitre de Zeidane et la contribution de Behnke sur la Namibie)[1].

Quand on parle de 'mobilité' il est possible de traiter des mouvements du bétail séparément des mouvements de la famille. La famille peut être sédentaire pour la plus grande partie de l'année, mais son bétail sera conduit quelquefois sur de grandes distances par des membres de la famille, du village, du clan, ou toute autre unité sociale. Cet ouvrage se concentre sur les pasteurs mobiles ou partiellement mobiles.

Les pasteurs transhumants forment une minorité significative dans le monde (Map 1.1, p. 2). Ils constituent environ 16 pour cent de la population des régions

---

[1] Il serait impossible de décrire toutes les variations qui existent parmis les systèmes de production et mode de vie pastoraux en Afrique. Nous n'avons pas un document qui contien une analyse complète du sujet, mais les réferences qui suivent fournissent un bon aperçu: Galaty et al 1981, Sandford 1983, Niamir 1990, Bonfiglioli et Watson 1992, Pastoral Development Network (ODI), et *Nomadic Peoples*.

sahéliennes (Bonfigioli & Watson, 1992). Dans quelques pays comme la Somalie ou la Mauritanie, ils sont en majorité. Les pasteurs mobiles ont presque invariablement établi des relations symbiotiques socio-économiques avec les populations sédentaires ou moins mobiles.

Les politiques de développement des zones pastorales se sont appuyées dans un premier temps sur la privatisation ('ranching') pour ensuite reconnaitre la viabilité et accepter les systèmes de gestion des biens communs, particulièrement ceux basés sur des techniques traditionnelles et des savoirs pastoraux indigènes. Mais dans la pratique, les problèmes écologiques et politiques sont rarement pris en considération lors de la conception de régimes de propriété communautaire dans les projets de développement, ce qui rend d'autant plus difficile la recherche d'un usage durable des biens communs.

Il est clair qu'une gestion locale réussie des ressources naturelles demande des institutions locales appropriées. Mais on s'est plus consacré à mettre sur pied les aspects institutionnels qu'à définir les rôles et les fonctions même de ces institutions. En conséquence, la plupart des efforts de renforcement des capacités au niveau local ont été statiques et incapables de s'adapter aux besoins nouveaux. La sécurisation foncière est certainement la meilleure incitation à une gestion durable des ressources. Toutefois les lois en vigueur ne répondent pas, en général, aux besoins de la mobilité pastorale et de la flexibilité dans l'usage des ressources. Les systèmes juridiques et administratifs dits modernes recèlent dès a priori contre la mobilité pastorale qui ne peuvent qu'accroître les concurrences et les conflits sur les ressources.

Au coeur du problème on trouve l'incapacité de la science dite 'normale'[2] à gérer correctement les hauts niveaux de complexité, d'incertitude et d'imprédictibilité inhérents aux terres arides et semi-arides. Ce livre part du principe d'un nouveau paradigme qui affirme qu'une certaine forme de mobilité de bétail est une nécessité écologique et économique dans les terres arides et semi-arides, et avance quelques idées sur la façon dont l'environnement social, politique et économique peut être façonné pour répondre à cette nécessité.

Même si les systèmes de production pastoraux peuvent varier énormément d'un écosystème et d'un environnement économique à un autre, les tendances et les problèmes rencontrés par les pasteurs dans la deuxième moitié de ce siècle apparaissent d'une similitude frappante dans le monde entier. Citons par exemple:

o l'aridification de l'environnement résultant des sécheresses répétées
o l'accroissement de la population et l'extension de l'agriculture
o les subventions et les politiques en faveur de l'agriculture
o l'expropriation par l'État et les individus de terres pastorales soi–disant 'vacantes'
o le déclin de la propriété du bétail chez les éleveurs
o l'effondrement des institutions traditionnelles et locales et des systèmes de gestion des ressources naturelles

---

[2] Opposé à la science dite 'post-normal'. Funtowitcz, S.O. & Ravetz, J.R. 1993. 'Science for the post-normal age', *Futures*, 25 (7): 739–755.

o la séparation entre la propriété légale des ressources naturelles et leurs usagers
o les stéréotypes hostiles aux pasteurs qui engendrent des politiques nationales erronées et des approches du développement inadequates
o la sédentarisation et la concentration de bétail qui entraînent surpâturage autour des villages, et souspâturage dans les zones éloignées
o les services sociaux inadaptés ou inexistants pour les populations éloignées et mobiles
o l'octroi d'une plus grande autorité politique aux populations sédentaires par les états coloniaux et post-coloniaux.

Bien que quelques unes de ces contraintes soient liées aux tendances environnementales, la plupart sont influencées directement ou indirectement par les politiques, les procédures administratives, et les mécanismes légaux.

Ce livre traite essentiellement des terres arides et semi-arides de l'Afrique dont la plupart sont régies par des systèmes de propriété communautaire. On estime que les terres arides du monde qui comprennent les écosystèmes arides, semi-arides et sub-humides secs, constituent 32 pour cent de la surface terrestre (UNEP/GRID, 1991). Mais en Afrique les terres arides couvrent 66 pour cent de la superficie totale (Botterweg, 1994). La carte 1.2 (p. 4)donne une description générale de la végétation de l'Afrique aride. Les terres arides sont souvent considérées comme marginales en terme de production économique, diversité écologique, poids politique et importance sociale (préjugés contre les groupes minoritaires, services sociaux médiocres). Cela explique que peu d'attentions et peu d'aides publiques au développement aille vers ces régions.

L'objet principal de ce livre est donc de porter à l'attention des scientifiques, des étudiants, des experts, des responables nationaux et des organisations de développement bilatérales et multilatérales, un 'état de l'art' sur les données et les analyses du développement pastoral en Afrique. Ce livre reprend les principaux éléments du nouveau paradigme du développement pastoral, suivant lequel la mobilité du bétail est encouragée plutôt que découragée, et la communauté locale est rendue responsable de la gestion de ses propres ressources naturelles. Il débouche sur des orientations pratiques. Il fait avancer le débat théorique lancé grâce aux efforts de l'Overseas Development Institute (ODI), du Commonwealth Secretariat, de l'IIED[3], de l'IDS[4], des 'ateliers de Woburn' ou des travaux du CIHEAM[5] et de CIRAD, efforts qui ont permis d'animer les débats lors des Consultations Techniques sur le Développement Pastoral organisés par l'UNSO[6] et hébergés par différentes institutions d'aide bilatérale (1993–98).

Le but principal de ce livre est de faire une synthèse dans les différents domaines scientifiques engagés ces dernières années dans le développement pastoral et de fournir des outils pratiques et des orientations pour adapter les

---

[3] International Institute for Environment and Development, London.
[4] Institute for Development Studies, University of Sussex, UK.
[5] Centre International de Hautes Études Agronomiques Mediterraneennes, Montpellier.
[6] United Nations Sudano-Sahelian Office, qui s'appelle maintenant l'Office to Combat Drought and Desertification, New York.

approches et les concepts au contexte. Dans un certain sens, les mots sont aussi des outils pratiques. Le politicien en use dans sa rhétorique, le chercheur en invente de nouveaux pour acquérir la renommée, et le développeur accède au financement de ces projets en usant des derniers termes à la mode. Dans ce livre, nous avons essayé de faire un tri parmi les nombreux 'mots clés' de la littérature pour trouver ceux qui présentent une valeur ajoutée, c'est à dire qui portent de véritables nouvelles idées ou concepts, et de montrer leur application pratique immédiate.

On espère ainsi que ce livre contribuera efficacement à mettre en lumière les obstacles qui freinent la gestion améliorée de la mobilité pastorale en Afrique aride, ainsi que les bénéfices qui peuvent en résulter.

Dans le chapitre 2 nous passons brièvement en revue la littérature qui retrace les changements historiques principaux des systèmes pastoraux et les effets le plus souvent destructeurs qu'ont eu ces changements sur les systèmes coutumiers de transhumance quand leur mobilité se trouvait réduite. Ensuite nous présentons les principaux paradigmes du développement pastoral et nous démontrons que l'évolution de ces paradigmes a été en grande partie à la fois la cause et l'effet des changements survenus dans les systèmes pastoraux.

Les chapitres 3 à 10 sont des études de cas choisies dans toute l'Afrique. Les études de cas vont des systèmes arides aux semi-arides, des systèmes de production semi-mobiles aux totalement mobiles, et parcourent la gamme des régimes de propriété communautaire de l'état viable à l'état le plus dégradé.

Le chapitre 3 de Mohamed Ould Zeidane tire des leçons de quatres études de cas soigneusement choisies en Mauritanie. Il montre qu'il existe une très grande variété non seulement des systèmes de production mais aussi des problèmes qu'ils rencontrent. Dans certains cas, les institutions coutumières et les relations symbiotiques persistent, mais dans les zones de plus haute valeur fourragère les systèmes s'effondrent. Les institutions modernes telles que les associations et les coopératives pastorales n'ont que partiellement comblé le vide. L'auteur souligne le parti pris des lois passées et présentes en faveur des activités agricoles, comparées aux activités pastorales, ce qui aggrave les conflits entre les deux systèmes.

Il relève la contradiction entre le besoin de réduire l'échelle géographique pour les besoins de l'administration et le besoin d'accroitre cette échelle, en particulier dans les zones arides, de façon à répondre à la variabilité écologique et aux exigences de la mobilité. Il suggère qu'un modèle où les petites unités administratives auraient des accords solides et des alliances avec d'autres unités pour des octrois réciproques de droits d'usage, pourrait contenir une solution. L'auteur affirme que plus l'institution est permanente, plus générale seront ses activités et ses responsabilités, tandis que plus une institution est temporaire plus spécifique seront ses fonctions. Il conclut en disant que la sécurisation foncière est pour le moment le problème le plus urgent auquel se heurtent les pasteurs.

Dans le chapitre 4, Pascal Legrosse concentre son attention sur un cas très connu et intensément étudié où un système traditionnel de contrôle politique et institutionnel a été appliqué à une ressource variable. L'auteur a une connaissance intime des complexités de la société Peul Macina du Delta du

Niger au Mali. En se basant sur des nouvelles données, il redéfinit, et par endroits détrône, de nombreux termes et concepts attachés à cet ancien empire pastoral. Il affirme que la Dina du milieu du dix-neuvième siècle, ce code imposé par Sheikh Sékou Amadou, exemple d'une gestion réussie d'une propriété pastorale communautaire, fut en fait une tentative de sédentarisation des pasteurs Peuls de façon à améliorer la stabilité politique de cette région.

Le code et ces institutions n'ont pas survécu au régime colonial français mais des vestiges du vieux système persiste. L'état post-colonial a tenté d'en reprendre le contrôle, mais ses statuts et ses règlements ont surtout généré des conflits sur les ressources naturelles, et du surpâturage. Depuis peu, quelques élites locales traitent les ressources naturelles comme leur propre propriété privée, imposant des droits pour leur usage aux autres Peuls ou étrangers. Ce chapitre débouche naturellement sur le chapitre suivant, le premier présentant le problème et le second proposant des orientations pour des solutions possibles.

Le chapitre 5 de Matt Turner compare les régimes de propriété communautaire des deux groupes Peuls, dans différentes parties du Sahel de l'Ouest, tous les deux transhumants, mais les institutions du groupe Macina plus formelles que celle du Peul du Niger. Il montre que l'impact du mariage de la théorie de 'la tragédie des communaux' (Hardin, 1968) avec l'*a priori* classique des pastoralistes sur les taux de stockage, 'une épistémologie de tour d'ivoire', a eu des conséquences tragiques. Dans leur hâte à suivre les exigences du paradigme de la tragédie des communaux, les chercheurs se sont fourvoyés dans les caractérisations du système Macina du Delta du Niger. Ils le décrivent de façon romantique comme un contrôle exclusif sur des ressources naturelles, alors qu'il n'est guère différent des régimes de propriété communautaire des agropasteurs sédentaires.

Selon l'auteur, il ne faudrait pas user du système foncier comme d'un instrument pour empêcher l'accumulation du bétail mais comme un mécanisme permettant de contrôler avec souplesse l'accès à des ressources variables. Les restrictions apportées à la mobilité du bétail ont créé un déficit artificiel de ressources fourragères dans la région. L'auteur souligne que si la nature non-légale du droit foncier est plus adaptée à un environnement variable et précaire, elle rend aussi le contrôle et la gestion des ressources plus vulnérable aux changements socio-économiques et aux manipulations politiques. Il suggère que dans ce contexte où le système mobile de transhumance ne fait plus partie d'une organisation sociale, un modèle 'hôte-transhumant' (d'accueil, d'accords de réciprocité) de régime de propriété communautaire, qui renforce les relations symbiotiques entre les agropasteurs sédentaires et les transhumants, pourrait être plus viable.

Le chapitre 6 nous conduit dans un écosystème semi-aride du sud Soudan chez les agropasteurs Nuer. Wal Duany décrit les Nuer comme une société 'sans État' à l'intérieur d'une nation plus grande. Il montre comment l'administration coloniale britannique, n'ayant pas trouvé de code légal formel chez les Nuer, a déduit que leur système légal traditionnel manquait de consistance. En imposant un système de loi britannique, elle effacerait sûrement les mécanismes coutumiers de contestation et de discussion dans les processus de résolution de conflits. L'auteur souligne que le droit de posséder quelque chose dépend de la reconnaissance de ce droit par les autres. Les conflits, comme la guerre civile entre le nord et le sud Soudan, surviennent quand ce droit n'est pas reconnu. Les Nuer voient l'État comme une institution centralisée et prédatrice, qui revendique la possession des terres Nuer, mais qui n'a pas d'institution capable de les gérer de façon durable.

L'auteur réussit cette tâche difficile d'interprèter les symboles traditionnels de la langue et des coutumes qui font partie de son propre bagage culturel, dans une langue moderne accessible aux étrangers. La négociation n'est possible que si des lignes de communication restent ouvertes entre les adversaires grâce à des arrangements institutionnels, et plus important encore en raison d'une interdépendance mutuelle. L'auteur fournit quelques exemples récents de conflits résolus de façon interne aux Nuer, pour montrer que l'auto-gouvernance est possible en dépit, ou peut-être à cause de l'hégémonie nord soudanaise et de la guerre civile.

Le chapitre 7 est un cas presenté par Maryam Niamir-Fuller sur les Karimajong en Ouganda. Il fait l'historique d'un passé de conflits entre les tribus et les sous-tribus dans cette partie de l'est Ouganda, et décrit les tentatives récentes des Karimajong et du gouvernement pour résoudre et prévenir ces conflits. L'auteur nous renvoit à la vue du monde par les Karimajong, vue basée sur une classification tertiaire de tous les peuples non-Karimajong, selon laquelle un individu peut-être un ami, un ennemi, ou 'l'autre', cette dernière catégorie étant un concept dérogatoire utilisé pour décrire celui qui n'a pas de bétail, c'est à dire de biens. Les animateurs du développement étant classés dans la dernière catégorie ont peu de chance d'être d'efficaces agents du changement, à moins qu'ils trouvent des moyens de devenir 'ami' des Karimajong.

L'auteur décrit les mécanismes traditionnels et modernes pour la prévention et la résolution des conflits. Elle fait état de tentatives récentes de l'intelligentsia Karimajong (parlementaires, chercheurs) de dévier le processus conflictuel vers des négociations pacifiques et le contrôle du banditisme. Elle souligne que le processus de gestion des conflits est tout aussi important que le but: accepter de s'asseoir à une table de négociation est une importante concession et un engagement. Elle présente plusieurs orientations sur la gestion de conflits empruntés à d'autres contextes, qui pourraient être adaptés à cette situation pastorale.

Dans le chapitre 8, Roy Behnke nous emmène dans une Namibie très aride où les pasteurs élèvent des chèvres et du bétail adaptés à un environnement très rude. Les familles sont presque totalement mobiles. Il montre que la première contrainte qui s'exerce sur le système de production est l'eau, la végétation ne venant qu'en second. Il montre aussi que les invités ou les étrangers sont souvent bienvenus en raison des pénuries de main d'oeuvre pour construire et entretenir les points d'eau. Il décrit un système de mobilité assez proche d'une forme irrégulière de rotation qui exagère l'hétérogénéité naturelle de la végétation. Il soutient que l'ajustement très précis de la charge en bétail à la variabilité de l'écosystème est permise, grâce à l'absence d'une autorité centrale qui essayerait d'imposer des règles rigides, par le recours à un système de production au niveau des familles basé sur la reciprocité. Toutefois, plus les ressources sont sensibles, plus elles sont controlées et gérées à des niveaux institutionnels supérieurs. Le système traditionnel, souligne l'auteur, est extrêmement professionnel et techniquement compétent. Le réformer ou l'améliorer ne serait pas facile, d'autant que les changements seraient dramatiques. Il n'est même pas sûr que le coût du changement serait récompensé par quelques bénéfices d'un modèle classique de type 'ranch'. Pour l'instant, le développement relève de la 'contestation bruyante', c'est à dire que ceux qui intriguent le plus activement obtiennent les plus gros bénéfices.

Ian Scoones, dans le chapitre 9, présente une étude de cas sur une zone sub-humide d'Afrique de Sud habité par des agropasteurs qui élèvent du bétail sur un mode à la fois extensif (les terres communales) et intensif (les résidus de récolte). Partant de l'éparpillement des ressources naturelles dans les terres semi-arides, l'auteur avance quelques réflexions sur les principes fondamentaux de la gestion 'opportuniste'. Le document montre bien comment la valeur relative des différentes ressources (en termes de critères économiques et politiques) détermine le type de régime de propriété lié à ces ressources, et comment ceci peut varier en fonction du temps et du contexte socio-politique. Les ressources de faible valeur sont gérées par les institutions informelles alors que les ressources de grande valeur sont gérées et controlées par des institutions formelles. La même ressource peut-être étroitement contrôlée si elle est rare (ex. par raison de sécheresse ou d'un accroissement de population), et peu contrôlée si elle est abondante (forte pluviosité, pression faible). En terme d'analyse, cette dichotomie est utile mais l'auteur souligne qu'en réalité il y a de nombreux degrés de valeur et que de nombreuses institutions qui se recouvrent et s'emboîtent gèrent des ressources communes. L'auteur suggère de porter attention à six domaines précis pour conçevoir de meilleures institutions et de meilleurs régimes de propriété communautaire.

Le chapitre 10, rédigé par Alain Bourbouze, décrit plusieurs études de cas dans le Maghreb, au Nord de l'Afrique. Les régions arides dont il est question ici, sont peuplées d'un mélange de pasteurs semi-nomades et transhumants. C'est un écosystème très sec et sujet à de fortes variations climatiques. L'auteur retrace les changements écologiques, économiques et sociaux qui ont affecté ces systèmes au siècle dernier et conclut qu'à notre époque la gestion des ressources naturelles, dans la plupart des cas, n'est plus le fruit d'un effort commun pour conserver les ressources, mais un moyen de contrôler les populations voisines de façon à définir la place de chacun dans les terrains communaux. Il relève la distinction entre l'exploitation commune, et la 'gestion commune de la concurrence' par les individus sur les ressources. Il décrit brièvement la nouvelle technique motorisée où le fourrage et l'eau sont apportés par camions au bétail plutôt que le bétail conduit à l'alimentation, technique qui s'est traduite par une dégradation des terres. Mais il souligne aussi que cette option, surtout pratiquée actuellement par les riches pasteurs, intéresse un espace considérable.

Il décrit un cas dans la région du Haut Atlas où une 'Charte de Trans-humance', conçue à l'origine dans un cadre coutumier, reprise ensuite par les autorités coloniales puis par le gouvernement et basée sur des principes légaux modernes, a été mise en application avec succès. Cette Charte englobe les droits de tous les usagers, mais elle n'apporte pas de réponse à la décroissance des espaces pastoraux et à l'accroissement des conflits en raison de sa généralité et de son manque de souplesse. Les pasteurs ont utilisé cette Charte comme un outil pour reconnaître des droits de propriété, alors que la gestion quotidienne des ressources était fondée sur des structures et des techniques plus traditionnelles. L'auteur montre que la plupart des conflits sont désamorcés par la recherche du consensus dans la prise de décision, ainsi que par une certaine tolérance vis-à-vis des infractions mineures. Seuls les conflits sérieux sont renvoyés devant les instances judiciaires, qu'elles soient coutumières ou modernes. L'auteur est d'avis que le formalisme du système judiciaire moderne est un point faible car il peut-être utilisé par les individus à leurs avantages. Par exemple, les amendes imposées aux auteurs d'un défrichement abusif sur collectif, renforcent leurs futures revendications sur les ressources communes, car ils ont déjà 'payé', le procès verbal valant titre. Ceci incite à la prudence dans l'usage des concepts de souplesse et de participation qui peuvent être utilisés à la fois à des fins bonnes ou mauvaises.

La plupart des études de cas portent sur des tentatives faites au niveau local pour gérer les conflits. Mais elles reconnaissent tous les liens très forts, en négatif ou en positif, qui s'éxercent à un niveau global par le biais des politiques nationales adaptées, des procédures administratives, des mécanismes légaux, ainsi que des marchés régionaux ou internationaux, et des guerres.

Dans son dernier chapitre, cet ouvrage explique que la recherche insiste de plus en plus sur l'argumentation économique et écologique en faveur de la création et du renforcement d'institutions qui facilitent la mobilité du bétail dans les environnements arides. C'est donc une nouvelle perspective qui s'ouvre n'ont seulement pour le développement pastoral, mais aussi pour les institutions sociales qui maîtrisent la gestion des ressources naturelles dans les zones arides. Le dernier chapitre intègre les nouvelles recherches sur la théorie du non-équilibre écologique, la gestion des communaux, la résolution des conflits, la planification participative, ainsi que les institutions informelles et formelles de gestion des ressources. C'est donc un cadre de travail qui fournit des outils pratiques, des solutions, des mots clés, et des principes sous-jacents à une création des conditions sociales, légales et institutionnelles qui peuvent légitimer la transhumance. Pour terminer, ce chapitre replace le cas africain dans une perspective plus vaste où l'on trouve la résurgence récente de la gestion des biens communs et de la transhumance dans d'autres continents.

2

A review of recent literature on pastoralism and transhumance in Africa

Etudes récentes sur le pastoralisme et la transhumance en Afrique

MARYAM NIAMIR-FULLER and MATTHEW D. TURNER

ABSTRACT

This chapter provides the contextual background and necessary definitions for the case studies and conclusions that follow, as well as a concise review for those readers not overly familiar with the recent literature on pastoralism. The importance of mobility as an adaptive tool to the variable ecosystem and as a risk-management mechanism is highlighted. The paper then traces the impact on pastoral societies of changes in the national political, social and economic systems of African countries, concluding that the majority of impacts have been negative (social and political marginalization, lowering of standards of living, resource shortages, environmental degradation, etc.), and have largely been due to external forces rather than internal forces within the pastoral societies.

The evolution of development paradigms is briefly discussed, showing how, in many cases, this evolution has itself been one of the causes of fundamental change in pastoral systems. In recent years, there has been a fortuitous convergence of different trends: since the 1980s there has been a concerted effort toward decentralization and popular participation in development projects and government actions at the local level; pastoralists are starting to demand their own share of the 'political space' by forming local, national, and even regional-level NGOs (non-governmental organizations); a new development paradigm has emerged, largely due to theoretical advances in several related scientific fields (ecology, institutional economics, common-property theory, integrated assessment), that recognizes the importance of livestock mobility for the sustainable management of arid ecosystems. The paper concludes by defining and describing the different terminologies, concepts, and key words important to the 'mobility paradigm', such as: non-equilibrium theory, reciprocity, social capital, opportunistic management, tracking, inclusive rights, and co-management.

RÉSUMÉ

Ce chapitre fournit le contexte et les définitions nécessaires pour les monographies et conclusions qui suivent, ainsi qu'une revue rapide des études publiées récemment sur le pastoralisme pour les lecteurs peu au courant de la question. Il met en lumière le rôle de la mobilité en tant que moyen d'adaptation à un écosystème variable et mécanisme de gestion des risques. Les auteurs exposer ensuite les incidences sur les sociétés pastorales des modifications des systèmes politiques, sociaux et économiques nationaux des pays africains et concluent que la plupart de ces incidences ont été négatives (marginalisation sociale et politique, abaissement du niveau de vie, pénuries de ressources, dégradation de l'environnement, etc.) et ont été plus souvent imputables à des forces externes qu'à des forces internes des sociétés pastorales.

Les auteurs étudient brièvement l'évolution des paradigmes de développement en montrant comment cette évolution a souvent constitué elle-même une des causes du changement fondamental des systèmes pastoraux. Au cours des dernières années, différentes tendances ont eu des effets convergents de façon fortuite: depuis les années 80, l'ensemble des projets de développement et des interventions des gouvernements au niveau local ont été orientés vers la décentralisation et la participation populaire; les pasteurs commencent à revendiquer leur propre place dans l'espace politique en constituant des organisations non-gouvernmentales (ONG) aux niveaux local et national, voire régional; un nouveau paradigme de développement reconnaissant l'importance de la mobilité du bétail pour la gestion durable des écosystèmes arides est apparu grâce surtout aux progrès théoriques réalisés dans plusieurs domaines scientifiques connexes (écologie, économie des institutions, théorie de la propriété commune, évaluation intégrée). Les auteurs terminent avec la définition des différents concepts, terminologies et mots clefs qui facilitent la comprehension du 'paradigme de mobilité', par exemple la théorie du non-équilibre écologique, la réciprocité, le capital social, la gestion opportuniste, les droits inclus et la cogestion.

IN THIS CHAPTER, we attempt to synthesize the extensive literature on pastoralism in Africa by first giving a brief summary of historic changes in pastoral management and livestock mobility, and, second, by summarizing the so-called 'mobility paradigm' that has emerged in the 1990s. By doing so, we hope to achieve two objectives:

1. to provide a concise review for those readers not familiar with the recent literature on pastoralism
2. to provide the contextual background and necessary definitions for the case studies and conclusions that follow.

Outsiders' recognition of the importance of mobility for pastoral systems is not new. Ethnographers catalogued its various manifestations and adaptive mechanisms early on. What *is* new, is the emergence of a more unified, multi-disciplinary rationale for livestock mobility in arid and semi-arid lands, accompanied by a more sophisticated understanding by researchers and development workers of the ecological and economic implications of its specific attributes, and a greater possibility for pastoral participation in this new development discourse. The 'mobility paradigm', therefore, should be seen as a historical evolution of ideas and actions, amalgamated into an approach quite different from what, for convenience sake, can be called the old or 'classical' paradigm.

Historic changes in pastoral management and livestock mobility

Are sedentarization and a village life, supported with modern education, health and communication facilities, the best development paradigms for pastoralists, as has been believed by governments and development workers since the colonial era? By advocating livestock mobility, are we not denying pastoralists the fruits of modern industry and of modern democracy? These questions are difficult because they touch the emotional fibre of both pastoralists and non-pastoralists. They implicate the moral values of each side, which are shaped by different cultural and environmental exigencies. Stakeholders often adopt extreme positions because they subscribe to different paradigms of development. Scientific literature, development reports, and village debates are characterized by strong words, accusations of romanticism, and of conspiracy.

A brief overview of the evolution of knowledge concerning pastoral development is, therefore, necessary to widen the emotional content of the discussion, and to show how the recognition that animal mobility is an essential ingredient for sustainable development in arid lands, has been arrived at.

Traditional risk management and livestock mobility

Dryland production systems are characterized by periodic 'boom and bust' – good years followed by bad years (droughts, epidemics, raids). The climatic cycles can be quite extreme. For example, a study in northern Mali covering 1984–1990 showed coefficients of variation of 28–37 per cent in mean annual rainfall, and 64–86 per cent in mean herbaceous biomass. As a result of the devastating 1970s drought in the Sahel, total livestock population in this area decreased by 44 per cent due to a combination of livestock mortality and net movements of livestock to the south (de Leeuw et al., 1993). Such periodic stress on the production system poses grave risks to the short-term viability of individual households. Pastoralists have developed strategies to reduce their vulnerability to these shocks.

These adaptive mechanisms and how they were managed by customary institutions have been catalogued quite extensively since the time of the early ethnographers of the 1940s and 1950s, such as Evans Pritchard on the Nilotes of East Africa, Cunnison on the Sudanese Baggara, Dupire and Stenning on the Peul, to name but a few. The book edited by Bovin and Manger (1990) gives an excellent analysis of the adaptive strategies of both pastoralists and agropastoralists in Africa. Since risks are often location specific, these strategies influence pastoral mobility either directly or indirectly. Here we choose to focus on several key strategies which especially influence pastoral mobility.

To begin with, pastoralists typically follow an *opportunistic stocking strategy*, accumulating livestock numbers that exceed subsistence demands during good years so that they can be assured of enough heifers and cows surviving the 'busts' for re-establishment of the herd after the bad years (Sandford, 1982). Depending on the aridity of the system, a minimum of 50 livestock units[1] per household in

---

[1] A livestock unit is defined here as follows: 1 cow, or 6 sheep or 6 goats.

20

arid systems, and 30 livestock units in semi-arid systems is the threshold below which a typical pure pastoral household cannot survive across drought cycles (Bremaud & Pagot, 1962; Dahl & Hjort, 1976; Sandford, 1982). An opportunistic stocking strategy requires that rangeland use patterns adapt to herd sizes. High primary productivity in good years provides an incentive to herders to reduce mobility, but they have to balance that with the needs of a larger herd. A smaller herd could be kept closer to home, but in bad years may need to be taken further afield to reach patches of good feed.

An important risk-management strategy of many pastoral groups is diversification, which is evident in many ways. A diverse mix of livestock species managed by the same household, not only better serves subsistence needs and more fully utilizes the range of fodder resources, but also reduces risk from different external shocks. A diverse mix of species requires access to a diverse, heterogeneous mix of resources, hence the need both for a biodiverse environment and for mobility. A household's livestock are often spread across a number of widely dispersed herds through herd splitting, which allows a manipulation of the mix of species according to ecological potential. Herd stratification, where the milch herd remains with the family (older men, women and children) in areas of higher milk demand while the rest of the livestock are managed in a more mobile fashion by younger men, is a common strategy across dryland Africa.

Livestock loans, gifts and entrustments represent the material ties of the 'moral economy', which facilitate herd re-establishment after bad years. In addition, the loans and entrustments also work to separate and distribute a household's livestock wealth, thereby lowering the exposure to location-specific risks (disease, raids, etc.) at minimal labour cost. Economic diversification and occupation switching also buffer a household from risk. Post-drought responses may include: investing in small stock as a rapid re-stocking measure, temporarily engaging in crop cultivation, temporarily engaging as migrant labour, etc. All these strategies affect the mobility patterns of the herds. Among the Fulani in West Africa, historically there has existed a gradation, driven by relative access to labour and livestock, and environmental variability, in a household's reliance on cropping versus raising livestock to meet subsistence needs (Toulmin, 1983; Bonfiglioli, 1990; Mace, 1993). Even 'pure pastoralists' often are involved in trade, hunting, gathering, warfare, or extraction activities that provide them an income source independent of the livestock economy.

Livestock mobility is the major way in which pastoralists have historically managed uncertainty and risk in arid lands (Scoones, 1995). Mobility is an adaptive tool for livestock production that simultaneously serves:

o the opportunistic usage of ecosystem productivity, which is spatially and temporally variable and to a large degree unpredictable
o the ability to use under-utilized pastures distant from settlements
o the use of pastures that are only seasonally available, and
o the provision of fodder to livestock at minimal labour and economic cost.

In addition, moving their capital stock (animals) provides pastoralists with effective ways to lower their vulnerability to disease outbreaks, droughts, potential raids, tsetse and other insect-borne health challenges (e.g. Bassett, 1986).

Livestock mobility has both spatial and temporal components. Factors affecting the distance, timing and location of livestock mobility include:

o the location and seasonal changes of water and pasture, disease outbreaks, localized drought, agricultural harvest residue, market location, ceremonies and periodic social gatherings
o economies of scale in certain cases (e.g. communal herding, drawing water from wells), and
o political factors such as relations with other pastoral or non-pastoral tribes, and needs for defence.

Pastoral mobility is not limitless. Limits of geographic knowledge, resource claims, social contacts, and security result in pastoralists revisiting encampment points on an annual or semi-annual basis. Travel corridors between particular encampment points are often well established. The romantic view of the pastoralist freely roaming the range, chasing rain clouds, diverges clearly from reality.

Livestock mobility does not necessarily correlate with human mobility. In cases such as the Maasina (see Chapters 4 and 5), the main transhumance herd (*garci*) moves almost continuously while the majority of the herding family remain with the milch herd near the home pastures. In other cases, the whole family moves with the herd during all or part of the year, such as among the North African Berbers (see Chapter 10). It is important to note as well that these features are not culturally fixed. Pastoralists often vary their human and livestock mobility patterns in response to political, economic, and ecological circumstances.

Mobility is possible because tribes and groups at any one point in time have access rights and ownership claims to large land areas. Each claim and right is mediated by internally cohesive social institutions, norms of reciprocity, and formal and informal regulations. Access to a wide range of potential pastures is facilitated through a combination of recurrent investments in social networks, alliances and force of arm. The herder's ability to respond to the high variability in ecosystem productivity is largely met through rules and accords governed by negotiation and mediation between tribes and sub-groups. Raids and wars are also potential tools for expanding one's pool of resources (pasture and livestock), or as a last resort when negotiations fail.

Pastoral ecosystems in the last century were relatively healthy despite several severe drought episodes (Cissoko, 1968; Waller, 1985; Bonfiglioli, 1988; Gritzner, 1988; Smith, 1992). A combination of important factors explains this phenomenon: lower human-population density (but not necessarily livestock density); land-tenure security vested in customary communal institutions; mobility of animals; and traditional natural resource management and improvement techniques. It is easy to romanticize, however, and the pastoral development literature has its own share of such examples. Pastoral systems are not a 'lost paradise'. Communities were not homogenous, nor peaceful, and competition between individuals and groups for natural resources, economic gain, and political power shaped the dynamic evolution of the societies. Customary authority was rarely egalitarian, and many hierarchical societies had highly unequal distribution of livestock wealth. However, what is important to sift out of the debate is the array of social, economic, political and ecological mechanisms that allowed pastoral production systems to remain viable for so long – and for some, well into our own time.

Changes and transitions

The events of the last century have modified these systems sometimes to the point of non-recognition. Livestock mobility has declined, as evidenced by high rates of sedentarization and a reduction in daily grazing radii around encampment points, movements among encampment points within a pastoral area, and the frequency and distance of historic transhumance movements. 'Spontaneous' sedentarization has been driven by a combination of factors that interact and reinforce each other:

o major droughts
o differential government support of agriculture
o lack of government support for transhumance
o the 'benign neglect' syndrome (Swift, 1993)
o population- and policy-driven extension of cultivation into rangelands
o increased individualization and disruption of political structures within pastoral societies
o state claims over 'vacant' pastoral land (state farms, national parks, forest reserves)
o rural violence
o increased ownership by investors outside the pastoral sector, and
o the growing economic vulnerability of pastoral groups.

Among pastoralists who have managed to remain mobile, the decline in the frequency and distance of livestock mobility is due in large part to agricultural encroachment onto rangelands, by both farmers and agropastoralists, leading to a general shortage in pasture area and blocking of traditional transhumance routes. As will be discussed later in this chapter, such a reduction in livestock mobility has increased the ecological and economic vulnerability of pastoral systems in dryland Africa.

European conquest and the imposition of colonial rule initiated a whole series of changes within pastoral societies. The continental-scale rinderpest epidemic that is estimated to have killed more than half of the herds in the late nineteenth century, differentially impoverished pastoral groups with respect to neighbouring farming groups at the outset of a period of political transition in rural areas. Pastoral groups, because of their mobility, offered particularly stiff resistance to colonial rule (e.g. Twaregs in West Africa, and Karamajong in Uganda).

One of the results of the colonial conquest of pastoralists was that they were 'frozen' (in an administrative sense) into land areas that they happened to occupy at the time of conquest (see Chapter 7). Although conflicts over land

23

were common in pre-colonial times, the formalization of pastoral territorial boundaries (during which process most mobile pastoralists lost significant portions of their ancestral territories) has added significantly to the sources of recent conflicts. Colonial military presence in some areas had the positive effect of reducing the threat of violence used by some pastoral groups (e.g. in East Africa) to maintain control over agricultural clients, but at the same time had the negative effect of reducing their control over their pastoral resources. Most colonial systems of indirect rule chose their indigenous leaders from sedentary populations, thereby affecting the balance of power between pastoral and farming communities, and further weakening pastoral groups politically.

In addition to these negative factors, colonialism also brought positive changes such as veterinary services, education, and greater access to regional and international markets. The political programmes of colonial and post-colonial governments, however, were very much directed against one feature of pastoral systems: their mobility. Colonial officials brought with them a cultural bias against mobile peoples, who were viewed as primitive, shiftless, and immoral. European ignorance of the nature of pastoral mobility resulted in it being seen as antithetical to good land husbandry because there were very few visible ties to the land[2]. The 'nomad' would destroy the land with his livestock and move onward to 'greener pastures'. Mobile populations were also less easy to administrate. Colonial governments often instituted measures to limit the mobility of their subjects and their subjects' animals. Once settled, a rural population could be medically treated, schooled, and taxed. These cultural and political biases against pastoral mobility have been continued in many cases into the post-colonial period, by government officials from agricultural backgrounds.

Early concerns by environmental scientists about regional-scale desertification also resonated with these political prerogatives (Stebbing, 1935; Lamprey, 1983). Ecological work during the colonial and immediate post-colonial periods was dominated by synchronic or, at best, diachronic studies. These studies were less focused on understanding ecological processes and the inter-annual dynamics of ecological change and more on describing savanna and steppe environments, mapping vegetation, or estimating carrying capacities for domestic livestock. Despite large inter-annual variations in rainfall, most environmental analyses consisted simply of comparing vegetative production with an assumed biological potential or 'natural' species composition (Homewood and Rogers, 1987; Turner, 1993; Dodd, 1994). Controlled grazing experiments were rare. Reductions in vegetative production combined with often anecdotal evidence of heavy livestock use were used to support arguments of widespread overgrazing in dryland Africa. Recurrent droughts since the 1970s fuelled long-standing concerns about

---

[2] Bias against pastoralists is not confined to Africa. Ranchers in the arid lands of the USA at the turn of the century had a difficult time gaining access to viable-sized land areas, since the philosophy of John Locke (1690) was applied to private ownership. This translated as a fixed allocation of 160 acres of land per homestead, no matter what production system it involved (Raymond, 1997).

desertification, with livestock seen by some as playing a major role in regionally scaled environmental decline (Lamprey, 1983; Sinclair & Frywell, 1985). Such assessments provided an environmental rationale for destocking and sedentarization programmes by colonial and post-colonial states.

Environmental concerns, cultural biases, and the political prerogatives of the State were reinforced by agronomists and range managers who viewed livestock mobility (along with pastoralists' accumulation of low weight-gaining livestock), as irrational, and who limited the potential for fixed-capital investments to improve livestock production. The tendency to evaluate the economic and biological productivity of pastoral systems through the viewpoint of European humid temperate biases has led to an underestimation of the economic contribution of these systems, and an overzealous attempt to change their way of life[3].

Government policies since colonial times, therefore, have favoured crops over livestock. Common agricultural polices included high import duties to protect domestic cereal prices, and subsidies on fertilizers and fuel (acting as an indirect subsidy on tractors). These policies have upset the economic balance that existed between crops and livestock, making the latter far less profitable, and discouraging investments into improving the range and livestock sector. They have also contributed to the expansion of crops into rangelands (Little et al., 1987; Lane, 1991; Niamir et al., 1994; Steinfeld et al., 1997). Because of their higher agronomic potential, the more productive pastures are often the first to go. The exclusion of pastoralists from these 'key pastoral resources' can lead to significant disruption of the annual transhumance cycle.

The earliest forms of livestock development in Africa involved incentives for settlement, such as ranching, destocking, and specialization (often called 'stratification') between reproduction and fattening (Ndagala, 1982; Oxby, 1982; Joof et al., 1988; Peluso, 1993; Neumann, 1995). Settlement schemes are justified by governments with a promise of socio-economic benefits (services, inputs, water, markets, infrastructure), however these services rarely materialize because they are implemented as public goods which are often proven to be too costly for most governments. Settlement schemes have failed on both ecological and economic grounds (Sandford, 1983; Homewood & Rogers, 1987; Thébaud, 1988; de Haan, 1994). In addition, local authorities take on the responsibility to administer the schemes, but rarely can enforce the rules and prevent transgressions by non-pastoralists (Okoh et al., 1988). De-stocking programmes have left households with less than viable herds and have contributed to their marginalization.

Sedentarization of pastoralists, whether forced or spontaneous, has resulted in severe land degradation in the semi-arid zones. Decreased mobility of animals means increased continuous grazing around the settlements resulting in reduced

---

[3] There are quite a few myths about pastoralists: they are backward, violent, and irrational; they overstock the range because they value cattle more than anything in the world; they do not want to sell their animals; they are averse to any kind of authority; they love to roam. Judging by recent debates in the development forum, by the type of workshops being organized by African scientists, and by the increasingly vocal Pastoral Organizations, the debunking of these myths is slowly but surely occurring.

25

vegetation cover and diversity, and soil degradation. At the same time, lower grazing pressure in distant pastures results in an invasion of unpalatable plants (Galaty, 1988; Warren & Rajasekaran, 1993; Niamir, 1997). Settlement also results in a loss of traditional knowledge and controls on range use, leading to less efficient management of the arid resources (Jacobs, 1980; Farah, 1993).

Socio-political systems have been subsumed under the hegemony of the central State (nation) leading to a weakening of the traditional leadership, and a fragmentation of authority. The customary judicial system has been relegated to deal with minor internal conflicts. Communally held land (the ownership of which was generally vested in a deity), has been abrogated by the Nation-State often under the pretence that it is not being put to productive use. In these ways, the authority of pastoral leaders has been seriously eroded.

In most cases, public works sponsored by the State (e.g. water, dips) have disturbed the traditional mobility patterns and created an 'open access' situation. They were intended to increase livestock productivity by 'opening up' new pastures, and combating parasitic diseases. However, their positive impacts have to be balanced with the negative. The permanent, high-output water points have been shown to contribute to a large radius of land degradation around them (Thébaud, 1988). Since no institution was left in place to maintain and manage the dips, most of the gains were relatively short-lived.

In addition, the growth of rural labour markets and commodification of agriculture have in part contributed to individualization, increased stress between elders and juniors or between the élite and other members of society, and reduced the cohesiveness of pastoral social formations (Dalli & Ezeomah, 1988; Chapter 10 in this volume). Since pastoral mobility requires management and political coordination, these changes have led to a reduction in higher-level coordination of transhumance movements, especially in areas of agricultural encroachment and political instability.

The major droughts of the 1970s and 1980s forced a mass movement of herders in the Sahel toward the south, but many have been able to return to their previous transhumance system since then. According to Thébaud (1998), if after ten years a sedentarized household has not been able to amass enough capital to reinvest in livestock, it will probably never be able to return to transhumance.

A lowering of standards of living and a de-capitalization of pastoral livestock wealth, are a few of the major results of these changes since the last century. For example, by the mid-1980s in northern Nigeria, half of a sample of pastoralists had less than the 'survival' threshold of 30 heads (Joof et al., 1988). Between 1900 and 1990 the average livestock holding per household in Karamoja decreased from 100 to 28 (Chapter 7, this volume). Part of the loss has been compensated with alternative income sources, especially crops and wage work. The high rate of out-migration of pastoralists towards smaller towns and cities (e.g. an estimated 30 per cent in Tanzania), or towards sedentarization (e.g. the almost complete disappearance of transhumance in Tunisia), is one manifestation of the continuing decline in living standards among most pastoralists. Another trend is the increasing concentration of wealth in the hands of a few

26

since the 1970s. It is common to find that about 15 per cent of the population control 80 per cent of the livestock (Little, 1985a; Sutter, 1987; Ndagala, 1991). Increasingly, the large owners are investors from outside the traditional pastoral sector, who entrust livestock to pastoralists or hire herders for a wage (Bonfiglioli, 1985; Little, 1985a; Turner, 1992; Bassett, 1994). Often these new owners, to maintain oversight of their wealth, place limits on the mobility of the herd.

There are clearly regional differences in the type of changes and transitions in the last century, but most of the changes are due essentially to a neglect of pastoralism, in favour of crop cultivation and forestry/wildlife, and to profound perturbations of socio-cultural interrelationships. Pastoral systems no longer have control over their socio-political environment, and are therefore not able to collectively adapt their mode of production and livelihood to the ecological changes (Digard et al., 1993). 'Pockets of resistance' where pastoralists still function according to traditional systems still exist – primarily in more economically and ecologically marginal areas untouched by market and administrative forces. The vast majority of pastoral systems, however, have undergone profound changes.

Evolution of development paradigms

A brief history of the evolution of development paradigms will help to provide a necessary perspective as well as show how each successive paradigm has learned lessons from its precedents (Figure 2.1). Pastoral development as such, is a relatively new paradigm that began in the mid-1980s in Africa with the advent of natural-resource management projects. Before that, and going back to the colonial era, the focus was on livestock-productivity development, rather than on the enhancement of livelihoods. There are exceptions to this rule. As early as the 1940s British colonialists introduced pastoral development in Kenya and Rhodesia, although the schemes were not very successful. The main objective of the livestock-development paradigm was to increase exports of products to urban centres and international markets. The main interventions were the application of the classical 'ranching' model from the United States, water point development, and vaccinations against contagious diseases and epidemics.

Classical range scientists advocated a conservative and sedentary approach to stocking rangelands. The carrying capacity of the land, in their opinion, should be calculated on the basis of the worst year. This conservative approach has been the justification for fenced ranches and de-stocking. However, it is now recognized that the greater the degree of variability in rainfall over time, the greater will be the opportunity cost of conservatism (Sandford, 1983). The resultant understocking in normal to good years, and reduced mobility, have been shown not only to constitute a waste of forage, but in many cases to lead to pasture degradation (see, for example, Seligman and Perevolotsky, 1994). As described earlier, policies on sedentarization, de-stocking and water development, were unsuccessful in increasing livestock productivity, at best, and in some cases were very destructive in the long run.

The success of veterinary interventions is a matter for debate. Some say that

27

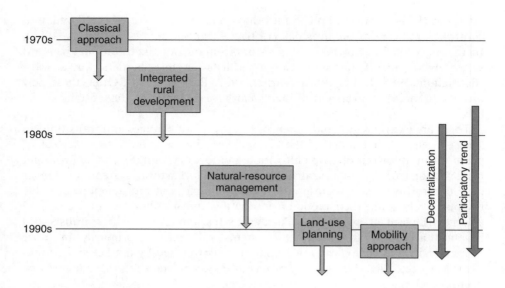

Figure 2.1 *Evolutionary diagram of different approaches to pastoral development in Africa*

these interventions were so successful that they resulted in an increase in livestock populations beyond the carrying capacity of the land, thus contributing to land degradation (e.g. Mamdani et al., 1992). Others believe that the interventions were neither so effective, nor so widespread as to make such an impact (Roeder, 1996). Today, most livestock dips have fallen into disrepair and veterinary medicines are hard to come by. Government policies towards privatization of the veterinary sector have reduced coverage of transhumance areas, because there are fewer economic incentives to reach remote and dispersed populations who can't pay much for the services. While privatization appears to have increased the supply of this service in rural centres, mobile pastoralists can only take advantage of them while passing by on their transhumance routes.

In recognition of these problems, the classical 'range and livestock development' projects were transformed into 'integrated rural development' projects in the early 1980s. Interventions in the health and education sectors, as well as roads and other infrastructure, were added to a blueprint for range and livestock development essentially similar to the classical approach. While less coercive and service-oriented than previous programmes, they continued an implicit sedentarization agenda with a nodding appreciation for local perspectives. Most of these projects were deemed only partially successful: water points, roads, schools and clinics were built, but livestock productivity did not increase, and the infrastructure fell into disrepair once the projects terminated.

By the mid-1980s a new generation of projects emerged, labelled 'natural-resource management projects'. These recognized the need to focus on the ever-increasing land degradation problem, and the difficulties of managing

28

multi-sectoral integrated development projects. The focus shifted away from livestock, to rangelands and all of their resources. The approach adopted by these projects was still a 'blueprint' approach; remote sensing was used to determine carrying capacities, agro-ecological zoning, and land use 'guidelines' which were discussed with pastoral and sedentary land users only after the blueprint was created. However, they did break new ground by attempting to modify institutional structures for natural-resource management. Many projects created and legally registered 'pastoral associations' to which the responsibility (but not ownership) to manage a defined land area was given. The main problems faced by this approach were that it was too 'top-down', and not subject to consensual agreement by land users; the relationship of new institutions to customary ones were left undefined, leading to a lack of effectiveness at best, or a further breakdown of customary institutions at worst.

The 'gestion de terroirs villageois' approach in West Africa, which is similar to land-use planning efforts in East and South Africa, was part of the next generation of projects. It followed the natural-resource management principles of previous projects, but at a more localized (village) scale, and strongly influenced by common property theory. The new community-based resource-management approach is novel in that it intends to devolve greater resource-management authority to the local level, acknowledges the role played by customary common-property management regimes (CPRs), attempts to allocate common property tenure to local institutions, and facilitates more participatory forms of development.

After about two decades of experience, this approach has enjoyed only partial success in building local-level institutions for natural-resource management. The following shortcomings have been identified in the literature (Turner, n.d.; Marty, 1993; Painter et al., 1994; Engberg-Pedersen, 1995):

o existing informal local institutions for decision-making are often overlooked
o significant differences between the interests of leaders and non-leaders are ignored
o inconsistencies exist between the approaches' goals for natural-resource management and villagers' goals for infrastructure and social development
o inability to provide adequate incentives for people to undertake labour intensive resource-conservation activities that have noticeable returns only in the long run
o mobile pastoralists are ignored or delegated to a secondary 'receptive' rather than pro-active position
o an under-appreciation of the high spatio-temporal variability of resource endowment in dryland areas, evidenced by its focus on promoting exclusionary mechanisms in land-tenure systems
o a spatially myopic focus on the village.

As this brief summary suggests, the evolution of pastoral development paradigms itself has been one of the causes of fundamental change in pastoral systems, although this impact has not been direct nor strong in all cases. This evolution has been largely led by factors exogenous to pastoral systems, rather than, as would be ideal, by the dynamics of pastoralism. Such factors include regional and international market forces, national political systems, trends in international donor funding, and changes in the perceptions of development workers and the theories of researchers. It is only since the mid-1990s that pastoralists have started to demand their own share of the 'political space'[4]. This can be facilitated through such mechanisms as local and national NGOs, election of pastoral parliamentarians, and the growth of a 'critical mass' of educated pastoralists who can effectively communicate at the macro-level about the people's problems. It is hoped that the evolution of pastoral development paradigms may, in the future, be led more by changes in pastoral systems than by exogenous factors.

Ever since the 1990s more and more donors and governments have begun to invest in projects that rely on traditional systems and knowledge of agro-pastoralists (e.g. World Bank in Mauritania, Mali and Chad, GTZ and Danida in Burkina Faso, UNDP in Sudan). But very few have tackled the 'problem' of mobility. One major problem has been that despite embracing the value of indigenous technical knowledge in the abstract, developers still often misunder-stand and under-appreciate the mobility prerogative of indigenous systems. Mobility is still often seen as a problem to be done away with, rather than a trump card to be strengthened.

The mobility paradigm: linking managed mobility to sustainable development

Several scientific trends coincided in the early 1990s to produce what was then called a 'new paradigm' or an 'alternative paradigm for pastoral development' (Behnke et al., 1993; Niamir, 1994b). Chaos theory, post-normal science, and new institutional economics are all part of this trend. But perhaps the most influential direction for the new pastoral paradigm has come from the new ecology. Ever since the early days of this century, the science of ecology has been based on the principles of equilibrium (balance, harmony) and functional order. However, the recently emerging 'new ecology' asks questions of the core assumptions of equilibrium ecological theory (Botkin, 1990; Allen & Hoekstra, 1992; Zimmerer, 1994). Classical equilibrium theory is unable to capture the uncertainty and variability in arid ecosystems, making such concepts as carrying capacity and stocking rate less effective in predicting ecosystem productivity and dynamics (Westoby et al., 1989; Behnke et al., 1993; Scoones, 1995; Behnke, 1997).

Many elements of the classical paradigm have been completely turned around by the new paradigm. Ecological studies undertaken in the arid lands show that climate appears to be a more significant factor in determining vegetation struc-ture, function and dynamics, than either grazing or internal ecological processes (e.g. O'Connor & Roux, 1995; Perevolotsky, 1995; Hiernaux, 1996; Huntley &

---

[4] We are indebted to Gregory Perrier for this term.

Walker, 1979). There is evidence that in some areas undergrazing is a more serious ecological threat to arid ecosystems than overgrazing (Seligman & Perevolotsky, 1994).

Environmental analysts have been prone to characterize the grazing of domestic livestock as destructive to the environment – in truth, the relationship in Africa is much more complex, given the significant differences in livestock management and grazing history from the North American situation. With the long history of co-evolution of livestock and the African environment, livestock should be seen as an integral part of both conservation and development (Steinfeld et al., 1997)[5]. The new paradigm not only argues that transhumant pastoralism is not an archaic remnant of the past, but even asserts that it is a necessary pre-condition to sustainable development in arid lands. The flexibility and adaptability of these systems, that have allowed them to survive the perturbations of this century, are points of considerable strength that could be useful in designing sustainable and adapted pastoral systems.

The mobility paradigm does not advocate turning the clock back, nor attempting to freeze pastoralists in their current state. Rather, it wants to ensure that the appropriate policies, legal mechanisms and support systems are in place, in order to allow self-evolution of pastoralism towards an economically, socially and environmentally sustainable livelihood system.

There is a need for a holistic and integrated analytical framework that can incorporate all the new developments in each of the contributing scientific fields (economics, sociology, anthropology, ecology, and political science), and provide a sound basis upon which development activities can be designed. The analytical framework should act as a check-list of issues that need to be covered and adapted to the specific local conditions, rather than a blueprint for action. Of primary importance is the recognition that both the ecological and the anthropic parameters are dynamic and constantly changing. Very often events will outpace the analytical effort, making an iterative approach indispensable.

The following section is an analysis of the salient aspects of pastoral natural-resource management systems, as seen through the lens of the mobility paradigm. The discussion revolves around selected key words, and is divided somewhat arbitrarily between the resource base, the resource users, their adaptive strategies, and their common property regimes. The latter touches upon property rights, management regimes, rules and their enforcement, and conflict management.

---

[5] The value of domestic livestock for environmental conservation can also be expressed in terms of global climate change. Preliminary calculations show that extensive, transhumant/mobile grazing has a positive carbon sequestration balance (i.e. amount sequestered is greater than the amount emitted), while both intensive grazing and commercial (feedlot) systems do not (Steinfeld et al., 1997).

KEY WORDS: *high variability and uncertainty; non-equilibrium theory; ecological resilience; socio-ecological pasture units*

The new paradigm recognizes three characteristics of warm arid ecosystems[6]: *ecological variability, unpredictability*, and *high resilience*. The arid ecosystem is constantly changing from one state or level to another, making it difficult or even impossible to define a stable, equilibrium state. Whether the system can be characterized as 'multiple equilibrium' or 'dynamic equilibrium' or *non-equilibrium* is still a matter of debate. Despite the variability, there are patterns and cycles in ecosystem change, and although the degree of predictability is low, it increases as the space and time-scales increase – the longer you have hindsight, the more you can predict; the larger the geographical scale, the more you can generalize *and* predict generalities. However, for the sake of convenience, the term 'non-equilibrium' is used as a generic term to mean any kind of system whose dynamics are not governed by a single equilibrium.

Arid and semi-arid ecosystems in Africa are characterized by low net primary productivity, and high variability in ecosystem structure and productivity. The lower the rainfall, the higher the spatial and temporal variability (IUCN, 1989). Temporal variability is manifested by seasonal and yearly changes in rainfall, which can have a coefficient of variability as high as 40 per cent. Spatial variability refers not only to different eco-zones and their transition zones, but also to heterogeneity at the micro-level, or 'patchiness' (de Angelis & Waterhouse, 1987; Scoones, 1991). The most limiting factor in the arid and hyper-arid zones is water, but as the ecosystem becomes more humid, the most limiting factor gradually becomes soil nutrient content (Penning de Vries & Djitièye, 1982). In the Mediterranean climates of northern Africa, there is the added factor of temperature (cool winters, hot and dry summers).

The contribution of grazing to changes in productive potential of dryland rangeland in Africa is a topic of considerable debate (Behnke et al., 1993; Behnke & Abel, 1996). More recent research has provided support for the following arguments:

1. the scale and magnitude of *persistent* environmental decline in dryland Africa has been overestimated
2. the role of livestock grazing in these changes has been overestimated, and
3. the pattern of anthropogenic land degradation is much more severe around permanent settlement sites than it is in open rangelands.

---

[6] The discussion is confined to warm arid ecosystems (tropical and Mediterranean). Empirical data on cold arid systems, such as in Patagonia or Mongolia, where extremely low temperatures for most of the year have the effect of 'slowing down' the system, are not yet conclusive on the issue of non-equilibrium.

An influential diachronic study in Sudan arguing for regionally scaled anthropogenic environmental degradation (Lamprey, 1983) has been shown to be fundamentally flawed, because more detailed analysis over the same area, has found only localized degradation, with little support for region-wide degradation (Hellden, 1991). Vegetation change may result in less rangeland productivity, as with bush encroachment in the Kalahari, but it does not necessarily affect the resilience of the ecosystem. Moreover, vegetation change may or may not be symptomatic of land degradation – the latter being defined as irreversible damage to the soil, and soil/water/nutrient relationships (Dougill & Cox, 1995).

Remote sensing work has found that inter-annual variations in rainfall and vegetation structure are so high that they require decades-long monitoring to detect expansion or contraction of the Sahara (Tucker et al., 1991). The reported 'loss of great trees' in the northern Sahel could either be a result of human-induced deforestation, or a manifestation of a functional adaptability and of a resilience to long-term desiccation of the environment. Longer-term monitoring and experimental research have found that dryland vegetation is more resilient than previously thought to episodic or seasonal grazing pressures typically produced by functioning transhumance systems (Hiernaux & Turner, 1996). It is heavy persistent grazing pressure that is most ecologically damaging, particularly that occurring during the rainy season when both vegetation and soils are most sensitive to livestock-related stresses. It is therefore not surprising that more persistent forms of land degradation are found in more heavily populated zones and localized around permanent settlements (Gorse & Steeds, 1987).

What is clear is that there *is* continual vegetation change in these zones, and that it is important to be able to distinguish the severity and permanence of this change, as well as its causes, in order to develop adequate responses. A growing appreciation of the resilience of dryland vegetation to drought and human pressure, is directing research toward the fertility and structure of soils as more appropriate indicators of land degradation and ecosystem resilience (Thomas, 1993; Dougill & Cox, 1995).

The drier the ecosystem, the more there is an incentive to manage the natural resource communally[7]. In arid lands, *uncertainty* is high, and the risks of production and survival are higher. The risk burden is too much for an individual to bear, therefore common property regimes are devised to share the risk and spread the burden. In addition, the relatively low returns from the arid resource do not warrant the costs of organizing and enforcing more exclusive forms of tenure (Behnke et al., 1993).

Both climatic and anthropic factors are important in determining ecosystem dynamics, and should be taken into account in all management decisions. An ecosystem that may be functioning according to a non-equilibrium theory requires a different management style to an equilibrium system. Land-use patterns must adapt to the variability and uncertainty of rainfall using 'opportunistic', flexible and mobile strategies. Pastoralists respond to the two factors of

---

[7] See for example, papers from the Property Rights and Performance of Natural Resource Systems Workshop, The Beijer International Institute of Ecological Economics, September 1993.

space and time, through the tools of mobility and tracking, not through a static average of primary productivity such as carrying capacity. Very often, it is the quality of the pasture, rather than its quantity that determines production capacity.

As ecological theory is slowly maturing toward a greater willingness to tackle complexity and uncertainty, it has been able to appreciate the pastoralist's informal, innate understanding of dynamic ecological processes. Pastoralists have operationalized this knowledge according to what can be called *socio-ecological pasture units*. These units are not based on production criteria alone, but are also distinguished according to both:

o *ecological variables*: (type of resource, heterogeneity in quality and quantity, variability in time, relative capacity, potential uses), and
o *social variables*: (tenure regime, political status of community holding the property, divisibility of the resource within the unit).

The pasture unit not only provides fodder and water, but also is a tradeable asset through which the group can obtain favours and access to another group's pastures. Of primary importance are the high-value units, or 'key sites', such as water points and immediately surrounding pastures, salt licks, riparian areas, etc. (Niamir, 1994a; Scoones, 1994). Mobility allows pastoralists to use these high-value patches at crucial steps along the transhumance route.

The resource users

KEY WORDS: heterogeneity; ITK (indigenous technical knowledge system); social capital; reciprocity; interdependence; political alliance

In considering development practice more focus should be placed on determining the diversity and *heterogeneity* of local systems, in order to design institutions, laws and policies that accommodate different needs. Normal science has a propensity for aggregation, which then washes over important differences that determine the socio-political dynamics of any society. The élite have different expectations from those of the 'peasants'. Men and women control different functions. There are always minorities within a community, whether they are of a different ethnic background, or whether they are newcomers. Greater economic mobility in and out of pastoral communities has contributed to a greater diversity in needs, expectations, and power relationships. These factors need to be taken into account by any participatory project or programme that wishes to develop a sustainable system. They also determine the contents of the 'reservoir' of *indigenous technical knowledge systems* (ITK) that can be depended upon to develop new and innovative solutions and activities.

Despite the emergence of 'individualistic behaviour' among pastoralists (see, for example, Chapters 7 and 10 in this volume), most pastoral groups, whether homogeneous or not, share a 'sense of community'. This encompasses four aspects (Ostrom, E., 1990):

o the sense that they must share a *common* future
o the capacity to develop mutual trust
o the capacity to communicate with one another
o a set of shared norms, that can be seen as social capital.

With the breakdown of the 'community', any one of these elements is liable to change. Recently, there has been greater attention paid to 'building *social capital*'[8] in the context of local capacity building for natural-resource management. Without social capital, it is argued, there is less collective discipline to engage in and respect communal decisions. Social capital can include cultural and religious mores and values, social norms, ITK, perceived duties and responsibilities related to kinship bonds, and conflict-management mechanisms.

Customary leadership provides one of the mechanisms with which the social capital and collective discipline are kept viable. The 'mass' of leaders available to perform different functions is usually quite extensive in pastoral systems because of the need for opportunism and relative independence. Leaders are found for political roles, scouting and monitoring, group herding, negotiating access to pastures, etc. Leaders and their supporting institutions provide the process through which ITK is translated into collective action.

In most pastoral systems today, social capital is maintained through another important mechanism, that of *reciprocity*. Reciprocity is a finely crafted institution that replaces monetary transactions in cases where the economic value of the material being exchanged is difficult to calculate, and/or where the benefits accruing from the exchange are enjoyed by more than one person. Rarely written, the knowledge of the frequency and quality of favours, their provenance and forms of disbursement, is an art in itself. Reciprocity is the medium through which interdependence among individuals and groups is established and maintained.

This sense of community, and its accompanying social capital, is extended to other groups, whether pastoral or not, in the form of kinship bonds, reciprocity of group favours, *interdependence*, and *political alliances*. Almost all pastoral communities have established symbiotic relationships with farming communities along their transhumance routes, in terms of economic, social and ecological interdependence. For example, post-harvest residue is used in exchange for taking villagers' herds on transhumance, as in Burkina Faso, or in Nigeria, where the Fulani maintain a patron/client relationship with non-Fulani communities, sometimes as dominants, and other times as equals within an economically dependent relationship (Wilson, W., 1995). With recent trends in spontaneous settlement (partial or complete settlement), most pastoral communities have more access to agricultural products than before. In the case of partial settlement, symbiotic relationships between the settled and mobile units are very strong. In areas of

---

[8] For example, see the series of papers presented at the Workshop on Participatory Natural Resource Management, April 6–8 1998 at Manchester College, Oxford, organized by Roger Jeffery of the University of Edinburgh (Centre for South Asian Studies) and Bhaskar Vira of the University of Cambridge (Department of Geography).

encroachment by non-pastoralists, symbiotic relationships have been eroded in many cases and replaced by competition and conflict.

In the past, when each tribe had greater autonomy than now, political alliances were important covenants that assured security and political strength. In some cases, such alliances were extremely fluid, such as among the Nilotes of East Africa (see Chapter 7, this volume), and in other cases, they were more permanent, such as among the Fulani of West Africa (Chapters 4 and 5, this volume). With the advent of post-colonial domination by a central state, such political alliances either became irrelevant (where the State was strong, such as in Mali and Senegal), or alternatively were able to function as before in cases where the State was unable to exert a strong influence on the periphery, such as in Morocco and Mauritania. The value today of political alliances as a socially binding mechanism rests on two conjunctive forces: the internal political strength and consistency of a group, and the degree to which the central State can exert its influence on the pastoral group. The more the internal political structure has broken down, and the stronger the central State, the less important is this mechanism in determining local politics. In Sudan, these political alliances have become a measure of defence by the periphery (pastoralists) against a weak central State.

Adaptive strategies

KEY WORDS: opportunistic mobility; tracking; micro-mobility;

macro-mobility; negotiation;

indigenous communication; safety nets

Recently, various studies are showing that mobile production systems in Africa appear to be more economically efficient than sedentary systems, even more so than commercial ranching. If flexible access to different habitats and resources is assured, higher populations of herbivores can be maintained in any given area (de Ridder & Wagenar, 1984; Westoby et al., 1989; Scoones, 1993). Studies in Zimbabwe, Botswana, Uganda, Mali and elsewhere show that overall returns per hectare (counting all products, not just meat) are higher in mobile pastoral systems than in agropastoral or commercial systems (see Sandford, 1983; Scoones, 1994, pp.12–13). However, productivity per animal is lower (primarily because of the lack of external supplementation and low veterinary input).

The mobility paradigm contends that relatively higher stocking rates can be achieved on arid rangelands if the animals are allowed full and *opportunistic mobility*[9] (Behnke et al., 1993). This argument runs counter to the established position that land degradation in Africa is due solely to overstocking. In many countries, overall livestock numbers have actually decreased since the last century, while pastoral populations have increased, thus reducing the economic

---

[9] The term 'opportunistic' is used in this context not in its negative sense (exploiting opportunities without regard to ethical or moral principles), but in the positive sense of taking advantage of opportunities as they present themselves.

36

viability of the individual household (e.g. Chapter 7). At the same time, total rangeland area has also decreased both in quality and quantity. The new mobility paradigm argues that the same or higher stocking rates than now can be maintained if the animals are allowed to be effectively mobile.

Resource-users have evolved adaptive mechanisms that allow opportunistic use of variable resources. One important mechanism is that herders scout and *track* ecological variability, both spatially and temporally, by constant monitoring and adjusting the behaviour of their animals (Scoones, 1994; Niamir, 1997). In a sample of Fulani herders in northern Nigeria, more than half of respondents said they depend on specialized scouts for news on pastures, the rest either on information passed around at markets or by other passing herds. Only a very few (3 per cent) discover this information by themselves (Adepetu et al., 1988). Fulani scouts in northern Nigeria always double check their information: the first visit evaluates the suitability of the pasture, in order to choose the itinerary, but a return visit at the last minute is necessary to warn of any changes (Ezeomah & Egbe, 1988). Tracking is possible if there is freedom of movement, and specialized labour and talent for tracking and evaluating ecological processes. Scouts must monitor indicators that are sensitive to ecological changes. The indigenous African indicators are more sensitive to the variability in the ecosystem, than to its condition at any point in time (Niamir, 1997). They must be supported by an institutional structure at the local level that can act upon the information received, and there must be free flow of information among different groups through scouts, herders, leaders, traders and other itinerants.

The control of *micro-mobility*, or daily movements around the village or camp, uses information derived from tracking, and results in a manipulation of the effective stocking rate over space and time. Micro-mobility is controlled by herders through four main variables:

o length of continuous grazing on the same pasture or patch
o frequency with which the same patch is re-grazed
o dispersion of animals and herds in the pasture and around the camp
o interval during which the patch is rested.

Management strategies and scenarios are developed based not on average-situation calculations (as with the scientific concept of 'carrying capacity'), but on how best to buffer production against variability and uncertainty. Specific strategies such as herd rotation, deferment, and seasonal or annual reserves, are incorporated through publicly transparent decisions (Niamir, 1990).

The variable environment necessitates continuous re-allocation of access rights that define *macro-mobility* (or long-distance routes and seasonal grazing areas). Traditional African processes of *negotiation* are used in three basic kinds of resource-management decisions: conservation, regulation and allocation. In Western resource-management situations, these three decisions are usually made by non-users for the benefit of users, but in traditional African systems, the two are the same. In the Western model, allocation (or re-allocation) decisions are usually taken only after a need arises (e.g. periods of physical or economic scarcity). By then it is a zero-sum game because user expectations are always higher than what is possible (Hanna, 1995). However, that is not the case in pastoral systems, because allocation decisions are being taken almost continuously within an atmosphere of negotiation and compromise.

Pastoral systems do not exist in isolated niches: they depend fundamentally upon interaction and communication between pastoral and non-pastoral systems. *Indigenous communication* systems, based on a relatively large set of local disseminators and non-electronic channels, are quite efficient at passing on useful information and weeding out 'junk mail'. The content of the communications is relevant to daily use (rangeland opportunities, scouting, marketing, epidemics), to decisions concerning coordinated movements between herding units and between tribes and larger groups, and to decisions concerning admissibility of outsiders.

Pastoralists have developed *safety nets* to buffer the effect of an unpredictable environment, devastating droughts and other stresses such as epidemics. Those safety nets that depend on cohesion of the larger group are fast disappearing, such as restocking schemes, in-kind credit, and village-level insurance customs. Those mechanisms that depend on cohesion among a smaller group (for example, in a group of households, or among kin) have largely survived the transitions – for example, reciprocal exchanges, shared labour, split herds, etc. In the absence of banks, credit unions, insurance companies, and other modern forms of risk management, traditional safety nets are important mechanisms for long term sustainability.

Common property regimes

> KEY WORDS: *common-pool resources; nested property; fluid boundaries; inclusive rights; transboundary resources; informal institutions; co-management; conflict management; popular enforcement*

The term 'common property regime' (CPR) is used here to refer to the management regime or institutional structure/function that is used to manage the natural resources, or *common pool resources*, to which the members of the group have common property rights. Management regimes can range from *laissez faire* (or individual), to informal common property institutions, and to formalized state ownership. The term 'institution' is used in the literature in two senses. Project reports and other 'practical' literature use it to mean an organization with a structure and function, while in academia it can be used in a wider sense, to include rules, social norms, and regulations. Marriage, for example, can be defined as an institution. It is in this latter sense that the term will be used here.

A discussion of CPRs is always preceded in the literature by a reference to the 'tragedy of the commons' (TOC), a term which was coined and described by Hardin (1968) but has its roots in Aristotelian philosophy (Ostrom, E., 1990). Hardin's tragedy represents a special case of the prisoners' dilemma, popular among game theorists for explaining the divergence between individual rationality and social optimality. The tragedy is seen as resulting from the fact that individuals enjoy the benefits of the commons property regime while the group pays the full cost of the individual's behaviour. As a result, there are strong incentives for free-riding. This argument has been applied to many diverse situations in natural-resource management. Although Hardin's example uses the term 'open to all' (i.e. open access), subsequent authors have applied his arguments to all forms of common property rights. It has come to symbolize the degradation of the environment to be expected whenever many individuals use a scarce resource held in common.

It is only recently that researchers have shown that managed commons do not follow this conservative dictum (Feeny et al., 1990; Ostrom, E. et al., 1992; Lane & Moorehead, 1993). In the African debate, critics point out that traditional common pool resources often were well regulated through common property institutions, but were mislabelled as open-access situations by colonial and post-colonial governments. Tragically, the disruption of indigenous political structures and the nationalization or privatization of these resources have in many cases led to situations that approximate open-access situations (Ostrom, E., 1990; Peters, 1987). In these cases, tragedy misdiagnoses have resulted in true tragedies of the commons (Peters, 1987).

The term CPR is particularly useful because it amalgamates the issues of property rights, with that of institutions for managing those rights. Experience to date from local-level natural-resource management projects shows that, in practice, projects tend to concentrate more on building the structure rather than the function of management regimes. This leads to overly static community organizations that rarely are able to achieve the project's self-imposed objectives, let alone leave behind a sustainable result. Projects usually are unable to tackle the issue of property rights, for several reasons. Very often, customary tenure systems are too complex, dynamic, and disaggregated to be easily comprehensible to the outsider, or worse still, have already broken down. Furthermore, the State is not able or ready to alter its legal instruments, and projects in isolation cannot hope to bear the necessary pressure to alter them. A clearer understanding of CPRs, the property rights and management regimes, would help design more appropriate short-term projects and long-term programmes in this regard.

Property rights The property rights regime called the 'commons' is defined as that which is 'owned by an identified group of people, which has the right to exclude non-owners and the duty to maintain the property through constraints placed on its use' (Hanna et al., 1995). The important notions are 'right' and 'duty', two sides of the coin of governance. Common-pool resources, because of the difficulty or high costs to divide, exclude, or bound them, are often considered as common property (Ostrom, E., 1990). The productivity of arid and semi-arid lands is both marginal and variable, and therefore these areas have a benefit–cost ratio that discourages investment in exclusionary, private, mechanisms (Dyson-Hudson & Smith, 1978; Bromley, 1989; Ostrom, E., 1990).

Another reflection of the relatively high exclusion costs associated with range-land resources, is that pastoral groups obtain access to a broader range of resources to fulfil their needs, through controlling key resources or access points. For example, controlling access to water points or other key resources along a transhumance route will confer *de facto* control over a broader area of pasture. Resource scarcity, outside interests, and the local geographic features influence the nature of the territoriality of pastoral tenures. Tenure in the Sahel is largely 'point-centred' (see Chapter 5), i.e. based on the ownership and access rights to resources in and around water points, salt licks, markets, and seasonal wetlands. In East and South Africa, other resources can also define the boundary and content of tenurial packages, such as rivers, hills, craters, etc. In North Africa, tenure is also defined by altitude, each group trying to own or negotiate rights to both low and highlands.

Institutions governing access to a particular pastoral resource therefore cannot be understood without an understanding of the production system through which it is accessed (Berry, 1989; Leach et al., 1997). Given the costs of exclusion, access to a particular resource may not be governed by institutions specific to it, but by institutions governing the resources that are necessary to reach and utilize it. These institutions may be nominally tied to different resources, sometimes distant from the resource in question.

Customary land tenure is often composed of a hierarchy of *nested property rights* (Peters, 1987; Ostrom, E., 1990; Vedeld, 1993b). A sovereign might grant land to a chief who then will allocate cultivation rights to individual households, while giving management control over pastoral areas to a sub-chief, clan head, or client. Individuals may gain more exclusionary rights by investing their labour in the development or maintenance of water points or other specific resources (e.g. beehives, salt licks). Therefore, most rangelands are mosaics of more private, common, and open-access resources as mediated and reinterpreted by local political systems.

Since pastoralists are often required to move over vast distances, they must negotiate the boundaries of authority over resources that are owned versus those that are used, much more so than sedentary cultivators. Because of their mobility, pastoralists leave much of their land seasonally 'empty'. This has given the excuse to individuals or governments to expropriate communal land either out of ignorance or through deliberate exploitation. Understanding these over-lapping authorities, paying careful attention to the distinction between sovereignty, ownership and usufruct, is indispensable for the understanding of pastoral tenure systems.

Unlike countries that follow the principle of nationhood, pastoral communities do not recognize unlimited territorial integrity and sovereignty (Casimir & Rao, 1992). The literature on western CPRs usually assumes that clearly defined boundaries are a necessary condition for long-enduring CPR institutions (e.g. Ostrom, E., 1993). However, that was not a necessity in arid pastoral lands before the advent of colonialism. Pastoral territorial boundaries today are still characterized by the flexibility of their boundaries (Moorehead, 1993; Salzman, 1994; Niamir, 1997; Chapter 5, this volume). *Fluid boundaries* are based on:

- mutual recognition
- acceptance of a system of priority users
- the concept of equitable utilization (as opposed to egalitarian use)
- inclusivity rather than exclusivity
- opportunistic use of boundary resources within certain agreed socio-cultural bounds
- a fluid geographical definition of the territory's boundary related both to ecological exigencies and political power plays.

Pastoral production systems require inter-territorial mobility, and must use transboundary resources, whether these cut across fluid customary boundaries, or fixed administrative and national ones. This requirement, which stems from the high variability of the arid environment, has not been addressed by conventional common property theory. There are important social, management, and economic limitations for territorial expansion, and today, political limitations can also be added to the list. Therefore, pastoral groups must depend on gaining access to resources controlled by others in a seasonal, annual or episodic fashion. Outsiders are routinely granted use rights by owners of pastoral resources (water or fodder). For this reason, pastoral tenure systems tend to be less exclusive than is considered ideal by common property theorists.

The *inclusive* (or porous) nature of pastoral tenure institutions has often been misread as evidence for the lack of institutions governing resource access – e.g. a resource open to all. Such conclusions confuse a lack of rigid exclusion (a defined membership) with the lack of exclusionary powers. In fact, outsiders can only use resources with the permission of the usufruct owning group. Even in pastoral situations where the ethic of hospitality is strong (see Chapter 5), costs are borne by the outsider in gaining access to the resource. Therefore, while outsiders may move in and out of a clan's territory, these movements are not free nor outside the control of the host clan. Access rights to both sojourn areas (camp sites) and passage routes are negotiated through appropriate customary institutions. Examples can be found among the Kababish of the Sudan (El-Arifi, 1979), the Twareg of Mali (Winter, 1984), the Zaghawa of Chad (Tubiana & Tubiana, 1977) and the Bedouins of Morocco (Chapter 10 of this volume).

In the case of *transboundary resources*, such as pasture units that straddle tribal lands, or national boundaries, there are only a few notable cases where there is a formal higher authority, such as a federation of tribes, or a regional government, that can control or manage the resource. Therefore, informal negotiation, or alternatively war, become the primary options available to resource managers. Although war has been the only option in certain notable cases (e.g. cross-border fights between Mauritania and Senegal, or Mauritania and Mali, and between Chad and Niger), in most cases, the institution of reciprocity is a powerful incentive that fosters inter-dependence and informal negotiation rather than war.

Imposition of a western paradigm on land tenure onto the pastoral system has greatly perturbed the 'macro' level relationships, although the 'micro' level or daily land-use patterns continue according to traditional informal rules and regulations (Niamir, 1997). At the macro level, the nationalization of land has separated the ownership of land from its users, has divested customary land-tenure of its religious connections to a deity, and has created a situation of 'open access' thus legitimizing uncontrolled use by others, whether pastoral or non-pastoral. It is rare to find generalizations that fit all situations, but in the case of common property in both developed and developing countries, two main trends have contributed to the breakdown of common property management regimes:

o those that increase demand for privatization of land at the expense of common land (population increase, government subsidies and policies, economic hard-ship, droughts)
o those that decrease the moral authority of local leadership and social cohe-sion (central government hegemony, redrawing of administrative boundaries, deliberate destruction of 'archaic' pastoral governance, fragmentation of authority).

Not only has this resulted in a disintegration of customary institutions respon-sible for natural-resource management, but it also has provided the legal mechanism by which proponents of privatization have been able to push through their agenda for land reform resulting in widespread alienation of land. For example, one can cite evidence from Uganda in Bazaara (1994), Kenya in Fratkin (1994), and Namibia in Cox et al. (1998). The rate of land expropriation is so severe now, it has been labelled a 'land crisis'. Most pastoral advocates, including pastoral organizations, are calling for an immediate moratorium on land titling until land rights can be equitably regularized (e.g. Niamir, 1994a; Salzman, 1994).

Management regimes   Pastoral management regimes are dependent on a variety of institutions that work to allocate resources, punish transgressors, and mediate conflict. In the pastoral context, organizations, rules and regulations that are used to manage common pool resources, can be divided into formal and informal ones. *Informal institutions* can be distinguished from the formal by their lack of delineation as laws, statutes, codes or formalized social organizations. Instead, they are often based on 'common-sense' codes of behaviour and social norms that are part and parcel of culture (Ostrom, E., 1993; Niamir, 1997). Informal institutions are continuously created and modified through political and judicial processes, and are based on fundamental shared social principles. They are therefore more open to negotiation and reinterpretation.

Most formal institutions in pastoral areas are identified with those created by the central State, such as private titles, administrative boundaries, or wood-cutting permits. However, much of customary law and society in Africa falls somewhere between the formal and informal (Rose, 1992; Shipton, 1994). Several pastoral groups have indeed developed formal institutions – for example, the Maasina Empire of Mali (see Chapter 4), the Turkana of Kenya (Gulliver, 1975), the Rufa'a al Hoi of the Sudan (Ahmed, n.d.) and the Il Chamus of Kenya

42

(Little et al., 1987). In northern Sudan, chiefs of neighbouring tribes and sub-tribes meet once a year at the end of the rainy season, and before the herders start on their dry season long-distance movements, to negotiate last year's violations and fines to be paid, as well as this year's transhumance routes, grazing area and grazing time. Such yearly meetings are continuing despite the perturbations from the civil war.

Many of the case studies in this book show, however, that it has been these formal customary institutions that are fast disappearing within pastoral communities, not the informal ones. This is partly due to the fact that formal institutions are dependent upon the very societal structures that have defined them, and if these start to disintegrate, then so will the formal institutions. Informal institutions by their very nature are able to change, adapt and survive, because they respond to specific current needs. In fact, many existing informal institutions may be viewed as remnants of formal institutions (e.g. see Artz et al., 1985, on Morocco). Therefore, even if formal institutions governing livestock mobility appear to be weakening, informal ones that define daily mobility patterns appear to have survived, if only for the sake of preventing conflicts between individuals. There is a striking similarity between informal rules across Africa in relation to livestock mobility. These include (Niamir, 1990):

o the concept of 'first come first served' as applied to specific resource patches
o passive coordination or 'choreography' of movements in a desire to avoid other groups or to occupy niches appropriate to one's mixture of animals
o rules on timing and intensity of use of pastures, and
o concentration or dispersion of animals depending on the ecosystem dynamics.

It may be useful to enshrine these informal institutions into formal Codes of Law, backed by State institutions, in order to prevent their disappearance. However, there is a danger of binding them into a too static structure, and therefore not allowing them to adapt to changing needs. Ensuring an appropriate legal and procedural mechanism that allows continual revision and update of these informal rules is not an easy task.

Western developers and analysts tend to look for formal, constitutional institutions in pastoral systems and when not found characterize informal institutions as weak and prone to free-riding. This preoccupation with formal institutions is driven in part by the guidelines provided by common property theory for successful management of common-pool resources (see e.g. Wade, 1987). These include: small and clearly delineated natural resources, small and fairly homogeneous group of users, high overlap between location of the resources and residence of users, high degree of internal social discipline, rapid detection of infringements, and high levels of devolution. The first three of these guidelines appear to be inconsistent not only with the informal nature of pastoral institutions but also with pastoral production itself. Pastoralists are not all homogeneous, and have to use a large area of land, often far from their residence.

Informal pastoral institutions are inherently malleable and as such are more consistent with the inclusive nature of pastoral tenure than more formalized institutions. In such systems, the vulnerability of pastoralists is diminished since

43

they are able to trade a wide range of present and future assets to gain access to a highly variable resource (see Chapter 5). Common pool resources therefore can be managed successfully in part through a reliance on informal institutions.

Ever since the Earth Summit of 1992, there has been a strong momentum towards CPR systems that combine government decentralization with community participation in what has often been called *co-management*. This term is more common in forestry and fisheries literature than pastoral studies (e.g. Berkes, 1995; Borrini-Feyerabend, 1996), but nevertheless is still as applicable. Co-management can be defined as an appropriate sharing of responsibility for natural-resource management between national and local governments, civic organizations, and local community (Leach *et al.,* 1997).

Community-based sustainable-development efforts in this decade have not quite achieved sustainable and functional institutions. Although they have benefited from many important principles, such as participation and decentralization, their efforts have not created the necessary diverse set of institutions (formal/informal, national/local) that could effectively cope with the dynamics of resource allocation and governance.

In the pastoral context, customary institutions rely on negotiation for their maintenance and evolution. However, negotiation is not always successful and conflicts do occur. Environmental variability results in a high degree of conflict and competition between groups of land users, particularly during years of drought and shortage (Cousins, 1996), or because of economic and political gain. Traditional pastoral societies, as mentioned earlier, were not always pacifist. 'Conflict management' is a term used to refer to both conflict prevention and conflict resolution (see Chapter 7). The principles upon which conflict-management are based include: dialogue, consensus, facilitation, reconciliation, arbitration, mediation and adjudication (Pendzich, 1994; Anderson et al., 1996). Conflict prevention is possible through development and enforcement of rules over natural-resource use, collective acceptance of such rules, and continuous negotiation of conflicting demands.

The main objective of conflict-resolution in a traditional system, is not so much to restore the patrimony of the individual, but to restore stability and social cohesion. External conflicts, i.e. inter-group conflicts, in the past were resolved either through mediation of a neutral ethnic group, or through the creation of an *ad hoc* 'parliamentary' body. For example, conflicts among the Afar and Issa of Djibouti are mediated by a group of men and women who have intermarried into each group, thus representing a relatively neutral body with a vested interest in keeping the peace. If external conflicts could not be resolved peacefully, then war was the last recourse.

In the case of internal conflicts, customary judges attempt to maintain a precarious balance between the interests of the individual and the needs of the community. They do not necessarily follow 'precedence' nor a host of detailed legal texts and rules, but enjoy considerable liberty in shaping each decision to the needs of the situation, using a few grand cultural principles or religious codes recognized by the social group (Ouedraogo & Rochette, 1996).

Although many observers are aware of the fact that a 'community' is not a homogeneous entity, it is often assumed that a public airing of conflict is sufficient to create social consensus and solidarity (Mosse, 1994). Formal structures involving representatives of heterogeneous groups do not automatically equalize the power relationships around the decision-making and negotiation tables. Approaches developed in the 1990s, particularly with the management of conflicts over forest resources, move beyond these blunt outside attempts to increase community cohesiveness and, in so doing, could provide interesting methods and guidelines for conflict management in a pastoral context.

Under situations of high intergroup trading of resource access, the question of enforcement of formal rules and informal codes arises. A few traditional pastoral systems have specialized law enforcement units, such as paid or voluntary sentries (Randhir & Lee, 1996), but most systems rely on *popular enforcement*. The key to this informal system is the existence of social norms and conventions that are shared by individual resource-holding groups. Members of the broader society are monitors and reporters of infringements. Enforcement of rules is based on several cultural mechanisms, which are presented here in order of informal to formal (Niamir, 1990; Swallow & Bromley, 1991):

o observance of an informal 'fairness' ethic
o the perceived interdependence between groups, which requires respect of mutual laws
o social ostracism
o prestige from being model members of the society
o the future possibility of reciprocated access rights
o other forms of incentives, such as benefits accruing from successful communal discipline or action
o accountability in political leadership
o public sanctions and punishments
o force on the part of the judiciary and police force.

The very fact that one individual or community exercises its rights, forces others to comply with those rights. Reciprocity is a very powerful tool and motive in pastoral systems for enforcing rules. Not only can access rights be reciprocated, but violation of rules can also be reciprocated (through revenge), thus acting as an incentive not to violate. In the spirit of co-management, conflict management and enforcement of rules should be carried out through a mixture of formal and informal institutions.

Conclusions

In the preceding review, we have shown how the 'classical' paradigm has led to misguided policies and projects. We have discussed the various aspects of the mobility paradigm, by focusing on several key issues: the resource base, the resource users, their adaptive strategies, and their common property regimes.

We have shown that the mobility paradigm presents a quite different framework for analysing these issues. By highlighting the key words associated with each set of factors, we have tried to explain some of the concepts and perspectives introduced by the mobility paradigm.

As yet the mobility paradigm has not moved out of the theoretical realm. Hardly any project in Africa has attempted to implement it in its entirety, although several can be mentioned that are working on parts of it[10]. Therefore, it may be premature to assert that the mobility paradigm is the best suited to contemporary pastoral situations. However, it is probably the only paradigm available to date that not only gives mobile livestock production systems a *raison d'être*, but also tries to redress the imbalance engendered by too much of a focus on intensive production.

The principal downfall of the mobility paradigm is that it is too difficult to implement. Many of its recommendations require profound changes in government, and sometimes customary, institutions. Such changes require time, a valuable commodity in short supply, both to most pastoralists who are under pressure to sedentarize, as well as to most donors under pressure to rapidly disburse and account for their budgets. The paradigm is still seen as the intellectual property of outsiders, although some argue that what it does is merely to translate indigenous pastoral knowledge into a language comprehensible to outsiders. It is probably both: an amalgamation of the ITK of pastoralists, with the scientific justifications and theories of outsiders.

---

[10] For example, the GTZ project PSB/Dori in northern Burkina Faso, is working on mapping overlapping land-tenure systems, on strengthening negotiation mechanisms, and conflict management. The UNSO project in Mali in the Bourgou, is developing a system of 'home bases' and eventually plans to work on strengthening institutions for negotiation of access rights. The GEF-funded project in northern Sudan (Gireigikh) has since 1993 been working with several pastoral communities to negotiate land management and access rights between sedentary populations and transhumants.

Le rôle des institutions provisoires dans la gestion des ressources à propriété commune des terres arides en Mauritanie

The role of transient institutions for the management of common property resources in the arid lands of Mauritania

MOHAMED OULD ZEIDANE

RÉSUMÉ

La Mauritanie est dotée d'importantes ressources pastorales. Depuis huit ans, les effectifs du cheptel ont fortement augmenté grâce à plusieurs années successives de bonnes pluies. Le pays compte cinq grandes zones agro-écologiques qui s'étendent de la zone sahélienne aride à la zone du fleuve Sénégal en traversant deux zones sahéliennes (est et ouest), et enfin une zone côtière. Les principaux systèmes de production traditionnels sont la transhumance pure, la transhumance associée à la culture et l'agriculture sédentaire. En outre, de nouveaux systèmes comme l'élevage 'moderne' et l'élevage extensif (ranching) sont apparus récemment.

Le problème des régimes fonciers n'a rien de nouveau, mais il a été aggravé par la pénurie relative de ressources et la constitution de rapports nouveaux à l'intérieur des communautés et entre elles. Les éleveurs modernes ont perturbé l'ordre établi (système coutumier), car ils n'ont aucun droit traditionnel sur les pâturages. L'existence de plusieurs types de droit (droit coutumier, droit moderne et droit coranique ou *Chariâ*) complique les arbitrages des conflits. Alors que le droit coutumier admet la primauté de la tribu, le droit moderne ne reconnaît à celle-ci la propriété d'aucun bien, sauf si les pouvoirs publics la lui ont accordée. La *Chariâ* ne reconnaît pas l'existence de droits exclusifs sur les ressources naturelles; les terres de pâturage et l'eau sont accessibles à tous et ne peuvent être expropriées. Les régimes coutumiers ont été appliqués dans le cadre de la *Chariâ* grâce à l'acceptation mutuelle des modes de tenure coutumiers sans que personne soit empêché d'avoir accès à la terre et à ses ressources.

La plupart des litiges sont réglés dans le cadre de la tribu. Les accords et les règlements sont réalisés de façon informelle suivant un jeu subtil d'alliances et d'intérêts. Chaque tribu a un conseil ou *Jemaa* qui regroupe les chefs patriarcaux les plus influents. Les décisions sont généralement prises par voie de consensus et respectent donc les intérêts de tous les membres de la tribu.

Les Associations pastorales (AP) qui ont été créées par les projets et les pouvoirs publics depuis quelques années ont pour mission de trouver un compromis entre les différents modes d'accès aux ressources. Elles ne donnent pas de résultats entièrement satisfaisants en raison de leurs grandes dimensions et de leurs responsabilités énormes, ambitieuses et multiples (santé animale, production, crédit, etc.). Il ressort de l'expérience des sept dernières années que les dimensions et le domaine d'intervention des AP doivent être réduits et que celles-ci devraient s'occuper uniquement de la sauvegarde et de la mise en valeur des ressources pastorales. A cette fin, les associations doivent jouir de droits réels sur les ressources, tenant en compte leur précaritées d'existence et leur volatilitées. Elles pourraient ainsi gérer les ressources, contrôler l'accès des tiers et percevoir des contributions en vue de la régénération et de l'entretien des ressources.

ABSTRACT

Mauritania possesses important pastoral resources. Livestock populations in the last eight years have increased considerably due to successively good annual rainfall. There are five major agro-ecological zones in the country, stretching from the arid saharian zone to the Senegal River zone, across two sahelian zones (east and west), and, finally, a coastal zone. The main traditional production systems are: pure transhumance, transhumance associated with agriculture, and sedentary farming. Other new systems have appeared, such as 'modern' breeding, and ranching.

The land-tenure problem is not a new one, but has been exacerbated by the relative scarcity of resources and the emergence of new relationships within and between communities. The modern breeders have disturbed the traditional order, since they have no traditional rights to the land that their animals use. The existence of several types of law (customary, modern, and Koranic or *Chariâ*) complicates the process of arbitration. While customary law acknowledges the tribal primacy, modern law does not recognize the tribe's ownership of any property except that granted by the public power. The *Chariâ* does not recognize exclusive rights to natural resources; grazing lands and water are open to all and cannot be expropriated. Customary systems have worked within the *Chariâ*, by mutually recognizing customary tenure arrangements, without preventing anyone from access to land and its resources.

Most conflicts are resolved within the tribe. Agreements and settlements are arrived at informally, according to a subtle game of alliances and interests. Each tribe has a council or *Jemaa*, which includes the most influential patriarchal chiefs. Decisions are usually taken through consensus, and therefore respect the interest of all members of the tribe.

The Pastoral Associations (PAs) that have been created by projects and government efforts in the past few years, are meant to find a compromise between the different modes of access to resources. They are not entirely satisfactory, due to their large size and their huge, ambitious, multi-sectoral responsibilities (animal health, production, credit, etc.). The experience of the last seven years shows that the size and field of implementation has to be reduced, and the focus of the PAs must be solely on the preservation and development of pastoral resources. For this to happen, the PAs must enjoy real, but not necessarily permanent rights to the resources. In this way they could manage the resources, control access by third parties, and exact contributions for the regeneration and maintenance of the resources.

Introduction

La MAURITANIE, pays sahélien, charnière entre l'Afrique noire et l'Afrique blanche, est à vocation pastorale évidente (Carte 3.1). Près de 70 pour cent de la population vit directement ou indirectement de l'élevage. En outre, la part des productions animales dans le Produit intérieur brut dépasse 15 pour cent et elles constituent près de 80 pour cent de l'apport total du secteur rural.

On estime le cheptel national à 1,3 million de bovins, 1,1 million de camelins et près de 9 millions de petits ruminants (ovins et caprins). Rapporté à la population, ce cheptel permet à la Mauritanie d'avoir le ratio UBT/habitant le plus élevé de la sous-région (environ 1,4). Les productions sont estimées à 400 000 tonnes de lait par an et 70 000 tonnes de viandes rouges, sans oublier les quantités importantes de cuirs, peaux, cornes, onglons, etc., dont la valorisation reste cependant relativement faible.

L'existence d'un nombre aussi important d'animaux d'élevage n'est pas sans

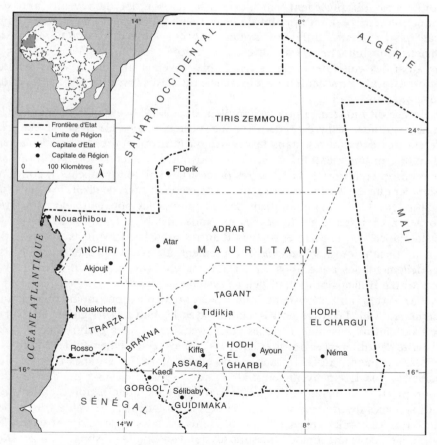

Carte 3.1 *Les régions pastorales de la Mauritanie*

poser certaines difficultés, dont l'une des plus délicates est le problème foncier. La tenure foncière pastorale, si elle pose la même problématique globale partout dans le pays et, probablement, dans tout le Sahel, n'en diffère pas moins suivant les systèmes et les modes d'élevage.

Pour bien appréhender la situation foncière, il importe donc de s'intéresser à ces différents systèmes et modes de production. Il y a en Mauritanie trois systèmes de production dominants: l'élevage traditionnel pur, l'élevage traditionnel associé à une activité agricole (qui se divise en transhumants et sédentaires) et l'élevage dit moderne, en ce sens qu'il adopte de nouvelles méthodes et le bétail y est propriété de certains opérateurs du secteur moderne (fonctionnaires, grands commerçants, etc.).

L'élevage traditionnel pur (transhumance pure) est celui pratiqué par des pasteurs qui n'ont d'autres sources de revenu que l'exploitation de leur troupeau. Ces groupes n'ont, le plus souvent, pas de revendications anciennes sur les terres de parcours exploitées.

L'élevage traditionnel associé à l'agriculture (agro-pastoralisme transhumant) se différencie du précédent par l'existence de sources de revenus annexes (production agricole), en appoint de ceux tirés de l'élevage. Les éleveurs dans ce cas appartiennent à différents groupes, d'origines très diverses. On rencontre d'anciens transhumants purs convertis en agro-éleveurs, suite à la sécheresse, mais aussi des cultivateurs déçus des faibles revenus tirés de la seule récolte. L'agro-pastoralisme sédentaire se différencie du dernier système par l'absence de mouvement du bétail.

L'élevage dit moderne a pris de l'ampleur depuis 1970, avec l'avènement de la première grande sécheresse de l'époque postcoloniale. Ce système est très différent des deux autres, dans la mesure où le propriétaire du bétail n'a pas une culture ni une mentalité d'éleveur et surtout, n'a aucune revendication (au sens du droit coutumier) sur les terres parcourues par son troupeau. Il convient de signaler que la *Chariâ* – loi coranique – ne reconnaît pas de droits exclusifs sur les pâturages. Ce que ne manquent pas d'exploiter les nouveaux éleveurs. Très innovateur, ce système a adopté des méthodes flexibles, tenant à la fois de la transhumance, du *ranching* et de bien d'autres modèles, suivant la nécessité du moment. Néanmoins, le *ranching* proprement dit n'a jamais existé, à cause essentiellement des conditions climatiques sévères (mais aussi du fait que le colonialisme français pratiquait peu ce système).

Enfin, divers autres systèmes agro-pastoraux plus ou moins marginaux existent: au Sud, l'élevage en association avec les cultures irriguées, au Nord, avec l'agriculture oasienne, dans les villes avec l'agriculture périurbaine.

Il existe des différences notoires entre les ethnies (essentiellement les Maures et les Peuls) et même, au sein d'une même ethnie, entre les groupes tribaux. Les spécificités locales peuvent engendrer des types d'élevage et des modes d'appropriation du bétail et des terres très différents d'une région à l'autre.

Les systèmes de production décrits ci-dessus exploitent et gèrent, à travers des institutions pastorales créées par les différents groupes d'éleveurs, les terres, les pâturages et, dans une moindre mesure, les points d'eau. On distingue, en général, deux types d'institutions qui diffèrent par leur finalité et surtout par leur durée:

50

○ les institutions pastorales permanentes dont l'activité est de portée générale. Ces organisations sont, en fait, le démembrement des structures tribales traditionnelles et ont pour rôle de gérer au mieux l'espace pastoral. Elles relèvent des *Jemaa* (assemblées, tribales), mais sont de plus en plus entre les mains des grands propriétaires de troupeaux, ce qui limite la concentration de leur sein

○ les institutions pastorales dites transitoires ou provisoires, structures *ad hoc*, créées le plus souvent pour faire face à une situation d'urgence. Leur rôle est, certes, circonscrit dans le temps, mais elles revêtent une importance particulière, en raison des arbitrages qu'elles mènent entre les différents groupes et parfois au sein du même groupe.

Ainsi, ces institutions sont mises en place quand des litiges fonciers pastoraux surgissent entre communautés utilisant le même espace ou les mêmes points d'eau. A côté des représentants des communautés en litige, des médiateurs sont désignés, avec l'accord de toutes les parties, en raison de leur sagesse, leur impartialité et de leur connaissance des sources de droit et des problèmes fonciers.

Depuis le milieu des années quatre-vingts, suite aux calamités naturelles (sécheresses, maladies du bétail) qui se sont abattues sur certaines régions, beaucoup de groupes ont désigné, de leur propre initiative, des organisations *ad hoc* en vue de négocier, avec les pouvoirs publics locaux et centraux, des sujets aussi importants que la réforme agraire, le code de l'eau, l'aide d'urgence, l'organisation des campagnes de vaccination, etc.

La crise récente des mécanismes d'appropriation et d'exploitation des terres (agricoles et de parcours) a engendré l'émergence de nouvelles institutions pastorales, sur l'initiative des pouvoirs politiques. C'est le cas, notamment, des Associations coopératives pastorales.

Après avoir fait un rappel de l'évolution historique et des enjeux actuels, dans la première partie, il sera procédé, dans la partie deux, à la description des ressources-clés, avant d'aborder, dans la partie suivante, les caractéristiques des usagers des terres pastorales. La partie quatre s'intéressera aux institutions locales et à la tenure foncière. Les deux dernières parties font état du devoir d'accès et des règles d'utilisation des terres pastorales, pour ce qui concerne la cinquième, et des mécanismes de résolution des conflits, pour ce qui est de la sixième partie.

Évolution historique et enjeux actuels

De tout temps, la terre a fait l'objet de convoitises et de conflits entre groupes humains vivant dans la même zone ou partageant le même espace socio-économique. Les terres à vocation pastorale n'échappent pas à ce constat. Alors que dans le passé, les relations au sein et entre communautés étaient régies par des normes et des codes acceptés par tous, on assiste, depuis la colonisation, à une remise en cause quasi totale du consensus social autour de l'usage et de l'appropriation des domaines pastoraux.

La tenure foncière (agricole comme pastorale) a traversé quatre phases distinctes, marquées par des événements socio-politiques majeurs, ci-dessous décrites.

La phase précoloniale

L'instabilité politique et les razzias durant cette période avaient poussé les propriétaires de bétail à faire allégeance aux émirs (suzerains issus de grandes tribus guerrières, ayant une emprise sur des territoires plus ou moins bien définis), qui avaient divisé le pays en sphères d'influence, dont les limites étaient toutefois très floues. L'ordre foncier, imposé par les armes, ne signifiait pourtant pas une quelconque appropriation des terres pastorales par les tribus guerrières (Hassane), peu enclines à l'élevage, activité jugée indigne des combattants et laissée aux mains des tribus maraboutiques (Zewaya) et des tributaires (Zenaga). Les conflits fonciers étaient réglés par le recours à la loi coranique, aux alliances traditionnelles et – ultime alternative – à la force armée. Les propriétaires de bétail devaient payer une dîme annuelle variable et participer aux éventuels efforts de guerre – suivant la décision de l'émir. Aujourd'hui, le pouvoir émirat n'existe plus de manière formelle, même si les descendants des émirs continuent de jouir d'un certain égard de leurs anciens vassaux.

L'époque coloniale

L'administration, qui avait alors une connaissance limitée des droits et usages coutumiers, avait fixé trois objectifs stratégiques aux activités pastorales – les seules pratiquées à grande échelle en ce moment: pourvoir en viande la zone arachidière du Nord Sénégal (à travers les réquisitions de bétail), asseoir l'autorité coloniale en assurant la promotion des groupes tribaux alliés de la France, et favoriser l'émergence de tendances individualistes, nécessaires au développement des échanges monétisés et à la pénétration capitaliste. La colonisation n'a cependant pas modifié fondamentalement les mécanismes traditionnels, auxquels les Gouverneurs de Cercles faisaient parfois recours pour régler des litiges.

Les années de l'indépendance

La loi 60.139 du 2 août 1960 était basée sur les droits fonciers coutumiers, dont l'emprise était particulièrement forte, à un moment où l'État était naissant et synonyme de persécution et de spoliation (car confondu avec le pouvoir colonial). Très vite, les pouvoirs publics, en dépit de déclarations favorables au développement du monde rural, s'étaient focalisées sur le seul secteur minier, qui assurait un excédent confortable de la balance des paiements. Durant les années soixante-dix, il était apparu, aux yeux de l'administration, que cette loi constituait une entrave au développement des régions, l'État ne pouvant pas accéder librement aux terres qu'il désirait mettre en valeur.

Les années 1980

La réforme foncière et domaniale de 1983 consacre l'appropriation unilatérale des terres non mises en valeur (dont, entre autres, toutes les terres à vocation pastorale) par l'État. Celui-ci est censé en distribuer, selon les impératifs du développement social et économique des régions et des zones concernées.

L'un des objectifs principaux de cette loi était de permettre de rentabiliser les investissements du secteur privé, ce qui n'était possible qu'à travers l'accès des promoteurs individuels aux terres situées au bord du fleuve Sénégal, seul source d'eau douce permanente du pays.

Jusqu'à la fin des années soixante-dix, les conflits n'avaient pas connu une grande ampleur. En effet, la pression des populations sur la terre était plus faible, d'une part, et les paysans croyaient que l'exode rural constituerait un recours durable au chômage des ruraux, d'autre part. La pression démographique s'est accentuée et l'exode s'est avéré moins prometteur que prévu en termes de création d'emplois et de relèvement du niveau de vie.

Depuis lors, les ressources pastorales ont été à la base d'un nombre croissant de conflits liés aux convoitises des uns et des autres et à la multiplicité des sources de droit. Pendant ce temps, ces mêmes ressources-clés subissaient une détérioration sans précédent.

Description des ressources-clés

Les ressources pastorales, tout comme le cheptel lui-même, ont particulièrement souffert des sécheresses récurrentes qu'a connues le pays ces deux dernières décennies.

Une étude pluridisciplinaire récente de la FAO[1] a défini cinq grandes zones agro-écologiques: la façade maritime, la zone du fleuve Sénégal (zone soudano-sahélienne), les zones sahéliennes Est et Ouest et la zone saharienne aride.

Cette dernière, la plus importante en terme de superficie (environ les 2/3 du territoire), est aussi la moins arrosée. La pluviométrie, très aléatoire, y dépasse rarement 100mm de pluie par an; il n'est pas rare qu'il ne pleuve pas durant deux ou trois années successives. Des pluies d'hiver tombent dans certains cas, avec des effets très positifs sur la qualité et la pérennité des pâturages (Carte 3.2).

Les oases, couvrant une superficie totale estimée à 5000 ha et dont plusieurs dizaines sont florissantes, jouent un rôle primordial dans la vie des villages environnants et le maintien des populations. Les palmeraies à base de *Phoenix dactylifera* sont dominants et couvrent des sous-étages d'arbres fruitiers, de légumes, de céréales, de cultures fourragères et d'arbustes tels que le henné (*Lawsonia alba*).

Les oueds présentent également une végétation aérienne plus ou moins éparse, composée de *Capparis decidua*, *Acacia raddiana* et plus rarement, de *Boscia senegalensis*. Dans les régions ensablées, on rencontre un tapis herbacé composé de *Panicum turgidum* et de *Stipagrostis pungens*.

La zone sahélienne Ouest couvre environ 7–8 pour cent du territoire. Les sols se caractérisent par une proportion relativement élevée d'humus et contiennent du fer en quantité suffisante (couleur rougeâtre). On distingue la sous-zone

---

[1] Définition d'une politique de l'élevage en Mauritanie, rapport principal, MDRE-FAO, 1993.

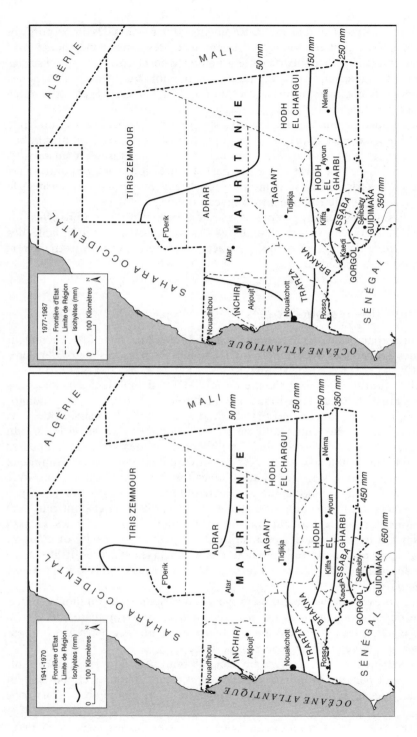

Carte 3.2 *Les isohyètes annuelles de la Mauritanie, 1941–87*

54

argileuse et la sous-zone sableuse, où la végétation est dominée par la savane arborée avec comme ligneux des Combretaceae et comme graminées *Cenchrus biflorus*. Au nord de cette zone, on rencontre *Caliotropis procera*, *Leptadania pyrotechnica* et, de moins en moins, *Balanites aegyptiaca*. Au sud, c'est le domaine traditionnel de *Acacia senegal*, en régression, associé à d'autres types de gommiers.

La zone sahélienne Est couvre environ 10 pour cent du territoire et renferme plus de la moitié des possibilités et ressources sylvicoles et pastorales du pays. C'est la zone de transhumance par excellence. L'agriculture sous-pluie est très présente, la superficie cultivée pouvant atteindre 50 pour cent du total national. La végétation y est semblable à celle de la zone Ouest, avec l'apparition, dans l'extrême Est, de populations importantes quoiqu'en recul suite au déficit pluviométrique, d'*Acacia seyal*.

La zone du fleuve Sénégal couvre près de 2 pour cent du territoire. Les sols hydromorphes sont dominants grâce surtout à la présence de l'eau temporaire ou permanente (dans ce dernier cas, les sols sont compacts et imperméables). Il y a seulement quelques années, c'était le domaine de forêts inondables et de prairies aquatiques, mais avec l'avènement des sécheresses, et surtout avec le défrichement massif en vue de l'agriculture de décrue, les possibilités fourragères ont considérablement diminué. Les cultures irriguées ont pris une grande ampleur depuis 1984, grâce à l'afflux des promoteurs privés dû aux incitations de l'État.

C'est dans cette zone que l'on rencontre encore les quelques forêts classées, que n'épargnent d'ailleurs pas les charbonniers, en dépit de la réglementation en vigueur. En vue de préserver la diversité éco-biologique de la vallée du fleuve, il a été créé un parc, le Parc national de Diawling (qui constitue, avec le Parc national du Banc d'Arguin et les forêts, les seules aires réservées). L'accès y est interdit aux éleveurs transhumants, aux chasseurs et autres coupeurs de bois.

La façade maritime, enfin, est composée d'une bande large de 50km et s'étendant le long de la côte mauritanienne. Les sols sont constitués en sebkhas pratiquement dépourvues de végétation et impropres à la culture. En revanche, la présence de salsolacés et de *Tamarix senegalensis* rend cette zone attractive pour les dromadaires. Ce territoire est le domaine traditionnel de certaines tribus de pasteurs (et pêcheurs occasionnels), qui se déplacent avec leurs chameaux le long de la façade, en parcourant le Trarza, la région de Nouakchott, et les villages Imraguen.

Le Parc national du Banc d'Arguin, véritable réserve naturelle d'oiseaux, situé entre Nouakchott et Nouadhibou, allie à la steppe désertique, la végétation côtière.

L'approche méthodologique a consisté à effectuer des études de cas sur les sites les plus représentatifs, à la fois sur le plan physique et sur le plan humain. Pour tenir compte de la diversité climatique et humaine, quatre sites ont été retenus (Carte 3.3):

o Khweiwira, dans la région du Hodh El Gharbi, au sud-est du pays, point de
  rencontre de groupes transhumants et de groupes sédentarisés depuis
  longtemps. Des liens de solidarité existent entre ces groupes, qui appartiennent

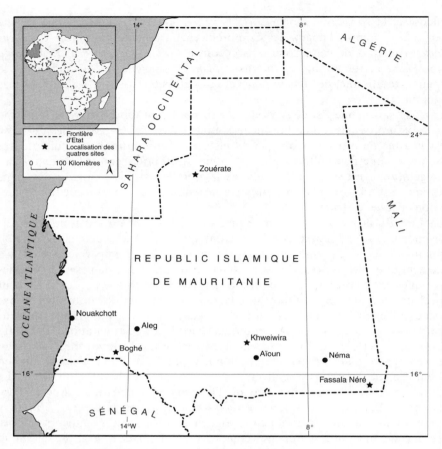

Carte 3.3 *Localisation des quatres sites (Mauritanie)*

parfois à la même tribu. La localité de Khweiwira appartient à la zone
sahélienne Est. Elle est entourée d'une savane sablonneuse; les dunes
mouvantes ont fait leur apparition ces dernières années, ce qui gêne
considérablement la croissance de la végétation et menace d'ensablement
les habitations. La zone, qui était située sur l'isohyète 300mm pour la
période 1941–70, est située entre les isohyètes 150 et 250mm (pour la période
1977–87). Les cinq dernières années ont été, néanmoins, plus favorables. Les
ligneux, assez vigoureux dans les ravines des collines, se sont régénérés à la
faveur des conditions favorables depuis 1987. *Acacia raddiana*, *Acacia
ehrenbergiana*, *Acacia senegal*, *Bauhinia rufescens Grewia bicolor* et *Ziziphus
mauritiana* se retrouvent en association, dans les ensablements des ravines,
avec des graminées annuelles telles que *Aristida adscensionis* et *Schoenefeldia
gracilis*.

o La zone de Fassala Néré, dans la région du Hodh El Charghi, à l'extrême
Sud-Est, située sur la frontière avec le Mali. Fréquentée traditionnellement

56

par des groupes transhumants, cette zone connaît une concurrence sévère entre ceux qui se sédentarisent (donc devenus cultivateurs) et ceux qui continuent à nomadiser. L'absence de règles traditionnelles et de mécanismes de régulation est forcément un frein à une gestion communautaire des terres. Cette situation se trouve compliquée par l'instabilité liée à la présence de camps de réfugiés maliens et au vol de bétail. Le problème de la sécurité en zone pastorale, que l'on déplore dans d'autres régions d'Afrique et du monde, trouve ici une parfaite illustration.

Fassala Néré, située sur l'isohyète 250mm, appartient également à la zone sahélienne Est. Elle présente dans sa partie sud une végétation pouvant être assez dense (avec des groupements d'*Acacia senegal* et *Camiphora africana* – *Adress* en hassania, dialecte local) et dans sa partie nord un paysage ligneux plus clairsemé à base de *Balanites aegyptiaca* et de *Stipagrostis pungens*. Le peuplement végétal comprend également des espèces vivaces, surtout *Panicum turgidum*.

o La zone de Zouérate, au Nord, où les conditions climatiques ne permettent que l'élevage de dromadaires. La particularité ici vient de l'extrême rareté des pluies (moins de 50mm en moyenne par an) et des points d'eau, qui, de ce fait, font l'objet d'une grande concurrence et constituent un moyen efficace de contrôle de l'espace. Durant les rares années de bonne pluie, cette zone est fréquentée par un nombre très important d'animaux, dont une partie appartient aux nouveaux éleveurs (dits modernes).

Mais pendant les années normales (pluviométrie inférieure à 70mm), les coûts liés à l'activité sont tellement élevés comparativement à la production, que seuls des éleveurs ayant des moyens importants peuvent espérer garder leurs troupeaux autour de Zouérate. Certains des pasteurs traditionnels demandent, jusqu'ici en vain, l'aide des pouvoirs publics, sous forme d'aliment pour bétail.

Zouérate et tout le Tiris Zemmour font partie de la zone saharienne. Dans la région 'pastorale' de Maqteir, on rencontre une végétation à *Stipagrostis pungens* (*Sbatt* en hassania), associée à des annuelles et plantes éphémères, comme *El hadh, Lehbalia* et autres plantes qui apparaissent principalement à la suite des pluies d'hiver. Ces plantes, dont certaines ont un goût salé, sont bien appréciées par le cheptel camelin.

o La zone de Boghé, au Sud-Ouest, au bord du fleuve Sénégal. Région d'agriculture irriguée, on y constate, à mesure que la riziculture progresse, des difficultés d'accès des transhumants aux terres de bas-fonds, traditionnel-lement exploitées par les animaux (présence de bourgouttières).

Boghé, avec en moyenne 200mm de pluie, appartient à la zone agro-écologique du fleuve. On y rencontre encore certains peuplements d'*Acacia wilotica* (*Amour* en hassania), présentant des conditions pastorales globalement satisfaisantes, en dépit d'une baisse pluviométrique enregistrée les trois ou quatre dernières saisons des pluies.

Les zones agro-pastorales ci-dessus décrites sont confrontées, à des degrés certes divers, au problème de surexploitation. Dans les estimations, pour 1989, a une quasi-égalité entre la taille du cheptel (2,7 millions d'UBT) et la charge maximale que peut supporter le couvert végétal (2,8 millions UBT): on risquait donc, à tout moment, une rupture de charge.

S'il est vrai que la situation est précaire, il convient, cependant, de considérer avec beaucoup de réserve la notion de capacité de charge dans les systèmes non équilibrés comme c'est le cas en Mauritanie, en raison de la forte variabilité du climat et des stratégies adoptées par les éleveurs. Ces derniers optent le plus souvent pour un usage 'opportuniste' de terres de pacage, probablement très efficace (Behnke et al., 1993), à cause de la forte dispersion des ressources pastorales dans l'espace.

Des études menées par Breman et de Wit en 1983 en Afrique de l'Ouest (cité dans Behnke et al., 1993) avaient montré que les stratégies des pasteurs transhumants pouvaient dépasser les rendements obtenus dans les ranchs commerciaux. La stratégie opportuniste consiste à s'adapter rapidement aux changements pouvant survenir, grâce à la mobilité et au destockage ou restockage, en fonction des conditions climatiques. La notion de capacité de charge (au sens écologique comme au sens économique) est peut-être valable dans les fermes-prairies humides, où la zone et la durée de pacage sont bien définies. Elle ne l'est sans doute pas s'agissant des régions non équilibrées.

Les caractéristiques des usagers des terres pastorales

En fonction des zones agro-écologiques, des activités économiques, des mutations sociales et des liens entre les groupes concernés, se fait l'usage des terres pastorales.

A Khweiwira vivent des agro-éleveurs (une centaine de familles) et des éleveurs transhumants (une trentaine), tous appartenant à la même tribu de Ideibussat. Les transhumants sont présents uniquement pendant quelques mois de l'année, entre mi-juillet et début novembre, ce calendrier pouvant varier suivant la pluviométrie. A côté de cette localité, créée en 1961 mais réellement habitée à partir de 1975, il existe de nombreuses autres localités de création plus récente et habitées par les tribus de Tenwajiw et Laghlal, sur des terres leur appartenant traditionnellement. Malgré la diversité tribale et les conflits ancestraux, chaque groupe reconnaît à l'autre son domaine, sur lequel il ne peut empiéter sans autorisation préalable de la *Jemaa* concernée, autorité traditionnelle.

Le village de Khweiwira a construit, il y a deux décennies, un barrage à vocation agricole. Chaque année, en fonction du nombre de ménages, on procède à la délimitation et au tirage au sort des parcelles à allouer aux différentes familles. Les transhumants, qui sont pourtant absents pendant la majeure partie de l'année, ont le même droit que les agro-éleveurs constamment présents, parce qu'issus de la même origine tribale, et fréquentaient, depuis des temps immémoriaux, le même territoire. La force des liens de solidarité donne tout son sens à la notion de communauté, qui dépasse le simple cadre du barrage: les 'lois' et règles coutumières valent pour tous et pour tous les domaines de la vie communautaire. Même le départ des familles transhumantes avait été concerté et admis par tous.

Fassala Néré, plus arrosée, attire un nombre très important de pasteurs, dont certains viennent du Mali et de régions éloignées de Mauritanie. Les populations habitant plus au nord de cette localité, chassées par l'avancée du désert et la pauvreté des sols, se sont déplacées à Fassala Néré, pour y pratiquer les cultures sous-pluie. Il s'en est suivi, sur quelques-unes des terres non physiquement occupées, une compétition entre activités agricoles et pastorales, que les mécanismes traditionnels de régulation ne sont pas toujours parvenus à régler.

A Khweiwira comme à Fassala Néré, le système socio-politique traditionnel est dominé par l'existence de pouvoirs locaux exercés à travers les chefs de tribus, sur la base des concertations faites par les *Jemaa*, assemblées traditionnelles, où se côtoient les représentants de toutes les fractions tribales. Au sein de chacune des fractions, les décisions sont prises en commun, chaque chef de famille donnant son point de vue: au bout des discussions, une position est adoptée, que tout le monde se doit de respecter. A titre d'exemple, l'époque de départ en transhumance, le lieu de destination ainsi que les personnes devant se déplacer avec les troupeaux, sont autant de sujets abordés au cours de ces réunions.

En dépit d'une organisation sociale différente, les éleveurs peuls, majoritaires dans la zone du fleuve Sénégal et à Boghé, adoptent des formes de concertation assez proches, les décisions étant prises au nom de la communauté par les plus âgés (en général, le chef de village est désigné parmi les plus âgés – les plus sages). A la différence des éleveurs maures, les peuls ne font pas l'élevage des dromadaires, bien qu'ils les utilisent comme montures. Depuis les dernières sécheresses, les pasteurs dans les alentours de Boghé associent de plus en plus l'agriculture à l'élevage pour disposer de sources supplémentaires de revenus. Dans cette zone riche en pâturages, les déplacements sont de faible amplitude, sauf année excessivement sèche. En revanche, de nombreux transhumants sont attirés par l'abondance et la multiplicité des espèces fourragères.

Dans la région du Tiris Zemmour à l'extrême nord du pays, et à Zouérate en particulier, les conditions climatiques sont particulièrement sévères, avec des pluies aléatoires et erratiques. Les populations autochtones sont traditionnellement des éleveurs de chameaux. L'espèce bovine ne peut vivre ici, en raison du manque d'eau (les dromadaires peuvent faire jusqu'à cinq jours sans boire en saison sèche, ce qu'aucune vache ne peut évidemment supporter) mais aussi de nourriture.

L'adaptation durant des siècles du pasteur mauritanien à une nature aussi hostile traduit sans doute une connaissance et une maîtrise assez appréciables de son environnement physique. Même si la symbiose n'est plus vraiment établie entre les éleveurs d'aujourd'hui (fils des nomades d'hier) et la nature, on ne peut que s'étonner devant le capital énorme de connaissances dont ils disposent. Cela commence par la toponymie: chaque endroit, chaque arbuste, chaque dépression, chaque versant de montagne, etc. a un nom, généralement emprunté à l'atonomie. Pour se repérer, l'éleveur se sert de son sens aiguisé d'observation, le jour, et des étoiles, la nuit. L'état du ciel permet également d'appréhender les conditions pastorales de la saison suivante et d'anticiper les comportements à adopter pour éviter une éventuelle catastrophe.

Dépendants étroitement des pâturages, les pasteurs en connaissent parfaitement les propriétés; chaque plante, chaque arbuste, chaque arbre se trouvant dans leur zone de pacage possède un nom local. C'est ainsi que les animaux sont guidés vers les bonnes terres, dont les pâturages sont les meilleurs: les bergers sont choisis parmi les plus attentifs à la nature. Les plantes comestibles sont également bien connues et servent d'appoint alimentaire (*Ziziphus mauritiana, Grewia bicolor, Maerua crassifolia, Boscia senegalensis,* etc.) et sanitaire (gomme arabique, tirée à partir d'Acacias).

Contrairement à beaucoup de préjugés, les vrais pasteurs ne détruisent pas la végétation, même si l'usage en est ouvert à tout le monde: les éleveurs traditionnels connaissent trop bien la nature pour se rendre compte qu'une exploitation irrationnelle de la part de quelques-uns seulement peut entraîner une réduction des possibilités fourragères et donc, une diminution de leur cheptel ou une augmentation insoutenable de l'amplitude de leur mouvement. Selon certains éleveurs interviewés dans les environs de Fassala Néré, 'Avant ces dernières années, nous adoptions une stratégie consistant à compenser la fragilité de notre écosystème par des déplacements plus lointains, pour éviter de trop solliciter la Nature. Malheureusement, l'émergence de nouveaux systèmes d'élevage tend à perturber l'équilibre qui s'est perpétué pendant des décennies dans cette région.'

Les éleveurs traditionnels des quatre sites retenus ont tous de grandes connaissances de leur milieu, même s'il existe de différences entre Peuls et Maures, par exemple dans le domaine de la conduite des troupeaux.

Des différences plus marquées existent quant à la solidité des liens communautaires traditionnels ou coutumiers. Autant la situation est restée quasiment inchangée à Khweiwira et Fassala Néré, autant les choses ont considérablement changé à Boghé et Zouérate. Dans les deux premières localités, le sens d'appartenir à une même tribu continue de constituer un fort ciment pour tous les pasteurs quelle que soit la stratégie adoptée par les différents sous-groupes tribaux (fractions). Des signes d'évolution sont toutefois perceptibles, notamment à Fassala Néré, avec l'arrivée massive de réfugiés Touaregs et de leur bétail.

Les éleveurs se trouvant dans la région de Boghé font face, quant à eux, à une rude concurrence sur les terres situées non loin du fleuve Sénégal. L'attrait exercé par les eaux de ce fleuve a profondément bouleversé les donnes foncières prédominantes, avec notamment le développement de la riziculture mécanisée. A cela s'ajoute l'installation de nombreux éleveurs venant du nord du pays, chassés par la sécheresse, ainsi que de nombreux Mauritaniens rapatriés suite aux événements malheureux survenus entre la Mauritanie et le Sénégal en 1989[2]. Il s'en est suivi un certain relâchement des liens entre autochtones, qui ne se traduit néanmoins pas par une dislocation du système traditionnel, mais par une

---

[2] Des incidents regrettables entre agriculteurs et éleveurs des deux pays ont entraîné des règlements de compte entre les communautés dans chacun des pays. De nombreuses personnes ont payé de leur vie et des dizaines de milliers expulsés de part et d'autre du fleuve Sénégal.

adaptation au nouvel environnement: pour rester en indivision, les membres de la communauté se constituent le plus souvent en coopératives, comme le prévoit la loi.

L'évolution a été plus lente mais plus radicale à Zouérate et dans le Tiris Zemmour dans son ensemble. Dans cette zone, seul l'élevage de camelins (et accessoirement de caprins) peut-être envisagé. C'est ici qu'on rencontre les plus grands troupeaux de dromadaires – quand les conditions climatiques le permettent. De puissants commerçants et fonctionnaires retraités disposent, depuis les dernières sécheresses, de l'essentiel du troupeau camelin de toute la Mauritanie. Ils font du Tiris Zemmour le lieu principal de pacage de leurs animaux à cause de la qualité des pâturages (les meilleurs de tout le pays, quand il y en a), et surtout à cause de l'absence de problèmes fonciers majeurs (pas d'agriculture, peu de points d'eau, etc.). Chaque grand propriétaire ou groupe de propriétaires de bétail construit ses puits et assure, d'une certaine manière, le contrôle des pâturages environnants. Ce contrôle, qui ne correspond pas à une appropriation des terres de pacage, est néanmoins reconnu par la *Chariâ*, qui autorise sous certaines conditions faciles à remplir, la possibilité d'un usage exclusif du *harim*, c'est-à-dire du voisinage immédiat du puits.

Les éleveurs traditionnels peuvent évidemment bénéficier de la même disposition; ils n'ont cependant pas le moyen de creuser des puits, dont la profondeur peut aller jusqu'à 50 mètres. Ces dernières années, une société d'État implantée à Zouérate, la SNIM[3], a régulièrement approvisionné en eau quelques petits propriétaires, le long d'une ligne de chemin de fer reliant Zouérate et Nouadhibou. Mais cela ne permet pas de régler le problème d'alimentation en eau du cheptel.

Les populations autochtones étaient jadis transhumantes, mais, avec la sécheresse et l'instabilité liée à la guerre du Sahara Occidental (entre 1975 et 1978), bon nombre ont émigré, avec ou sans leur troupeau, en Algérie ou vers d'autres régions de Mauritanie. Aujourd'hui encore, les chameliers évitent de se déplacer au nord de Zouérate, de peur d'éventuelles razzias par les troupes sahraouies, qui font de cette zone un lieu de pacage pour leurs animaux durant certains mois de l'année.

Les relations entre les administrations régionales et locales sont caractérisées par la méfiance du côté des populations et par la prudence du côté des pouvoirs publics. Les institutions administratives, qui ont le devoir de faire respecter les lois foncières tout en maintenant l'ordre, ont souvent tendance à ne pas arbitrer de manière radicale.

En cas de conflit, les chefs coutumiers sont sollicités pour trouver un arrangement 'à l'amiable' entre les parties en litige, sur la base des coutumes foncières établies bien avant la naissance de la Nation. Les autorités administratives reconnaissent elles-mêmes leur incapacité d'exercer un contrôle efficace sur des domaines pastoraux immenses et non aménagés.

---

[3] Société nationale industrielle et minière, qui exporte environ 12 millions de tonnes de fer par an et rapporte près de 40 pour cent des devises du pays. Elle apporte une contribution sociale importante, notamment aux éleveurs, pour lesquels elle transporte régulièrement l'eau dans le train jusque dans leurs villages ou campements.

Les sociétés pastorales, quelle que soit la zone considérée, ont subi de profonds bouleversements ces dernières décennies, bouleversements qui se prolongent encore au nord comme au sud du pays. La crise de ces sociétés est tellement évidente, qu'à l'examen on peut réellement se poser la question si elles continuent encore d'exister.

Alors que les transhumants purs représentaient 78 pour cent en 1965, ils n'étaient plus que 12 pour cent en 1988[4] et probablement moins de 10 pour cent actuellement. L'exode rural s'est fait à une telle ampleur, que la Mauritanie a connu, entre 1979 et 1990, l'un des taux d'urbanisation les plus élevés au monde – près de 8 pour cent par an! Les femmes ont désormais en charge l'essentiel des travaux ruraux, les hommes s'étant absentés pour de longues périodes, souvent même pour toujours.

Les rares pasteurs qui continuent à vivre au gré des saisons, se sont regroupés autour de points d'eau, loin des zones enclavées. Pour éviter les longs déplacements, ils recrutent des bergers pour assurer le gardiennage et accompagner les troupeaux en transhumance.

Sur un plan social, des mutations importantes ont été enregistrées: évolution des rapports de forces internes au sein des entités pastorales, nouvelle division du travail, relâchement des liens de solidarité traditionnels, émancipation de la femme, changement des habitudes alimentaires et type d'habitat, etc. Pour les régions à grandes potentialités agricoles, la tendance à la fixation (sédentarité) et la création de ce qu'on pourrait appeler 'les nouveaux villages', a souvent entraîné des frictions entre les populations anciennement installées et les nouveaux arrivants. Ces frictions peuvent se transformer en véritables conflits si ces derniers réclament l'accès aux terres cultivables.

A côté de cette évolution sociale des pasteurs transhumants, s'est greffé un changement de propriétaires de bétail – les éleveurs ayant vendu une part importante de leurs troupeaux aux grands commerçants à des prix bradés, au moment des sécheresses. Ces derniers thésaurisaient dans le bétail qui n'était pas imposé (il l'est depuis 1995, mais faiblement) et qui bénéficiait d'une bonne couverture sanitaire et vaccinale. Le phénomène de la pauvreté s'est alors particulièrement développé en milieu pastoral, les éleveurs n'ayant aucune autre connaissance ni aucun autre métier que l'élevage.

Les problèmes fonciers ont également connu une ampleur sans précédent – 'les nouveaux éleveurs' n'ayant aucune prétention (au sens du droit coutumier) sur les terres qu'exploitent leurs animaux. La conduite et la gestion des troupeaux en a reçu en coup, les vrais éleveurs, qui avaient acquis un savoir et des techniques millénaires, ayant été dépossédés et poussés à l'exode.

Institutions locales et tenure foncière

Pendant longtemps, l'accès aux ressources pastorales s'est fait à travers l'appartenance à des institutions pastorales informelles, basées sur les liens de

---

[4] ONS (Office National des Statistiques)

parenté ou sur les solidarités géopolitiques. Les principaux acteurs de la tenure foncière traditionnelle étaient les émirs, les *Jemaa* tribales, les chefs de tribus et de fractions.

L'administration coloniale a également joué un rôle considérable, dont les répercussions se font encore sentir. Indépendamment des institutions locales traditionnelles, l'enjeu foncier dans toutes les régions de Mauritanie et pour tous les systèmes de production, intègre la dimension '*Chariâ*' (loi coranique), que personne ne conteste, mais que certains groupes évitent parfois d'appliquer.

En Mauritanie, toutes les institutions transitoires pastorales et tous les accords *ad hoc* se font dans le cadre de la tribu ou de l'ethnie, par le biais desquelles se font – de manière informelle – les arrangements et les alliances.

Le rôle de l'émir consistait à assurer la protection des animaux pâturant sur son territoire; toute agression contre eux signifiait une déclaration de guerre et entraînait une riposte et des représailles – à moins que les agresseurs ne remboursassent les dégâts et ne présentassent des excuses aux propriétaires des animaux et à l'émir lui-même.

Pendant la saison des transhumances, des accords sont passés entre émirs alliés pour permettre aux animaux des populations relevant des uns et des autres de pâturer en toute quiétude et d'utiliser les points d'eau comme les autochtones. Le bétail et les bergers sont alors sous la protection de l'émir hôte. De très nombreux conflits ont surgi suite au non-respect par certaines tribus (sous allégeance ou non) des accords passés entre émirs.

Depuis l'indépendance en 1960, les émirs n'ont pratiquement plus d'existence politique, même s'ils conservent un statut social particulier au sein de leur communauté. Ils n'ont également plus d'autorité sur les terres pastorales et agricoles.

Les *Jemaa* gardent un certain rôle de nos jours. Dans le temps, les problèmes de sécurité l'emportaient sur les autres préoccupations, le *Jemaa* décidait essentiellement de la guerre, de la paix et des accords à passer avec les autres ensembles tribaux. Dans ces conditions, chacun était libre de la gestion et de la conduite de son troupeau. Les grands propriétaires divisaient souvent leur cheptel en plusieurs lots, qu'ils faisaient paître séparément – pour ne pas tout perdre en cas de razzia.

La situation a évolué avec l'apparition d'un État moderne: le rôle des *Jemaa* a diminué au profit de nouveaux centres de décision, y compris au sein même de ces *Jemaa*, constitués par les hommes les plus riches et les plus influents de chaque campement. Ils décident aussi bien du déplacement du bétail que de l'octroi de l'autorisation de transhumance à accorder aux tiers. Le 'Conseil' se tient dans la demeure d'un chef qu'on choisit implicitement, sans qu'il y ait recours à un vote, pour ses qualités propres, son importance économique mais aussi son rang social.

Les seuls investissements réalisés dans les domaines pastoraux traditionnels consistent en la construction de puits, destinés à l'abreuvement du bétail. Les puits peuvent-être collectifs ou individuels (mais dans tous les cas, les droits du propriétaire ne sont pas absolus mais seulement prioritaires).

Le schéma décrit ci-dessus est valable essentiellement pour les tribus maures, qui sont toutes organisées suivant un système hiérarchisé comprenant au sommet la caste des guerriers (*hassanes*) et celle des marabouts (*zewayas*), et à la base les tributaires (*lahma*), ensemble hétéroclites de bergers-contremaîtres (*z'naga*), d'artisans (*m'almin*), de griots (*igawen*), d'affranchis (*h'ratines*) et d'esclaves (*abid*). Sans pouvoir parler d'une réelle spécialisation de la société maure traditionnelle, on peut néanmoins dire que chaque caste jouait un rôle social et économique bien déterminé – le plus souvent lié aux activités agro-pastorales.

L'organisation sociale chez les peuls n'est pas fondamentalement différente, en ce sens que la société est ici encore pastorale et fortement hiérarchisée. En revanche, les peuls avaient un mode de vie transhumant qui leur permettait de pratiquer l'agriculture sous pluie. Les événements de 1989 leur avaient fait beaucoup de tort, plusieurs d'entre eux ayant été victimes de vol ou d'expropriation de bétail, avant de se voir expulsés. En retour, certains d'entre eux ont fait régner l'instabilité sur la frontière avec le Mali (vol de bétail, attaque de villages maures, etc.).

Comme déjà souligné, la '*Chariâ*' continue de jouer un rôle important dans la tenure foncière traditionnelle. La loi coranique autorise des emprises non exclusives sur les terres à vocation pastorale et sur les autres ressources pastorales. Les personnes privées ou collectives qui construisent et entretiennent des points d'eau jouissent d'un droit de priorité d'usage, qui ne peut en aucun cas se muer en droit absolu. Toutefois, cette loi permet un contrôle des pâturages par les points d'eau, en cas de dégradation plus ou moins manifeste. Elle autorise également toute pratique coutumière (*orf*) qui ne viendrait pas en contradiction avec les dispositions du droit musulman.

La logique guidant la *Chariâ* est tout à fait cohérente avec le milieu naturel et les pratiques d'élevage, le nomadisme et la transhumance étant les principaux modes de production animale dans le pays. Une appropriation absolue des terres et des points d'eau par des communautés ou des individus aurait conduit à des conflits interminables et, vraisemblablement, à la disparition de l'élevage à terme.

Si personne ne remet en cause cette loi – somme toute divine (car explicitement tirée du Coran) – il arrive que les pratiques communautaires s'en écartent de manière plus ou moins subtile. En fait, les terres à vocation pastorale sont toutes revendiquées par les tribus. L'accès d'autres groupes aux ressources pastorales d'une tribu ne peut se faire sans l'assentiment de cette dernière – même si la *Chariâ* les y autorise dans une certaine mesure.

La conjugaison des droits traditionnels ou coutumiers et de la *Chariâ* permettait une certaine régulation et adéquation entre charge de bétail et potentialités pastorales. La bonne connaissance des éleveurs de leur milieu physique, ajoutée à l'existence de liens séculaires unissant les groupes utilisant le même espace pastoral, avaient permis de perpétuer l'équilibre écologique, en dépit des sécheresses et des conflits.

Les perturbations du milieu rural liées à la dégradation des terres et au relâchement des liens traditionnels ont amené le gouvernement à adopter une politique associant davantage les pasteurs à la gestion de l'espace pastoral.

L'organisation des éleveurs en associations coopératives pastorales s'est appuyée sur trois textes importants:

o l'ordonnance n° 83.127 du 5.6.1983 portant Réorganisation foncière et domaniale, et ses décrets d'application
o le décret n° 67.171 du 18.7.1967 portant Statut de la coopération, complété par la loi n° 93.15 du 21.1.1993
o l'ordonnance n° 82.171 portant Code forestier.

L'idée principale qui a présidé à la création des Associations coopératives pastorales (ACP) consiste à réunir au sein d'une même institution, les utilisateurs d'un même espace pastoral, en leur confiant la gestion des ressources-clés. Des équipes pluridisciplinaires ont été mises en place lors de la phase d'identification du projet; elles étaient sorties avec la conclusion que l'on pouvait créer des ACP viables, à condition:

o que le territoire alloué soit suffisamment riche en pâturages et en potentialités hydriques, pour permettre d'envisager le développement de l'élevage à moyen et long termes
o que la taille du cheptel total des associés potentiels soit suffisante
o que les ensembles tribaux et ethniques présents sur le territoire soient homogènes et acceptent clairement d'oeuvrer ensemble et de respecter les choix démocratiques qui seraient opérés
o que le droit des tiers (transhumants, agriculteurs, etc.) soit scrupuleusement respecté par les ACP
o que l'élection des bureaux des ACP soit faite de manière démocratique et transparente, chaque localité et chaque campement fréquentant traditionnellement le domaine pastoral concerné devant être invité à se faire représenter à l'élection.

C'est ainsi qu'à partir de 1987–88, ont été créées une quarantaine d'ACP, dans le cadre d'un projet cofinancé par la Banque Mondiale et la Banque Africaine de Développement (BAD). Ce projet comprenait en outre des volets d'appui aux nouvelles associations pastorales mises en place: crédit pour l'amélioration de la production, formation des membres du bureau aux techniques de gestion et d'organisation et aux statuts des coopératives, hydraulique pastorale (chaque ACP constituée devant bénéficier d'un point d'eau).

Sur un plan purement foncier, les ACP devaient bénéficier d'un droit sur les terres, sur une base contractuelle avec l'État, les conditions restant à définir en accord avec la loi foncière. Cette dernière, il convient de le signaler, fait peu cas des terres pastorales. Le contexte dans lequel elle a été élaborée était plutôt marqué par des tensions sur les terres de cultures, que le gouvernement cherchait à individualiser, pour inciter les exploitants et les investisseurs privés. Toutefois, une disposition du décret d'application de la réforme foncière permet une mise en valeur collective des terres: 'Toute collectivité qui exprime le désir de conserver ses terres indivises peut s'organiser en coopérative régulièrement constituée, dont les membres sont égaux en droits et en devoirs. Il en est de même pour les collectivités dont les terres ne peuvent être individualisées, pour des causes d'ordre économique ou social' (article 15).

La constitution des ACP repose sur cette disposition réglementaire. Ainsi, les groupes d'éleveurs exploitant habituellement le même domaine pastoral, peuvent avoir des droits juridiquement garantis, à condition que l'objectif recherché soit l'amélioration et la rationalisation de la gestion de l'espace pastoral commun.

Une circulaire du Ministre de l'intérieur, datée du 14 août 1990, destinée aux autorités administratives régionales, définissait les droits légalement reconnus, accordés aux ACP en matière de gestion pastorale et d'accès aux points d'eau. Les ACP peuvent bénéficier, sur leur demande, de locations portant sur des zones de pâturages ou sur des points d'eau réalisés par l'État et à vocation pastorale qu'elles désirent mettre en défens. Elles peuvent également jouir de la qualité d'auxiliaires des pouvoirs publics, en matière de police rurale: lutte contre les feux de brousse, contre les coupes d'arbres, etc.

Il est toutefois manifeste que la réforme foncière et ses textes d'application ne sont pas adaptés aux terres à vocation pastorale, qui appartiennent pourtant dans leur quasi-totalité à l'État. A titre d'exemple, la mise en valeur pastorale ne donne pas droit à la terre, contrairement à la mise en valeur agricole. Par ailleurs, les droits fonciers que l'État s'était engagé à accorder aux ACP tardent à être octroyés. La prudence affichée par les pouvoirs publics s'explique sans doute par la multiplicité des sources de conflits potentiels, liés aux contradictions présumées entre les différentes sources de droit (droit moderne, *Chariâ*, droit coutumier et traditionnel, etc.).

Pour l'instant, les compétences des ACP se limitent à des activités de production, de commercialisation et de sensibilisation. Elles ne gèrent donc pas l'espace et n'ont donc comme mécanismes de résolution des conflits, aussi bien en leur sein qu'avec les autres ACP, que le mécanisme traditionnel et communautaire.

Il est encore tôt pour avoir un jugement définitif sur les ACP, mais des leçons peuvent d'ores et déjà être tirées des expériences vécues par chacune des associations:

o Il apparaît de plus en plus clairement que la taille des domaines pastoraux doit être revue à la baisse, pour permettre aux membres des bureaux constitués d'exercer un contrôle et un suivi réels des activités. On a remarqué, en effet, que les organes des associations ne pouvaient pas exercer convenablement. On peut certes rétorquer à cela l'intérêt qu'il y a de disposer de domaines vastes, pour permettre une grande mobilité du cheptel, que rendent nécessaires des conditions climatiques aléatoires. Mais c'est oublier que quelle que soit la taille de l'association, la mobilité continuera d'exister, puisque l'élevage extensif ne peut pas être fait autrement. La solution consisterait à développer davantage les liens entre associations, pour permettre aux transhumants d'avoir accès, comme par le passé, aux zones qu'ils fréquentaient. Si le but recherché doit être d'améliorer la gestion de l'espace en assurant un minimum d'adéquation entre les besoins des utilisateurs et les ressources pastorales, il est impératif que les modèles préconisés soient adaptés à la stratégie de mobilité.

o Il convient de distinguer entre les activités d'élevage liées à la production, qui pourraient être confiées à des coopératives (de production, de commercialisation, de crédit, etc.), et les activités liées à la gestion et à l'amélioration de l'espace pastoral, qui devraient être confiées à des associations d'intérêt public, que l'État, dont elles constitueraient un démembrement, se doit d'appuyer financièrement et techniquement. L'expérience a montré, en effet, que les ACP avaient tendance à s'intéresser surtout aux avantages économiques qu'elles pouvaient tirer de la production ou de la commercialisation, au détriment de la préservation des ressources pastorales. Le vrai problème qui se pose à ce niveau est de savoir les liens qui devraient exister entre les coopératives et les associations, étant donné qu'elles visent les mêmes individus. On peut, *a priori*, suggérer des organes de concertation entre les deux types d'organisation, qui viseraient à prendre en considération leurs préoccupations réciproques, à harmoniser les approches et éviter les doubles emplois. Le type d'investissement à promouvoir dans ces zones pourrait aider grandement à concilier les objectifs des deux genres d'organisation. A titre d'exemple, l'aménagement d'une zone pastorale comportant la fauche et la conservation du foin, l'ouverture de pare-feux et de pistes rurales (mettons dans le cadre de l'appui à une association d'intérêt public) permettrait non seulement de préserver les ressources pastorales, mais aussi de faciliter les activités de production et de commercialisation. Il en va de même s'agissant de la réalisation de points d'eau là où l'eau manque et où il n'y a pas de risque de surpâturage – avec la participation pleine et entière des populations bénéficiaires. La construction de pistes à bétail et d'aires d'abattage, ainsi que la réalisation de programmes de formation et de vulgarisation au bénéfice tant des associations que des coopératives constitueraient autant d'actions qui devraient avoir un effet positif.

o Il est souhaitable d'encourager fortement l'initiative privée, notamment en matière d'investissement, au sein de ces organisations pastorales: il en résulterait vraisemblablement un effet d'entraînement important, pourvu que des assurances soient accordées aux auteurs de telles initiatives. Les investissements privés qu'on pourrait réaliser dans le contexte d'un usage communautaire des terres, dépendent beaucoup du degré d'éveil des communautés concernées et des garanties qu'elles pourraient offrir à d'éventuels investisseurs. Il est clair que ces dernières n'ont ni les moyens ni, la plupart du temps, la volonté d'investir dans leur terroir. La seule zone où les investissements auraient une rentabilité certaine est la zone environnante du fleuve Sénégal. A supposer que le régime communautaire reste appliqué dans cette région, on peut envisager que les éleveurs mettent en commun une partie de leur bétail. A partir de là, on peut supposer que des investisseurs privés, intéressés par la production laitière ou carnée, financent la réalisation d'infrastructures pour développer cet élevage, quitte à récupérer leurs fonds une fois la production améliorée.

Devoir d'accès et règles d'utilisation des terres pastorales

Les règles d'utilisation des terres pastorales restent dominées par les droits coutumiers et la référence au droit musulman, mais s'inspirent également des lois coloniales et, plus récemment, de la réforme foncière et domaniale (Tableau 3.1).

L'accès aux ressources varie en fonction de la nature de ces dernières. Les pâturages aériens et les herbacés (y compris les savanes et les steppes à Balanites) sont théoriquement libres; c'est du moins ce que prévoit la loi coranique. Les points d'eau peuvent faire l'objet d'un droit d'usage prioritaire, mais non exclusif. Dans les faits, les communautés locales (sédentaires ou transhumantes) ont une emprise, plus ou moins forte, en fonction de leur ancienneté dans les lieux et de leur perception des enjeux pastoraux et agricoles (concurrence autour des ressources).

Les oases constituent les seules terres dont l'appropriation, sous forme individuelle, était clairement établie depuis plusieurs décennies, surtout dans l'Adrar et le Tagant septentrional (au nord et au centre de la Mauritanie). Les

Tableau 3.1. Décrivant le mode de tenure en fonction de la zone

Zone/Tenure	Traditionnelle/ tribale/ethnique	Moderne (État)	Droit d'usage
Khweiwira	Dominante dans les faits	Quasi absent	Néant
Fassala Néré	Dominante	Les quelques litiges sont le plus souvent réglés par le mode traditionnel. Il arrive qu'on ait recours aux lois modernes, pour décourager les tendances en conflit.	Les éleveurs touaregs qui fréquentent la zone évoquent leur droit d'usage.
Zouérate	Plus faible qu'ailleurs	Les éleveurs 'modernes' font prévaloir la loi moderne pour accéder aux ressources pastorales.	Bien que n'ayant pas de droit coutumier sur les terres, les éleveurs modernes réclament le droit d'usage, car présents depuis plusieurs décennies.
Boghé	Assez important	La zone étant surtout agricole, de nombreuses terres de pacage sont perdues, au profit de l'agriculture irriguée. Cette dernière a été largement encouragée par la réforme foncière.	Les transhumants de plus en plus nombreux à s'installer réclament de nouveaux droits sur les terres.

oases des régions Est (Assaba et deux Hodhs) sont régies, dans la pratique, par les droits collectifs tribaux, au même titre que les autres terres. Mais la tendance est à l'individualisation de ces oases.

La situation, pour toutes les terres non oasiennes, est marquée par un accès quasiment libre aux terres marginales se trouvant au Nord, peu riches en pâturages et en points d'eau. En revanche, dans les régions à vocation pastorales, où la pression et la concurrence sont plus importantes, personne n'ose transgresser le domaine d'une tribu ou d'une fraction, sans avoir été clairement autorisé. Ce droit qui se veut exclusif rentre en contradiction à la fois avec la loi foncière et la *Chariâ*. En outre, cette forme d'appropriation des terres par les communautés ne s'est pas accompagnée d'une amélioration sensible de la gestion des ressources pastorales. En revanche, plus l'emprise communautaire est forte et reconnue par autrui, plus les pratiques d'élevage sont respectueuses de l'environnement.

Dans les régions septentrionales, quasi désertiques, l'emprise tribale est moins prononcée: l'accès des terres est ouvert à tous les pasteurs, qui sont cependant limités par le manque de points d'eau. Certains éleveurs de Zouérate ont creusé, il y a quelques années, des puits sur des terres individuelles ou au voisinage (en arabe, *harim*) d'une localité, en vue d'en priver nomades et transhumants fréquentant les mêmes pâturages. La *Chariâ* permet en effet, suivant le principe du '*harim*', d'en interdire l'accès à autrui, sous certaines conditions.

Mais qu'il s'agisse du droit coutumier (que l'État ne reconnaît plus, ce qui ne sécurise pas les populations) ou du droit moderne (la terre appartenant à l'État, donc à personne, selon l'entendement des populations), le résultat, en terme de valorisation et d'amélioration de la gestion pastorale, reste le même: les parcours se dégradent puisque l'appartenance de la terre est hypothéquée entre l'État et les communautés, les éleveurs achètent (presque par dépit) davantage d'animaux pour augmenter leur gain et ainsi on assiste à un surpâturage excessif, la désertification, les pertes massives d'animaux, et ainsi de suite. Il n'y a d'ailleurs pas que le problème de la dégradation des terres, les investissements sont également dérisoires, en raison toujours de l'insécurité foncière.

La solution consistant à individualiser les terres pastorales n'est pourtant pas envisageable à grande échelle en Mauritanie. D'une part, le mode extensif de l'élevage s'accommode mal de la sédentarisation des animaux, pour qui la mobilité reste le seul moyen d'assurer la rentabilité.

Même la seule société d'État chargée de la commercialisation du bétail, aujourd'hui liquidée, ne s'est pas vue octroyer un domaine pastoral exclusif – cela n'aurait servi à rien, les animaux étant la plupart du temps en déplacement lointain.

Les mécanismes de résolution des conflits

En période normale (pluviométrie acceptable), les pasteurs empruntent les mêmes axes de transhumance vers des terres fréquentées habituellement depuis des décennies. La nécessité d'établir des contacts ou de négocier l'accès aux ressources pastorales est exceptionnelle, sauf pour des communautés qui n'avaient pas l'habitude de fréquenter les terres. En général, il suffit simplement d'avertir les chefs des communautés locales et les habitants des localités concernées pour éviter d'éventuels conflits avec les agriculteurs.

En d'autres termes, le processus de négociation est intimement lié à la situation et au niveau des pâturages: en cas d'abondance, les négociations sont menées avec facilité et courtoisie, en signe de bonne volonté à l'égard des autres communautés et de peur de représailles antérieures.

Par ailleurs, la tradition islamique et les règles d'hospitalité dans les milieux arides militent en faveur du libre accès aux ressources, quand elles peuvent suffire à tous. Il y a cependant lieu de remarquer qu'en cas de feux de brousse particulièrement dévastateurs, et en dépit de la bonne pluviométrie, les communautés autochtones peuvent engager les négociations avec les transhumants présents en vue de leur départ et peuvent aussi limiter à d'autres éleveurs l'accès aux pâturages.

Toutefois, depuis la fin des années 80, avec les événements regrettables entre la Mauritanie et le Sénégal et avec les mouvements de dissidence des populations du nord du Mali, la transhumance trans-frontalière a été profondément bouleversée. Du coup, les déplacements du bétail à l'intérieur du pays en ont ressenti les effets restrictifs: ceux qui transhumaient traditionnellement vers le Mali et le Sénégal ont dû se rabattre sur des terres pastorales que d'autres utilisaient habituellement – d'où conflit. Fort heureusement, il n'y a pas eu depuis lors une année vraiment sèche, et les solidarités face à l'étranger ont jusqu'ici permis de contenir les litiges apparaissant çà et là.

En période de mauvaise pluviométrie, par contre, les mécanismes traditionnels et les droits coutumiers ne sont plus entièrement valables. La tension sur les terres pastorales est, en effet, telle que des conflits entre les populations locales, les transhumants habituels et les transhumants occasionnels, sont inéluctables.

Les sources de conflits (toujours latents) sont très nombreuses et variées. Elles dépendent d'un nombre important de variables et de paramètres, qui tiennent de la nature des relations liant les groupes humains, des types de pâturages, des espèces élevées, de la pluviométrie enregistrée, etc.

Il n'y a pas à proprement parler de juridiction chargée de statuer sur les conflits. Dans le temps, le chef tribal assurait aussi bien l'exécutif que le juridique, après avoir pris l'avis d'un Cadi. De nos jours, le chef tribal, qui existe encore comme symbole, n'a plus de poids réel, et l'avis du Cadi n'est plus recherché.

Les litiges les plus faciles à résoudre sont ceux opposant des fractions d'une même tribu, par exemple une fraction sédentarisée (pratiquant l'agriculture) et une autre qui continue de se déplacer. En général, les chefs de famille, en particulier ceux qui disposent d'un nombre assez important de têtes de bétail, se réunissent pour examiner les litiges, sur demande d'un ou de plusieurs membres de la tribu, dont les intérêts auraient été touchés. La sentence est prononcée par le Chef de tribu, que tout le monde se doit de respecter.

Il en va autrement des conflits entre agriculteurs sédentaires de longue date et transhumants issus de tribus ou d'ethnies différentes. Il arrive que des cultivateurs, pour éviter l'arrivée de transhumants, clôturent les mares et autres points d'eau habituellement fréquentés par ces derniers et se mettent à mettre les terres avoisinantes en valeur. La réforme foncière leur permet, en effet, d'avoir un droit d'usage exclusif (certes non définitif). Les conflits surgissant ainsi sont d'autant plus graves que l'année est mauvaise.

L'apparition de nouveaux villages, créés suite à la sécheresse et à la fixation des populations pastorales, constitue un piétinement sur les terres de pacage des transhumants et nomades, terres qu'ils considèrent comme étant la leur, au même titre que les villageois nouvellement fixés. La situation est aggravée par le fait que ces villages se délimitent des espaces vitaux, parfois démesurément grands, qu'ils interdisent aux pasteurs étrangers.

Ces types de situations conflictuelles se règlent par les assemblées tribales ou ethniques, en fonction des liens de solidarité, des rapports de force ou de l'intermédiation d'une troisième partie. Le processus de négociation se fait sur l'initiative de l'une ou l'autre partie (ou du médiateur, le cas échéant), mais en respectant des formes à convenir. Après plusieurs rencontres plus ou moins informelles, on convient de certains accords de principe, que les chefs de tribus ou de fractions entérinent au cours d'une cérémonie se déroulant en principe dans la demeure de celui qui a eu un rôle prépondérant dans la négociation, ou qui a joué les bons offices.

Les conflits peuvent avoir des racines profondes et lointaines, liées à des stratégies d'exploitation de l'espace différentes. C'est le cas de la région de Djenké, au Hodh el Gharbi, au sud-ouest de Khweiwira. Dans cette région vivent deux communautés pastorales différentes: les Ladoum, semi-sédentaires, installés depuis plus d'un siècle, et les Peuls, également semi-sédentaires, venus plus récemment. Les Ladoum, puisque les terres pastorales leur appartiennent, interdisent aux Peuls, mais aussi aux autres tribus, de creuser des puits au sein même de la Djenké. Cette interdiction ne vise pas à contrarier leurs nouveaux voisins, mais à dissuader les éventuels transhumants, dont l'afflux entraînerait inévitablement la dégradation de toute la région pastorale.

Mais les situations les plus conflictuelles sont apparues dans la vallée du fleuve Sénégal, suite à la réalisation des barrages communs entre les trois États riverains: Mali, Sénégal, Mauritanie. Outre les investisseurs privés attirés par les incitations accordées par le gouvernement depuis le milieu des années 1980, les populations haratines (esclaves affranchis) se sont ruées vers les zones aménagées, en concurrence parfois avec les populations toucouleurs autochtones. Les éleveurs traditionnels et leurs animaux n'ont plus accès à l'eau du fleuve à cause des aménagements hydro-agricoles (il n'existe aucun couloir de passage au fleuve). Les zones de bas-fonds, habituellement réservées à la pâture, sont de plus en plus cultivées et les animaux refoulés.

Tous ces conflits se règlent au travers d'arrangements informels certes, mais tirés le plus souvent des réalités socio-politiques des populations concernées. L'État, administrations comme instances juridictionnelles, n'intervient qu'en dernier recours, pour mettre fin à des situations qui deviennent insupportables – mais sans trancher de manière définitive sur le fond des conflits eux-mêmes. Les accords entre les transhumants et les communautés sont passés dans un cadre tribal: c'est la tribu qui, le cas échéant, arbitre les conflits nés d'un malentendu.

La réforme foncière est trop récente pour constituer une référence suffisante pour régler les conflits. Elle n'aborde d'ailleurs pas suffisamment les aspects liés à la tenure foncière pastorale. La *Chariâ*, quoique constituant une référence explicite sur plusieurs points, laisse une marge importante au *orf* (coutumes et traditions locales non contraires aux prescriptions du Coran).

Les ACP, en dépit des espoirs qu'elles suscitent, ne sont pas non plus outillées pour faire face aux conflits liés à l'accès aux ressources pastorales.

Remarques finales et conclusion

Les investigations réalisées sur le terrain montrent une diversité des approches adoptées par les communautés en matière de gestion de l'espace, en fonction des contraintes climatiques et des déterminants socio-culturels propres.

Aujourd'hui, l'élevage n'est plus un mode de vie qui ponctue le rythme de vie des populations comme il y avait 30 ans de cela. L'évolution socio-politique et économique du pays en a fait une activité principale de l'économie nationale, dont le développement est lié à la réalisation d'infrastructures et à la définition de politiques, qui cherchent toutes, consciemment ou non, à limiter les mouvements des pasteurs et des animaux, en vue de mieux planifier la croissance du secteur de l'élevage.

La dégradation des parcours et les sécheresses récurrentes ont été paradoxalement les catalyseurs essentiels de cette politique: elles ont poussé les populations à abandonner leur mode de vie pour se fixer (se sédentariser), remettant en cause l'organisation séculaire de l'espace et les droits coutumiers qui le régissaient.

Indépendamment de la diversité écologique et culturelle, il y a un fond théorique d'approche commune édicté par la *Chariâ*, qui demeure la référence ultime en cas de conflits majeurs autour des terres à vocation pastorale. Mais la complexité des problèmes posés et leur variabilité amènent à penser qu'il y a lieu de chercher une solution (non nécessairement unique), tenant compte des droits traditionnels, coutumiers, mais aussi des droits d'usage apparus à la faveur des changements apparus ces dernières décennies.

A côté de l'impératif de développement des ressources pastorales, il apparaît nécessaire de trouver une réponse aux inquiétudes grandissantes des pasteurs et agro-pasteurs transhumants qui, à l'inverse des nomades en voie d'extinction, sont de plus en plus nombreux. La sécurisation foncière est un besoin pressant qu'expriment de plus en plus les éleveurs confrontés à une concurrence sévère autour des terres pastorales qu'ils exploitaient traditionnellement.

L'ampleur des changements sociologiques et les migrations internes intervenues ces dernières années ont créé des nouveaux 'droits d'usage de fait', qui perturbent partiellement les droits coutumiers et traditionnels[5]. La sécurisation foncière doit

---

[5] Dans une note destinée à l'atelier sur la 'Tenure foncière pastorale et le développement au Sahel', tenu en novembre 1993 à Nouakchott grâce à l'appui financier de l'IIED, Pierre Bonte parle également de ces droits d'usage, avec lesquels il reconnaît qu'il convient sans doute de compter, si on ne désire pas créer des situations conflictuelles d'un genre nouveau.

convenir non seulement aux éleveurs traditionnels, mais également aux populations nomades qui se sont fixées et aux nouveaux éleveurs, qui possèdent désormais une part non négligeable du cheptel national. Une organisation pastorale basée uniquement sur les droits traditionnels ne peut être préconisée car elle méconnaîtrait les changements récents et engendrerait de nouveaux conflits.

La Mauritanie, à l'instar d'autres pays africains, a exploré la solution basée sur la création d'associations coopératives pastorales, respectant à la fois l'organisation traditionnelle de l'espace, les droits d'usage et la réglementation (loi foncière et domaniale).

Il est cependant clair que les ACP, dans leur forme actuelle, ne donnent pas satisfaction aux usagers de l'espace pastoral et ne contribuent en rien à la préservation de l'environnement. Il conviendrait de tenir compte des leçons tirées des sept dernières années, pour recentrer les ACP, notamment en revoyant à la baisse leur dimension, en engageant davantage de négociations entre usagers et surtout en définissant des plans d'aménagement réalistes respectant les droits coutumiers. L'appui technique et financier de l'État devra être apporté.

Ces plans d'aménagement doivent rester souples et être négociés avec tous les utilisateurs (y compris les cultivateurs et les éleveurs n'appartenant pas à l'association concernée). En particulier, il faudrait s'inspirer des accords traditionnels *ad hoc* existants (et sans cesse renouvelés), pour définir des règles d'accès aux ressources-clés que les ACP devraient faire respecter.

Les droits à octroyer aux ACP ne doivent en aucun cas être permanents, au risque de pénaliser les tiers (transhumants, agriculteurs), surtout en années de sécheresse. Si la solution à court terme est difficile à imaginer, en revanche, à moyen terme, elle consisterait peut-être à différencier les types de ressources, de sorte que les tiers n'aient accès qu'au vital. Ainsi, les ACP auraient la possibilité d'interdire l'accès de certaines zones, certains types de pâturages et de points d'eau. Elles pourraient exiger des tiers une contribution en échange des types de ressources exploités. Cela demande une connaissance approfondie de l'écologie et des ressources-clés et une maîtrise certaine de leur territoire par les ACP. L'apport des ONG en ce sens pourrait être déterminant, s'agissant aussi bien de la formation des éleveurs dans le domaine de l'écologie et de l'amélioration de la gestion des points d'eau, de la terre et des pâturages.

Enfin, la variabilité du climat et la fragilité de l'environnement tant physique que social, exigent que le cadre institutionnel et juridique soit révisé périodiquement, afin de permettre son adaptation aux évolutions actuelles et de garantir la paix sociale dans les zones pastorales sans laquelle aucun développement n'est envisageable.

Cette révision doit partir des populations rurales concernées elles-mêmes, qui connaissent évidemment mieux la nature des problèmes posés et ont souvent une idée assez précise des solutions à apporter.

L'atelier de Woburn, tenu en 1993 dans la banlieue de Londres sur l'écologie et le développement rural en Afrique, avait conclu à la nécessité d'intégrer les techniques et institutions traditionnelles locales à toute approche visant à améliorer les conditions de vie et de production dans ces milieux.

73

Dans ce cadre, les institutions provisoires semblent devoir jouer un rôle de premier plan, notamment dans le domaine de la gestion des ressources naturelles.

Ces institutions présentent, en effet, des avantages indéniables:

o Elles sont l'émanation des entités traditionnelles et ne donnent pas lieu, contrairement aux organisations greffées de l'extérieur, à un rejet de la part des populations. Qu'il s'agisse d'un conflit au sein d'une même communauté ou entre plusieurs groupes, la 'sentence' est le plus souvent acceptée sans heurt.

o Les institutions provisoires sont à la fois simples dans leur organisation et souples dans leurs méthodes, ce qui convient bien à un milieu traditionnel assez faiblement instruit et dont les connaissances en matière de droit moderne sont plutôt limitées.

o Ces institutions, en raison de leur caractère informel, sont, maintenant, les seules à pouvoir faire face aux conflits liés à la transhumance, difficile à codifier dans des textes législatifs ou réglementaires. Ce constat est particulièrement vrai durant les années à faible pluviométrie, où les animaux se déplacent pratiquement sans contrôle, très loin de leurs lieux habituels de pacage, ce qui n'est pas sans causer des tensions dans les zones fréquentées.

o Les institutions provisoires ont prouvé leur capacité d'adaptation aux changements sociaux et écologiques. Cette adaptabilité, observée dans toutes les zones pastorales de Mauritanie, est en fait le résultat des autres avantages déjà cités.

o Enfin, elles ne coûtent pratiquement rien, car n'occasionnent pas de frais, contrairement aux organisations formalisées, qui exigent des moyens liés à leur fonctionnement.

Bien sûr, les institutions provisoires présentent des inconvénients et des faiblesses: leur précarité dans le temps et leur caractère informel impliquent un éternel recommencement des processus de conciliation et d'arbitrage. En outre, ces institutions constituent surtout des sortes de comités de crise, et ne s'intéressent donc pratiquement pas aux problèmes de développement des zones pastorales.

D'où, sans doute, la nécessité d'encourager les populations rurales à créer, parallèlement aux institutions provisoires, des institutions plus durables, chargées plus spécifiquement des problèmes de développement et d'aménagement rural. Les associations coopératives pastorales (ACP), créées sur l'initiative des pouvoirs publics, constituent, certes, une ébauche de réponse, mais présentent certains inconvénients qu'il y a lieu de corriger. Tout d'abord, n'étant pas une émanation directe des populations, les ACP ne sont pas toujours 'internalisées'. L'effort doit donc porter sur leur adaptation en vue de leur appropriation adéquate par les groupes concernés.

Ensuite, les ACP se sont octroyées un champ de compétence trop large, qu'elles ne pourront jamais couvrir entièrement. Elles doivent donc se focaliser sur les problèmes de gestion et d'aménagement de l'espace rural, en laissant aux coopératives le soin de s'occuper des aspects de production et de commercialisation. Le règlement des conflits et les arbitrages, pour leur part, doivent rester du domaine des institutions provisoires qui ont incontestablement l'avantage comparatif dans ce domaine.

4

Les règles d'accès des troupeaux peuls aux pâturages du Delta central du Niger (Mali)

The rules of access by Fulani herds to the pastures of the central delta of the Niger River (Mali)

PASCAL LEGROSSE

RÉSUMÉ

Dans une analyse des règles d'accés des troupeaux bovins à la plaine inondée du Delta central du Niger, l'auteur présente le rôle essentiel du *Jowru* dans la gestion des ressources pastorales et met en évidence les enjeux que soulèvent les itinéraires de transhumance dans l'organisation des maîtrises foncières propre à cette zone. Ces itinéraires sont les premières appropriations lignagères des Peuls au cours de leur conquête des riches pâturages du Delta. Les formations politiques qui se sont succédées ont, pour leur part, découpées le Delta en territoires limités alors que la 'maîtrise-propriété' de l'herbe passe par le contrôle de gîte, formant comme des îles au milieu de l'étendue herbeuse. La préséance et les étapes que les *Jowru* ordonnent le long des itinéraires de transhumance, la redevance qu'ils perçoivent sur les étrangers, constituent les principaux éléments du contrôle locale des terres de pâture. En théorie, les pâturages ne pouvaient être déclassés pour l'agriculture sans l'accord du *Jowru*. Avec la fin de l'hégémonie peule, la politique des États, l'évolution démographique, l'insécurité alimentaire et la sécheresse, les champs ont étendu leurs limites, barrant la route aux pistes de pâture, et provoquant de nombreux conflits sans fin. Une nouvelle politique pastorale est nécessaire.

ABSTRACT

An analysis of the rules of access of cattle to the flood plains of the central delta of the Niger River shows the indispensable role of the *Jowru* in the management of the pastoral resources. The organization of tenure rights, unique to this zone, is a function of the claims imposed by transhumance itineraries. These itineraries were appropriated by each Fulani lineage during the historical conquest of the rich pastures of the Central Delta. Subsequent political domination of the zone by a central State has divided the Delta into administrative divisions; however, *de facto* ownership and management of the pasture continues to operate through the control of camp sites and transhumance itineraries. This local-level control over rangelands is exercised by the *Jowru*, primarily through the spatial and temporal organization of transhumance stages, assigning rights of precedence to the different lineages, and imposing taxes on outsiders. In theory, the rangelands cannot be converted to crop cultivation without the agreement of the *Jowru*. In practice, with the end of the Fulani political hegemony, the imposition of various State policies, the demographic growth, and drought and food insecurity, farms have expanded into rangelands, cutting off transhumance routes to the Delta and provoking numerous and serious conflicts. A new pastoral policy is necessary.

Introduction

DANS LE DELTA central du Niger, l'accès et l'exploitation des pâturages propres à chaque territoire agro-pastoral (*leydi*)[1] par les troupeaux bovins se font sous la direction de *Jowru* – maîtres de pâturages et organisateurs de parcours de trans-humance – depuis des gîtes d'étape et de repos (*winnde*) auxquels ils accèdent par des itinéraires précis (*burtol*) selon un ordre de préséance entre les troupeaux (Carte 4.1).

En octobre–novembre, les troupeaux se concentrent en bordure du Delta et se regroupent en *eggirgol* sous la direction d'un *Jowru* à l'entrée de leur *leydi* respectif. A l'intérieur d'un même *eggirgol*, les troupeaux sont regroupés suivant les *Jowru* dont ils relèvent. Les troupeaux 'citoyens' du *leydi* prennent la tête, suivis des troupeaux 'forains' d'autres *leydi* (du centre du Delta) qui doivent emprunter les mêmes pistes pour rejoindre leur propre pâturage. Derrière eux, fermant la marche, les troupeaux des pasteurs 'allochtones' des terres sèches (ceux qui n'appartiennent pas aux ayants droit sur des pâturages du Delta) pénètrent à leur tour (Gallais, 1967: 375–89). L'entrée autorisée aux pâturages de la plaine pour les troupeaux des pasteurs 'allochtones' (*SeenonkooBe, WuwarBe, FarimakenkooBe,* etc.) est subordonnée au versement d'une redevance à un ou plusieurs *Jowru* le long de leur parcours. Cette redevance manifeste d'une certaine façon la reconnaissance, par le pasteur allochtone, des droits de la communauté locale sur son territoire et sur les conditions de l'accès à ses ressources. De même, les troupeaux 'forains' ne pourront emprunter la piste de transhumance d'un *Jowru* et traverser son pâturage sans que des accords de réciprocité les lient.

Ce schéma, qui fonde l'accès régulier des troupeaux bovins aux pâturages du Delta, est décisif dans le contrôle foncier des ressources pastorales. Dans le débat sur la gestion des ressources communes, il est nécessaire de montrer l'importance et l'enjeu des itinéraires de transhumance dans l'organisation des 'maîtrises-propriétés'[2]

---

[1] La transcription phonétique utilisée ici pour le *fulfulde* (la langue des Peuls) est celle qui a été élaborée par le 'Congrès pour l'unification des alphabets des langues nationales de l'Ouest africain' (UNESCO, Bamako, 1966). Tous les termes *fulfulde* sont transcrits et mis en italique, à l'exception de certains mots, essentiellement des toponymes qui ont été depuis longtemps écrits à l'aide de l'alphabet français par les services administratifs et géographiques et qui sont connus sous cette orthographe. De même, les règles d'accord en nombre et en genre seront respectées pour les hétéronymes d'ethnies qui ont été francisés (ex. les Peul*s*, les Bozo*s*). Pour faciliter la lecture, les termes *fulfulde* seront systématiquement écrits à la même personne, soit du singulier: *winnde* (pl. *bille*) soit du pluriel: *RiimayBe* (sg. *Diimaajo*) selon l'emploi le plus courant dans le texte (B et D dans un mot indiquent des consonnes glottalisées).

[2] Au *Maasina*, on utilise le terme *jeyal* (du verbe *jeyude*, être propriétaire, posséder) pour indiquer les attributs de la maîtise foncière, de la propriété (et de la servitude). Claude Fay écrit très justement que 'Pasteurs, pêcheurs et agriculteurs libres détiennent les maîtrises-propriétés (*jeyal*) liées à leurs éléments de prédilection (terre, eau, herbe) et à leurs territoires, maîtrises incluant les prérogatives sacrificielles du lignage maître-propriétaire des eaux, des terres ou des herbes' (1995: 35).

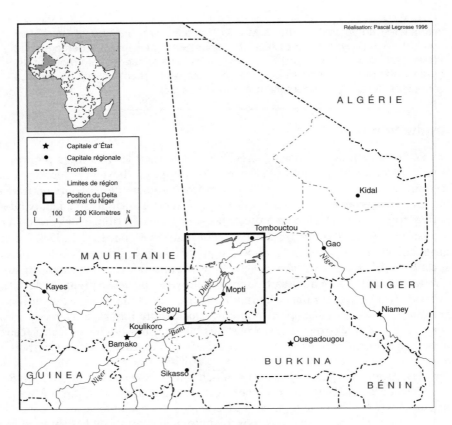

Carte 4.1 *Position du Delta Central du Niger (Mali)*

pastorales propres à la plaine inondée. Les politiques foncières ont souvent tendance
à dissocier le contrôle du sol dans les frontières du terroir (Vedeld, 1994) de
l'exploitation des ressources qui y sont attachées. Dans le Delta, les pâturages n'ont
pas de frontières mais ce sont les gîtes (*winnde*) qui déterminent les contours des
pâturages exploités (qui se juxtaposent aux pâturages d'autres gîtes). Ensuite, la place
d'un troupeau dans la série linéaire de la file des troupeaux transhumants (*eggirgol*)
renvoie directement à la localisation de son pâturage. Comme la préséance gouverne
le déplacement de l'*eggirgol*, la position qu'occupe un troupeau dans l'ordre de la
préséance lui assigne son statut et ses droits d'accès aux pâturages comme usager
primaire, secondaire ou tertiaire.

Historiquement, il apparaît que la conquête des pâturages de la plaine inondée
est d'abord passée par l'établissement de pistes reconnues par les autres pasteurs.
C'est sous la pression extérieure, guerrière des *ArDo*, puis légaliste de la *Diina* et
enfin administrative de la colonisation, que les maîtrises pastorales ont fixé leurs
contours à l'intérieur d'un découpage en *leydi*, en cantons et, aujourd'hui, en
arrondissements. Il existerait trente et un *leydi* dans le Delta aujourd'hui
(CIPEA, 1983, contre trente-sept pour Gallais en 1967). Mais les changements
socio-économiques et pluviométriques ont fait que les cultures se sont largement

étendues hors des zones propices qui leur étaient réservées et que les systèmes de production ruraux se sont profondément transformés. L'agriculture et l'élevage se trouvent en situation de concurrence et de conflit pour l'accès, le contrôle de la terre et de ses ressources, et la captation de la redevance prélevée sur cet accès. Une analyse de ces changements doit distinguer l'organisation linéaire de la transhumance visant à la réactualisation des maîtrises de l'accès et de la gestion de la ressource, de l'organisation spatiale du contrôle de la terre dans un *leydi* (et de l'histoire des pouvoirs qui ont contrôlé les différents systèmes de production, élevage, pêche, agriculture).

Le *Maasina* est au cœur de l'histoire des Peuls

L'essentiel ici est de comprendre la logique des règles d'accès aux pâturages du Delta des troupeaux transhumants peuls et les grandes lignes de ses évolutions et crises jusqu'à nos jours. Pour cela, on décrira l'institution du *Jowru* et l'organisation socio-territoriale de la transhumance autour des pistes de transhumance à travers les territoires pastoraux et les redevances de pâturage. Les exemples seront pris dans le *Maasina* où nous avons séjourné 15 mois entre octobre 1992 et mars 1995[3].

Précisons que le toponyme de *Maasina* désigne, suivant ses acceptions historiques et spatiales, une région plus ou moins vaste et se rapporte à différentes formations politiques au cours de l'histoire (voir Carte 4.2). Au quinzième siècle, des conquérants Peuls venant du *Fuuta Tooro* au Sénégal sous la conduite de l'*ArDo* Magan Diallo s'installent sur 'une colline appelée Mâsina' (Es Sadi, 1900: 283) dont la localisation est toujours discutée. Depuis cette colline, ils étendent leur autorité sur une grande partie de la rive gauche du marigot de Dia (le Diaka) – un défluent du Niger – zone d'influence limitée au sud par la ville de Dia, à l'ouest par les Markas du *Kubay* et au nord par l'influence des Touaregs. Le pouvoir des chefferies guerrières *ArDo* était d'abord un pouvoir sur les hommes, établi par la force. La maîtrise de lignages sur des territoires (*Jowru*, maître d'eaux, chef de terre) était directement attachée au 'bon vouloir' de ces chefs guerriers (les *ArDo*) qui levaient des tributs, désappropriaient, réappropriaient les maîtrises lignagères, tranchaient les conflits et maintenaient l'ordre entre lignages dominants[4]. Les descendants de Magan Diallo (les *JalluBe*) demeurent le lignage dominant jusqu'au dix-neuvième siècle, aussi bien des populations de la région (Bozo, Marka) que des autres lignages peuls nomades qui dépendent d'eux et leur versent des tributs; l'ensemble étant sous la pression des guerriers du Royaume bambara de Ségou qui les dominent depuis le dix-septième siècle (Gallais, 1967: 91; Imperato, 1986: 210–11).

---

[3] Ces recherches ont été effectuées en vue d'une thèse de doctorat en anthropologie sociale et ethnologie qui devrait être soutenue fin 1996 à l'EHESS, 54, Bd Raspail, 75006 Paris.
[4] Les questions liées à la définition d'un État *ArDo* où les *ArDo* seraient les détenteurs du monopole de la violence légitime et les fondateurs en droit des territoires (Fay 1995: 37), ainsi qu'à l'instauration par le pouvoir *ArDo* de plusieurs formations politiques lignagères sans réelles hiérarchies politiques institutionnalisées (cf. Irons, 1979), mériteraient un long développement qui n'a pas sa place ici (cf. Azarya, 1988; Kassibo, 1994; Fay, 1995).

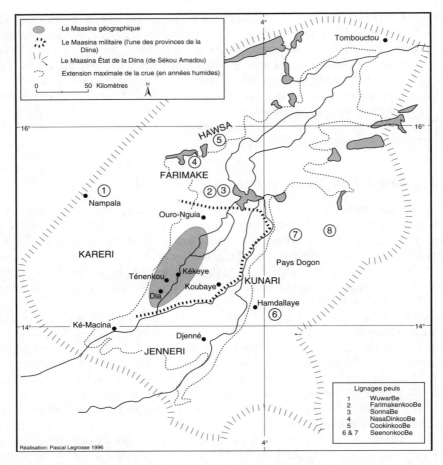

Carte 4.2 *Les différents Maasina et localisation des lignages Peuls allochtones (Mali)*
(Source: *Sanankoua, 1990; Fay, 1994*)

Lorsque le marabout peul Sékou Amadou soumet la région à son autorité en substituant la dynastie des Bari à celle des Diallo vers 1818 (Azarya, 1988: 112), la nouvelle formation politique, surnommée la *Diina* (littéralement 'religion'), prend par extension le nom de *Maasina*. L'État peul centralisé de la *Diina*, autour de sa capitale Hamdallaye (Brown, 1969: 20–21), est divisé en cinq provinces. Parmi elles, le *Maasina* s'étend sur la rive gauche du Niger, de Diafarabé au lac Débo, avec pour capitale Ténenkou, où Sékou Amadou installe son neveu Bori Hamsala comme chef[5] (Ba & Daget, 1984: 68). Chaque province regroupe plusieurs cantons (*lefol*

[5] Les quatre autres provinces sont le *Jenneri*, le *Fakala-Kunari*, le *Hayre-Seeno* et le *Nabe-Dunde*. Chaque province était dirigée par un chef militaire assisté d'un conseil religieux, d'un conseil judiciaire et d'un conseil technique.

leydi), héritage des chefferies *ArDo*. Ce que les *ArDo* n'avaient pu réaliser, la *Diina* l'a imposé: la sédentarisation des Peuls afin d'établir des hiérarchies politiques stables. En fait, la sédentarisation n'est réalisée qu'au milieu du vingtième siècle mais les institutions politiques de la *Diina*, mobilisées par les valeurs sociales et religieuses de l'Islam, vont légitimer, voire renforcer, l'organisation foncière peule élaborée au cours des siècles précédents.

La carte des *leydi* (voir Carte 4.3) est légèrement modifiée afin de récompenser ceux qui ont contribué à l'avènement de la *Diina* et taire les velléités de rébellion. Sans que cela soit systématique, diverses maîtrises foncières, principalement celles des Peuls, sont fixées en arabe sur des *tarikhs* par les marabouts afin d'asseoir le nouvel ordre politique; ce que certains ont appelé abusivement le 'code de la *Diina*', à l'image d'un code foncier édicté par un État moderne. Ce règne de la *Diina* se termine avec l'annexion du Delta à l'empire Toucouleur de El Hadj Omar (1864–93). La volonté des Toucouleurs de vaincre la résistance acharnée des Peuls va vider le *Maasina* de ses habitants. La stratégie de la 'terre brûlée' utilisée par El Hadj Omar provoque une émigration forcée vers l'Est. Incendier les greniers à mil et à riz était son moyen habituel de coercition. A partir de 1880, les grandes familles du *Maasina* sont déportées dans le *Kunari* et Tidiani peut structurer l'ensemble de son empire[6]. C'est seulement sous la colonisation française que les Peuls vont revenir de l'exode et repeupler le *Maasina*, mais ils ne pourront que partiellement rétablir leur maîtrise sur leurs territoires et sur leurs anciens captifs qui prennent leur autonomie. Les agriculteurs étendent de plus en plus efficacement leurs champs aux dépens de l'organisation pastorale. Une nouvelle gestion de l'espace se met en place avec l'appui de l'administration coloniale. Celle-ci recompose une véritable carte des divisions administratives du Delta. Le *Maasina* reçoit une nouvelle extension géographique avec la création du 'cercle de Macina'.

Le terme *Maasina* a donc recouvert différentes extensions géographiques au cours de l'histoire. Aujourd'hui, la première acception de *Maasina* forgée par l'histoire des *ArDo* (Diallo-Dicko) est toujours utilisée localement pour indiquer globalement la région de Ténenkou traversée par le Diaka et comprise entre Diafarabé et Ouro-Nguia. C'est dans ce sens que le terme sera employé ici.

Avec la décolonisation, les cantons sont supprimés et une nouvelle division administrative est mise en place avec des commandants de cercle et des chefs d'arrondissement. Les découpages ne tiennent plus compte des territoires lignagers, des *leydi* ni des anciens cantons. L'État malien proclame la domanialité des terres et des eaux et interdit certaines prérogatives aux *Jowru*, aux anciens chefs de canton, aux maîtres d'eaux (notamment les tributs, prébendes et redevances)[7]. L'État est alors perçu comme s'attaquant à l'ordre traditionnel des maîtrises foncières. Les conflits se multiplient et se résolvent peu.

---

[6] Les différentes provinces étaient les suivantes: le *Jenneri Pondori*, le *Maasina*, le *Gimballa*, Tombouctou, le *Kunari*, les pays de Douentza-Dalla-Hombori, le *Jelgooji*, le plateau central dogon, le *Seeno* Bankass, le *Gondo* et Louta. Le *Maasina* était gouverné par Famori Sissoko qui résidait à Toguéré-Koumbé dont il fit sa capitale (Barry, 1993).

[7] Il faut attendre 1985 pour que les *Jowru* soient reconnus officiellement, à la suite d'une intervention du Ministre de l'Intérieur à la Conférence annuelle des bourgoutières de Mopti.

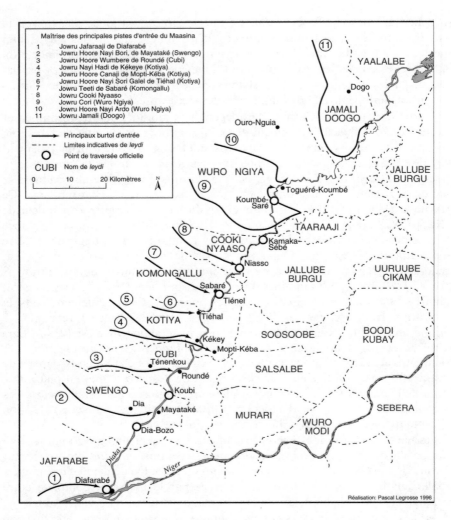

Carte 4.3 *Les principaux intinéraires d'accès au Maasina et le découpage en leydi (Mali)*
(Source: *CIPEA, 1983*)

Les responsables administratifs les entretiennent plus ou moins à leur avantage financier, volontairement ou du simple fait de leur fréquentes mutations, prétexte à soulever à nouveau les problèmes. Comme le fait remarquer Claude Fay (1995), les pouvoirs coloniaux et postcoloniaux n'ont-ils pas été facteurs de désordre? Et l'on peut se demander avec Moorehead s'ils n'ont pas hypothéqué une gestion durable des ressources du Delta? (1991: 251).

L'ensemble des problèmes (fonciers notamment) dans le Delta s'est amplifié avec l'avènement de la démocratie et le sentiment de vacance du pouvoir qui s'en est suivi (Fay, 1995: 46). Après l'inquiétude née des attaques régulières de petites bandes de pillards armés venant de la frontière mauritanienne depuis 1991, provoquant une désorganisation des retours de transhumance, le *Maasina* est aujourd'hui en effervescence avec les débuts du programme de décentralisation visant à un nouveau découpage administratif du pays: 'on déterre toutes les anciennes querelles'. Ainsi, la zone se voit durablement déstabilisé par la contradiction qui s'est amplifiée entre les familles 'propriétaires' traditionnelles des territoires et de nombreuses familles n'ayant qu'un droit d'usage sur des terres, des pâturages ou des eaux. Ces dernières tentent de s'approprier des maîtrises foncières à la faveur des faiblesses de l'administration (Fay, 1995: 40–51; Moorehead, 1991: 216–26)[8].

Le *burgu* et l'inondation de la plaine deltaïque

Le Delta central du Niger est aussi appelé '*burgu*' par les Peuls. Ce terme *burgu* désigne à la fois la plaine inondée, ses pâturages, les prairies pastorales appropriées et certaines herbes qui les couvrent. Le rythme et la nature des activités dans le Delta intérieur du Niger, son attrait au cours de l'histoire pour différentes populations et pour des empires successifs, viennent de l'eau qui inonde une plaine immense au milieu des sables et des grès. Ainsi, la submersion temporaire des plaines du Delta donne une impression d'unité due à la domination des savanes herbeuses et des prairies aquatiques. C'est cette unité 'd'exploitation' que désigne le terme de *burgu* en peul. Si l'ensemble de la plaine inondée entre le Diaka et le Niger se désigne sous le nom générique de *burgu*, le terme a surtout un sens spécifique et qualifie les pâturages, libérés lors du retrait des eaux; c'est dans cette acception qu'il est le plus souvent employé. De façon restrictive, le *burgu* désigne aussi l'association floristique de haute valeur fourragère qui compose ces pâturages aquatiques (*Echinochloa, Panicum* et *Polygonum lanigerum*) dont la principale est l'*Echinochloa stagnina* (*gamaraawo* en *fulfulde*). L'importance du *burgu* pour l'élevage fait que ce terme désigne surtout les pâturages dont un lignage peul a traditionnellement la jouissance: on dira le *burgu YaalalBe*, pour le *leydi* des *YaalalBe*, ou le *burgu Cufuuji* pour un des *sous-leydi* de *Wuro Ngiya*. Le terme *burgu* sera employé ici dans ce sens, pour désigner le pâturage approprié sous la responsabilité du *Jowru*.

Les meilleurs *burgu* sont ceux où pousse en abondance l'*Echinochloa stagnina* qui se présente sous forme d'une longue tige rouge pouvant atteindre deux ou trois, voire quatre mètres de haut selon la lame d'eau. La tige est souvent ramifiée dans sa partie terminale et présente des nuds radicaux formant de nombreux coudes, laissant dans la bouche un arrière goût sucré. L'*Echinochloa* produit en touffe une petite graine, semblable à celle du fonio[9], dont la cueillette

---

[8] Depuis l'avènement de la démocratie, les conflits fonciers sont systématiquement traduits devant la justice. Comme il n'y a qu'un juge par cercle, les procès traînent, la concussion n'a pas disparu et la région a encore connu plusieurs règlements de compte sanglants.

[9] *Digitaria exilis*. Graminée cultivée ou non, dont les grains sont utilisés pour l'alimentation, consommés crus, en bouillie ou en couscous.

apporte un complément alimentaire en période de soudure. Le cycle végétatif de l'*Echinochloa* est directement lié au phénomène de crue et de décrue dans la plaine (Figure 4.1). Durant la période d'exondation, les tiges sont rampantes. A la montée des eaux, elles se redressent, flottent en émettant un feuillage abondant. A la décrue, les tiges se marcottent en donnant autant de repousses qu'il y a de nœuds en contact avec le sol (Laine, 1987: 9–10). En arrivant dans le Delta, les troupeaux trouvent une biomasse importante (de 5–20 tonnes de matières sèches par hectare), formée d'herbes mûres de qualité moyenne à médiocre et d'accessibilité relativement difficile, ce qui entraîne un gaspillage considérable (mais stimule le recrû et la repousse). Après la première pâture, le recrû est consommé en même temps que le reste des herbes sèches dont l'accès est facilité par le brûlage (Wilson *et al.*, 1983: 44).

La modification de la situation hydrographique a été très néfaste au *burgu*: les vingt à vingt-cinq années de pluies médiocres, que le Delta a connu depuis la fin des années soixante, ont provoqué une réduction importante de l'alimentation fluviale par le Niger et par le Bani[10]. De même, l'extension actuelle de la crue, observée depuis une dizaine d'années, est infiniment moindre: au lieu d'immenses superficies d'un seul tenant, correspondant à un vaste ensemble de plaines inondées et de grands lacs, on ne trouve plus que des aires discontinues, inégalement réparties dans l'espace et variables dans le temps. Les superficies inondées actuellement peuvent être considérées comme moitiés moindres par rapport à celles figurées sur les cartes topographiques de l'IGN du début des années soixante[11]. Par contre, les crues exceptionnelles depuis celles de 1994–5 permettent de retrouver l'extension de l'inondation connue avant 1982.

Les études sur le taux et la capacité de restauration naturelle des pâturages du Delta se trouvent confirmées. Comme l'avait fait remarquer Wilson *et al.*, les herbages pérennes du Delta présentent peu de risques 'qu'un surpâturage qui compromette la productivité soutenue et à long terme de l'écosystème . . . se produise à grande échelle, en raison de la restauration de la fertilité du sol et de sa non-utilisation pendant les six mois que dure l'inondation, ce qui permet des taux d'exploitation élevés sur de courtes périodes' (1983: 182). Ce que confirme Pierre Hiernaux par ses analyses de la 'résilience' surprenante des pâturages du Delta (1993; CIPEA actualités, 1995)[12].

Par contre, le riz remplace dans de nombreuses zones les ressources fourragères leur substituant des chaumes ou des jachères. Les animaux en bénéficient après les récoltes mais n'ont plus de recrû de saison sèche. Quand le riz a remplacé l'*Echinochloa*, la perte alimentaire est considérable. Après

---

[10] Le cumul pluviométrique, qui était d'environ 550mm entre 1930 et 1974, est devenu inférieur à 400mm depuis 1974. En 1992, le cumul pluviométrique à Ténenkou est seulement de 284mm, et de 382mm à Mopti.

[11] La limite de la zone d'inondation portée sur les cartes présentées ici est issue de ces cartes de l'IGN.

[12] Au *Maasina*, il n'y a pas de terme *fulfulde* traduisant littéralement le surpâturage. On dira qu'il y a beaucoup (*heewi*) d'animaux ou on désignera les espaces dénudés par l'érosion ou l'assèchement (*karal*) formant, par exemple, comme une tâche au milieu d'un champ de mil.

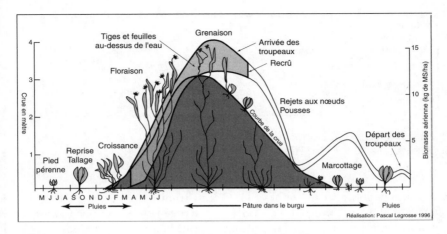

Figure 4.1 *Le cycle de l'*Echinochloa stagnina *(Mali)*
(Source: *Wilson* et al., 1983; Laine, 1987)

l'abandon de la culture, le pâturage peut cependant se régénérer rapidement (de trois à cinq ans).

La représentation peule du milieu

Dans le *burgu*, au-dessus du niveau de la crue se trouve la végétation des *toggere,* tertres à l'abri des eaux, souvent boisées (tamariniers, palmiers, etc.), où se sont édifiés les villages. Au fur et à mesure du retrait des eaux, les premiers pâturages à apparaître sont ceux des *ruunde* (levées, îles) (voir Figure 4.2). Ces levées sont des endroits relativement élevés. Selon la représentation peule du milieu, la tiédeur de leur sol, la proximité de l'eau et leur pâturage presque exclusivement composé d'*Echinochloa* en font des lieux très recherchés. Rapidement épuisées, les levées sont abandonnées pour des pâturages moins riches mais plus nombreux. Les gîtes sont généralement sur ces lieux stratégiques dans la plaine inondée (et sur les tertres en bordure de la plaine). Le temps d'occupation du gîte varie suivant la richesse des pâturages des alentours, c'est-à-dire sa localisation dans une plaine d'inondation (*feeyo*) ou à proximité d'une cuvette (*deebaare*). Ainsi, ces cuvettes (*deebaare*, littéralement cuvette, basse plaine herbeuse et herbe de ce lieu), sont moins élevées que les levées (*ruunde*), mais s'étendent généralement tout autour d'elles. Ce sont des bas-fonds, où l'eau stagne longtemps, où il fait plus frais, où le pâturage est plus maigre mais subsiste plus longtemps. En saison froide, les animaux n'y demeurent pas la nuit mais remontent sur une levée voisine, où la terre, très sèche, demeure plus chaude. Sur la rive gauche du Diaka, toute la zone basse est appelée *ponnga*, plutôt que *burgu*. Levées et cuvettes sont étroitement liées pour l'élevage et appartiennent généralement à un lignage ou à un segment de lignage. Un pâturage approprié comprend une ou plusieurs levées, entourées d'une ou plusieurs cuvettes. Entre eux, la plaine d'inondation (*feeyo*) est un vaste espace, plus bas que la levée, plus

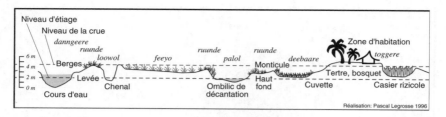

Figure 4.2 *Coupe du* Burgu *(Mali)*
(Source: *Gallais, 1958: 120; 1967: 138–139*)

élevé et moins humide que la cuvette. C'était traditionnellement le seul lieu de pacage qui pouvait être cultivé. Mais les changements socio-économiques et pluviométriques ont fait que les cultures se sont largement étendues.

Les parcours suivent le 'micro-relief' de la plaine. Les gîtes sont sur les zones plus hautes et constituent les points d'accès au pâturage. On plante les cordes à veaux sur ce lieu stratégique. De là, les animaux partent pâturer et reviennent se reposer et se faire traire. Ainsi, la maîtrise d'un pâturage ne se fait pas aux frontières mais de l'intérieur. On pourrait la comparer à ce que font les pasteurs des zones arides qui maîtrisent l'usage de l'herbe par l'appropriation d'un puits. Le CIPEA (1983) dénombrait environ 400 gîtes dans le *Maasina*, déservant entre 700ha et 2000ha en moyenne selon la région[13]. Les gîtes sont les points de replis des pâturages mais aussi les relais s'échelonnant le long des pistes de transhumance. Le temps d'arrêt prescrit à chacune de ces étapes (un, deux, cinq, voire dix jours) varie en fonction de la distance séparant deux gîtes consécutifs (afin de donner le temps aux bovins exténués de reprendre des forces) et des règles de préséance fixant l'enchaînement des accès aux pâturages.

Établi sur un écosystème dynamique, aux ressources très importantes, l'organisation de l'exploitation des pâturages du Delta tient compte du 'non-équilibre' des ressources (Turner, 1992) en s'adaptant dans le temps mais relève d'un système strict et rigide dans l'espace: des itinéraires précis de transhumance à travers le *burgu* approprié et des règles de préséance à l'intérieur de l'*eggirgol* entre troupeaux. L'usage et le contrôle des parcours dans le Delta sont la charpente d'un espace socialement maîtrisé, reliant les pâturages entre eux; pâturages où un 'lignage s'est assis', où il a marqué son emprise par un ou plusieurs gîtes, souvent par le sang versé à la suite d'une bataille ou d'une compensation. La priorité de l'accès au pâturage par le lignage s'exerce en complément du principe de réciprocité entre patrilignages alliés et permet la mobilité pastorale.

---

[13] En 1975, le Service de l'élevage dénombrait 385 gîtes dans le cercle de Ténenkou, dont moins d'une centaine n'étaient plus utilisés à cause de la sécheresse ou étaient cultivés. Des bornes en ciment ont été posées sur certains des gîtes utilisés afin de les protéger d'une mise en culture sur une aire de 200m de diamètre.

Le système pastoral du Delta et ses rythmes

Les espèces animales élevées par les Peuls du Delta sont essentiellement et en priorité les bovins (40,7 pour cent du cheptel en 1992); puis viennent les caprins (30,2 pour cent) et les ovins (25,3 pour cent). Les grands troupeaux transhumants sont uniquement composés de bovins. Ils varient de quelques têtes à 200 ou 300 têtes pour un ensemble de 500 000 bovins appartenant aux habitants du Delta alors que plus d'un million de bovins fréquentent la zone inondée en saison sèche (voir Figure 4.5)[14]. Seuls les pasteurs allochtones mélangent le petit bétail et les bovins qui transhument ensemble et fréquentent les mêmes lieux, ce qui est inacceptable pour un berger peul du Delta. Une partie seulement des petits ruminants quitte la zone en période des hautes eaux et reste sur les bordures. Quant aux bovins, ils sont regroupés en troupeaux différents au cours de l'année selon plusieurs critères de gestion, ce qui a pour effet de décaler les départs en transhumance du cheptel jusqu'en septembre. Par contre, tous reviennent ensemble en novembre. La stratégie est d'optimiser l'exploitation des vaches en lactation (*benndi*) tout en assurant une sortie du Delta à l'essentiel du cheptel (*garci*) pendant la saison des pluies. Cette stratégie consistant à diviser le cheptel du Delta préexiste au 'code de la *Diina*' et les Peuls allochtones font la même division chez eux. Les *WuwarBe* de Nampala séparent aussi le gros du troupeau transhumant (*garci*) des vaches allaitantes (*bireeteedi*) et tous les Peuls du Farimaké divisent le cheptel entre le *garci*, les vaches en lactation (*dabitooji*) – qui rejoignent le *garci* dans le Delta au mois de mai – et quelques vaches allaitantes (*ceetooji*) – qui restent au village toute la saison sèche.

Parmi les pasteurs des terres sèches venant transhumer dans le Delta, on distingue les *WuwarBe*, les *SeenonkooBe*, les *FarimakenkooBe*, les *SonnaBe*, les *CookinkooBe*, et les *NasaDinkooBe*. Ce sont ces lignages, pris dans leur ensemble, que nous désignons par 'allochtones'. En novembre, la plupart de ces pasteurs, ayant achevé la récolte de leurs petits champs de mil, barricadent d'épineux leurs paillotes, ou ferment leur maison, chargent leur hutte en pièces détachées sur des ânes et s'ébranlent en familles à la suite des troupeaux, ne laissant que quelques vieillards, *RiimayBe*, jeunes enfants et vaches allaitantes dans les villages. Puis les troupeaux se dirigent en ordre lâche vers le Delta. Tous ces pasteurs vont se retrouver, souvent plusieurs fois au cours de leur parcours, en relation avec l'un ou l'autre des nombreux *Jowru* du Delta. A chaque fois, ils auront à traiter avec eux, ou avec leurs représentants, des conditions de leur passage ou de leur stationnement dans un *burgu*.

Pendant ce temps, les troupeaux du Delta se rassemblent en bordure du *Maasina* et se mettent en 'file indienne' selon un ordre précis pour emprunter les pistes qui les conduiront à leurs pâturages (*burgu*) respectifs (voir Figure 4.3). Chaque *eggirgol* en file ordonnée de troupeaux porte un nom, et à chaque *eggirgol* correspond un itinéraire précis, la 'piste de transhumance' (*burtol*). On dénombre onze portes d'entrée principales pour l'accès au *Maasina* (voir

---

[14] *Recensement du cheptel national*, Direction Nationale de la Statistique et de l'Informatique, Bamako, mai 1992; chiffres cumulés pour les cercles de Mopti, Djenné, Ténenkou et de Youvarou, qui couvrent l'ensemble du Delta.

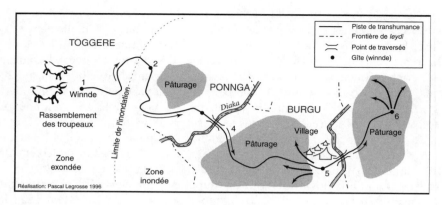

Figure 4.3 *Le parcours de transhumance (Mali)*

Notes

1. *Les troupeaux se rassemblent et forment un eggirgol face à leur piste (burtol, piste aménagée pour le passage des animaux) sur les bordures exondées (toggere) à la limite de la zone inondée et attendent le signal de l'entrée donné par le Jowru en tête de l'eggirgol.*
2. *L'arrivée dans un des principaux villages de la zone haute est souvent accompagnée de festivités (nabbere, montée du village par les animaux).*
3. *Le parcours se fait selon des règles strictes de préséance et d'étape sur des gîtes de repos et de traite (winnde), tout en permettant aux animaux de brouter suffisamment la nouvelle herbe libérée par la décrue. Les pistes que suit l'eggirgol à travers le ponnga (zone entre le burgu et les toggere) sont souvent dans l'eau (gumpol, voie tracée dans une étendue d'eau couverte d'herbes).*
4. *La traversée du Diaka donne toujours lieu à d'importantes festivités (lumbal, traversée d'un cours d'eau) où les meilleurs bergers sont récompensés pour l'embonpoint de leurs animaux.*
5. *Retour des bergers et de leur troupeau au village d'origine. L'eggirgol est arrivé dans les pâturages de ses lignages (burgu), ses troupeaux se dispersent.*
6. *Les troupeaux suivistes ('forains') continuent vers les pâturages de leurs propres lignages.*

Carte 4.3). Avec la nationalisation des terres dites 'vacantes', le service de l'élevage entreprend de contrôler la transhumance. Ainsi sept points de traversée du Diaka (voir Carte 4.3) sont surveillés par l'administration. La 'Conférence des bourgoutières' fixe annuellement les dates de traversée et organisait, jusqu'en 1991, une perception sur les troupeaux qui traversaient[15]. Dans le couci de préserver les pâturages du Delta, une politique hydraulique ambitieuse a été mise en place dans les zones sèches afin de retarder l'entrée des animaux dans la plaine inondée. Mal adaptée, cette politique eut pour conséquence d'empêcher la régénération des pâturages et multiplier les causes de conflits (Moorehead & Lane, 1995: 432).

Le rang qu'occupe chaque troupeau dans l'*eggirgol* et l'ordre d'accès de chacun dans les pâturages appropriés est extrêmement précis et strictement respecté. Le *Jowru* veille attentivement à ce que tous respectent cette discipline

---

[15] Son montant était de 5000 FCFA par troupeau aux points de traversée officiels. Les recettes ont été estimées à 100 millions de FCFA par an. Ces sommes devaient être investies par la Coopérative des éleveurs dans la promotion de l'élevage. Elles ont le plus souvent été détournées par l'administration ou utilisées pour l'accueil des personnalités lors des fêtes de retour de transhumance.

et suivent exactement la piste de transhumance et ses étapes: les lieux de regroupement nocturne, les gués, qui peuvent être taxés, ou encore les lieux de passage en terre sèche. D'ailleurs, son propre troupeau passe en tête; viennent ensuite ceux du patrilignage, puis des familles alliées et enfin ceux des autres familles du *leydi*. Les allochtones, au fur et à mesure de leur intégration dans l'*eggirgol*, sont maintenus à l'arrière. Cet ordre est défendu pour plusieurs raisons:

o 'des raisons de gestion économique qui font rechercher les pâturages vierges – les animaux s'empressant de brouter les meilleures herbes – dont seuls les troupeaux de tête peuvent bénéficier

o 'des raisons liées au 'droit de propriété' du lignage maître qui peut seul 'ouvrir la porte' de son *burgu*

o 'des raisons de prestige attachées au berger du premier troupeau (*Jowru*), qui a l'honneur et la responsabilité d'ouvrir la marche, de s'engager en tête du long *eggirgol*, de connaître où sont les passages, où les marigots se traversent, où les bas-fonds s'abordent

o 'des raisons mystiques, car le *Jowru* posséderait le don de nuire à l'élevage (faire tarir ou rendre stérile les vaches allaitantes) et pratiquerait diverses cérémonies propitiatoires (Gallais, 1958: 132; 1975: 360–1; 1984: 200).

S'il le faut, le *Jowru* fera respecter cet ordre de la transhumance et son intérêt (lié aux redevances) avec le bâton et l'aide de ses alliés (Gallais en donne plusieurs exemples sanglants 1967: 382; 1984: 206–207).

Les règles de la transhumance, l'origine du *Jowru*

Avant la *Diina*, l'existence ou la permanence des *Jowru* semble liée à la bonne volonté des *ArDo*. Certains aspects du rôle spécifique et de la compétence des *Jowru* justifient leur institution et leur reconduction ensuite par la *Diina*, qui les renforce dans leur fonction pastorale. Aujourd'hui, le titre de *Jowru* est intimement lié à la perception de redevances, *conngi*[16], en espèces ou en nature pour l'exploitation du *burgu*. Il en reverse une partie aux hommes de sa famille (le patrilignage). C'est également le *Jowru* (entouré des membres de son patrilignage) qui établit la durée des étapes le long de l'itinéraire suivi en retour de transhumance. Il veille aux champs non récoltés et à la hauteur de l'eau dans certains passages profonds (*deebaare*). Dans son *burgu*, il a toute latitude de déclasser quelques parties au profit de l'agriculture: la gestion foncière de son *burgu* lui incombe directement si son autorité est reconnue par les agriculteurs et l'administration.

Le Delta est découpé en territoires agro-pastoraux, les *leydi*. A la chefferie des

---

[16] Dans le *Maasina*, la redevance que versent les allochtones pour l'exploitation du *burgu* n'est pas dénommée du terme *'tolo'* retenu par la plupart des auteurs qui ont abordé ce sujet, mais par *'conngooji'* ou simplement *'conngi'*, un ampliatif de *coggu*, 'prix d'achat ou de vente d'une marchandise'. Ce terme technique de *conngi* désigne 'la redevance pour profiter du *tolo*', le *tolo* étant un terme désignant le pâturage, l'herbe verte (du verbe *talnude*, 'verdir'), la pâture étant soumise à redevance.

leydi correspondent des lignages ou des segments de lignage et un espace géographiquement donné et historiquement limité. Le terme *leydi* désigne à la fois le sol, la terre et le territoire, et suppose une organisation lignagère. Lorsque l'on parle du *leydi Wuro Ngiya*, il s'agit du territoire dépendant politiquement, administrativement et du point de vue foncier du chef du village de Ouro-Nguia (ancien chef de canton). Le titre de *Jowru* est détenu par un segment de lignage lié à un pâturage particulier qui a ses gîtes et ses règles d'accès pour les animaux, fixées par la préséance à l'intérieur du *leydi*. Dans certains *leydi*, il existe une différenciation entre *Jowru* suivant le type de *burgu*, ou sa localisation, fruit de l'histoire: il est *Jowru Jom huDo, Jom tolo* ou *Jom togge* (Ba et Daget, 1984: 73–4). On pourrait y retrouver une sorte d' 'écologie politique' (Gallais, 1984: 37 et 125–6) qui ferait correspondre une configuration naturelle à un titre et au statut d'un lignage (voir Figure 4.4).

Dans la pratique, ces différenciations signifient que l'utilisation des espaces agro-pastoraux des *leydi*, à quelque fin et par quelque individu que ce soit, n'est jamais soumise à l'approbation des lignages, mais laissée à la discrétion du *Jowru Jom huDo* pour le *gumpol*, du *Jowru Jom tolo* pour le *burtol* et le *burgu*, et du *Jowru Wuro* pour le *harima* (pâturage réservé aux vaches allaitantes du village). Dans l'organisation des maîtrises foncières, ces titres sont donc attachés à l'usufruit permanent et héréditaire de parcelles du *leydi*[17] dont les plus visibles sont les 'sous-leydi'. Dans notre exemple, les *sous-leydi* de *Wuro Ngiya* correspondent à des troupeaux de transhumance, tels *Hoore Cori, Nayi Aamadu, Cufuuji, Lawsi* (voir Carte 4.4). Chaque *sous-leydi* est sous les ordres d'un *Jowru*. Cet ensemble est administré par *Jowru* de tête de l'*eggirgol* et par un chef berger dans certains domaines. Néanmoins, les différents *sous-leydi* qui le constituent ont une autonomie assez forte. C'est ainsi que le *sous-leydi Taaraaji* a pu se détacher en 1979 de l'ensemble du *leydi* et gérer seul ses terres.

Entre l'étude des *leydi* du Delta de Jean Gallais faite en 1967 (1967: 141–143) et celle du CIPEA effectuée en 1983, on constate une bonne concordance avec cependant des différences significatives. Dans le *Maasina* en particulier, là où Jean Gallais relevait onze *leydi*, le CIPEA n'en trouve plus que cinq. En vingt-cinq ans seulement, on assiste à des naissances, comme *Taaraaji*, et à des regroupements. En fait, les frontières des *leydi*, leur nombre et leur répartition n'ont pas cessé de se modifier ou d'être modifiés. Cette mobilité des *leydi* apparaît forte et semble contradictoire avec ce qu'affirment facilement les Peuls quant au caractère intangible des frontières pastorales. Étant le jouet des changements politiques et des rapports de force, leur plasticité est aussi directement liée aux modifications des itinéraires de transhumance, mais surtout à la création ou à la suppression de gîtes. Au cours de la même période, un seul parcours a été créé, semble-t-il, celui du nouveau *leydi Murari*. À cela s'ajoutent plusieurs modifications dont la plus importante est celle du parcours qu'un *Jowru* du *Maasina* a changé, par la force et au détriment d'un autre *eggirgol*

---

[17] Ces droits sont suspendus lorsqu'il s'agit d'une mise en culture: indéfiniment si le tenant est un autochtone du *leydi* et s'il exploite de manière continue l'espace qu'on lui a concédé; temporairement si, au contraire, ce tenant est un étranger.

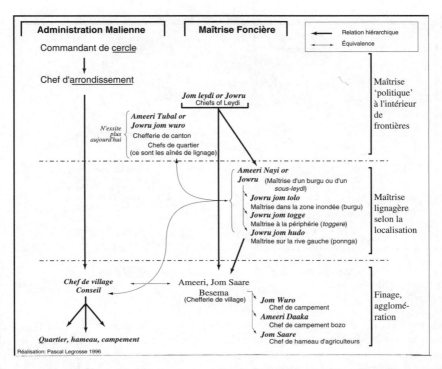

Figure 4.4 *Schéma de la structure institutionnelle et de ses découpages spaciaux (Mali)*

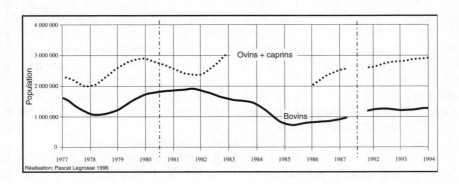

Figure 4.5 *Evolution des effectifs du cheptel dans la Delta central du Niger (Mali)*
(Source: *DNE (1977–80), Estimations ODEM (1981–7), Recensement du cheptel national (1992–4)*)

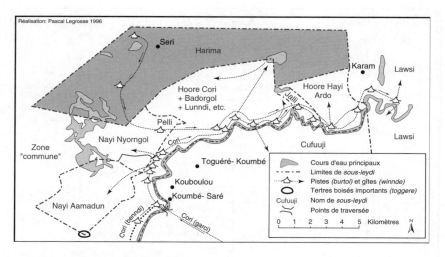

Carte 4.4 *Exemple de division en* leydi *et* sous-leydi *(Mali)*
(Source: *CIPEA, 1983*)

bien trop faible numériquement, pour emprunter un itinéraire plus court, rendu possible par la sécheresse. Cette modification permet à ce *Jowru* d'attirer plus de troupeaux 'forains' (pressés de rejoindre leur *burgu*), et de recevoir ainsi en retour des redevances importantes.

La collecte du *conngi*

L'accès autorisé au *burgu* se conclut par le versement d'une redevance au *Jowro*. Ces prestations sanctionnent la distance sociale existante entre les droits des troupeaux 'citoyens' du *leydi,* les 'forains' ressortissant d'un autre *leydi* (Schmitz, 1986: 378) et les 'allochtones'. Selon un *Jowru* du *Maasina*: 'si tu t'es installé le premier sur une zone et que tu te l'appropries, si quelqu'un vient sur cette zone, il faut qu'il te donne quelque chose. Il faut que tu parles et que tu t'entendes avec lui. Il enlève quelque chose et te le donne.' Le paiement de redevances est directement lié à la reconnaissance du droit du premier occupant qui implique le geste de reconnaissance du propriétaire par les allochtones qui viennent paître sur ses terres. Le versement du 'prix de l'herbe' marque la différence de statut entre autochtones et allochtones, et maintient la distance sociale.

Le *conngi* donne un droit de camper et de pâturer limité dans le temps: l'accès aux pâturages ne peut se faire qu'après l'entrée des troupeaux des autochtones (règle de préséance) et jusqu'à l'arrivée des premières pluies, signal du départ de tous les troupeaux hors du *Maasina*. Son montant va être l'aboutissement d'une négociation jouant sur la date d'accès, les circonstances, la qualité du *burgu*, la reconnaissance de l'unité territoriale, de l'autorité qui la domine et des liens de clientèle. Son institution est liée à celle du *Jowru* 'au temps des *ArDo*'. Dès lors, sans être jamais supprimé, le *conngi* a évolué avec les tributs versés au suzerain

92

pour devenir une importante source de profit pour le *Jowru* et sa famille[18]. Les politiques étatiques successives n'ont pu ensuite modifier cette institution.

La première collecte se fait lors d'une assemblée en brousse au retour des animaux de transhumance. Il s'agit de 'l'assemblée de celui qui est venu et que l'on doit prendre', disent les Peuls. La réunion se fait près de la corde à veau à côté du lieu où le *Jowru* dépose ses gourdes de lait sur le gîte. Le versement du *conngi* à cette étape du parcours permet à l'allochtone d'intégrer son troupeau à l'*eggirgol*. Le *Jowru* lui donne une place, généralement à la queue de la file. En étant dans l'*eggirgol*, les animaux de l'allochtone vont profiter des meilleurs pâturages quasiment comme les animaux des autochtones. Ils camperont sur le même gîte et pâtureront les mêmes herbes (*burgu*). L'avantage est appréciable pour la prospérité des animaux et pour le prestige qu'il confère.

Sinon, pour les allochtones qui entrent dans un *burgu* après le passage de l'*eggirgol*, les animaux ne vont bénéficier que d'un pâturage déjà en partie brouté ou piétiné. Ils ont 'l'herbe de la boue'. Ils s'installent assez loin des villages ou sur le lieu habituel qui leur avait été indiqué par le *Jowru*. Ils vont soit s'adresser au *Jowru*, chez lui, soit ils attendent son passage ou celui de son représentant. Le patrilignage propriétaire peut déléguer deux ou trois personnes pour se rendre auprès des allochtones afin de prendre ce qu'ils doivent verser. Les délégués reviennent rendre compte de leur gain et le résultat est partagé dans le patrilignage. Les fils et les frères du *Jowru* vont aussi chercher le *conngi* pour le patrilignage.

Une réciprocité lie aussi les propriétaires de *burgu*. De fait, entre *Jowru* de terrains complémentaires, il est bon d'avoir de petites attentions: le *Jowru jom togge* ne fait pas payer de redevance au *Jowru jom tolo* au moment de son passage, et réciproquement, ce dernier s'abstient de demander quoi que ce soit au maître du *togge* (collines à la limite de l'inondation) ou du *gumpol* (piste dans l'eau entre les *togge* et le *burgu*).

La crise de l'organisation spatiale de la transhumance

L'ordre politique apporté par la *Diina* et fondé sur l'hégémonie peule a fait des *Jowru* les seuls et uniques propriétaires coutumiers des pâturages. Ils géraient très lucrativement l'exploitation même des *burgu*. Intransigeants dans cette entreprise, ils accordaient le droit de pâture, de culture, quand et à qui ils voulaient. Toute transgression, irrégularité, ou non-observation des coutumes étaient passibles de représailles souvent très sanglantes. Si les pistes, les gîtes, et leur propriétaire respectifs sont bien connus de tous encore aujourd'hui, en cas de conflit, cela devient chose particulièrement confuse car la coutume, le droit

---

[18] Le principal critère pour évaluer le *conngi* est la taille du troupeau. Si le troupeau est grand, on peut demander à l'allochtone un taureau de deux ou trois ans ou entre 5 000 et 25 000 FCFA (pour certain *burgu* du Delta, le montant peut-être bien supérieur. Des redevances de 100 000 FCFA ont été observées). Avec l'augmentation du cheptel, la réduction des pâturages, le développement de l'économie marchande, le *conngi* a pris de l'importance pour devenir une 'rente' se payant de plus en plus souvent en argent. De plus, son montant a connu une hausse importante depuis la *Diina*, qui avait tenté d'en fixer les règles, pour atteindre des sommets vertigineux (Gallais, 1984: 198).

coranique, les traditions, le droit de conquête et l'organisation imposée par Sékou Amadou, s'interpénètrent ou s'opposent. Le problème le plus immédiat auquel sont confrontées les sociétés pastorales est bien la réduction de leur domaine sous la poussée pionnière des paysans. Ici, les causes de conflit sont nombreuses puisque les cultivateurs dont les cultures sont limitrophes des pistes ont tendance à agrandir leurs champs au détriment de celles-ci. Ceci constitue un embarras pour les pasteurs et une source importante de dégâts dans ces champs lorsqu'ils ne sont pas récoltés à temps. D'autre part, les agriculteurs vont cultiver certains gîtes proches du village car, ayant servi au stationnement d'un bétail nombreux, leur sol abondamment fumé est de ce fait plus riche que les autres. Soit cette pratique relève du fait accompli, soit l'agriculteur demande l'autorisation de cultiver une partie du gîte au *Jowru* en échange d'une redevance en nature (le *conngi leydi*): la remise d'une fraction de sa récolte. Cette location informelle de terres peut le mettre à l'abri de plaintes éventuelles.

La fin de l'hégémonie peule, l'administration défavorable ou méconnaissant le pastoralisme, l'évolution démographique, l'insécurité alimentaire et la sécheresse ont abouti à une situation conflictuelle entre 'l'ordre du parcours' et 'l'ordre du champ', entre la pâture et la récolte, entre maîtrise pastorale et maîtrise agricole. Du point de vue pastoral, les conflits fonciers actuels ont le plus souvent pour origine la captation de la rente sur l'exploitation de champs aux confins de deux pâturages relevant de maîtrises lignagères différentes. Les deux patrilignages vont se quereller pour qu'une frontière soit définie entre eux (celle-ci peut n'avoir jamais existé ou être connue implicitement par les usagers). Ils font appel au *Jom leydi* (voir Figure 4.4), s'ils relèvent du même *leydi*, ou à l'administration pour que les limites de leur pâturage soient reconnues, que leur préséance sur l'autre partie soit acceptée. Si leur prétention est légitimée, les champs sont alors de leur ressort. Il n'y a plus d'instance supérieure au *Jom leydi* et seule l'administration ou la justice sont compétentes. Leur intérêt pécuniaire est d'ailleurs trop important pour laisser au *Jom leydi*, au *Jowru*, ou au *Jom Saare*, le soin de régler un conflit.

D'autres phénomènes jouent dans le déclin de l'organisation traditionnelle de la transhumance: le changeant de propriétaire du bétail (riches *RiimaayBe*, commerçants *JaawanDo*, anciens combattants de l'armée coloniale, fonctionnaires de l'État malien) conduit une bonne partie des pasteurs à devenir des bergers salariés. Pour des raisons d'insécurité dans le Sahel et le Gourma, on assiste à un raccourcissement de la longueur des parcours de transhumance. Cette pratique est aussi le fait d' 'éleveurs absentéistes', de propriétaires de troupeaux dont l'activité principale n'est pas l'élevage. Elle est donc liée à l'émergence de nouveaux systèmes d'élevage dans la zone. Inversement, les retours de transhumance sont effectués souvent de manière plus précoce qu'auparavant. Ceci s'explique par l'insécurité, la succession de mauvaises pluviométries d'hivernage, mais aussi par la concurrence entre les troupeaux pour l'accès à un *burgu* appauvri. Par conséquent, les occasions de conflits entre agriculteurs (cultivateurs de mil ou de riz) et pasteurs se trouvent accrues, car quand les troupeaux reviennent trop tôt, les champs ne sont pas encore récoltés.

Si les traditions de l'élevage et de l'agriculture demeurent, ce n'est plus avec le même contenu: les anciens pasteurs préfèrent cultiver leurs terres à *burgu* au lieu de les livrer à des troupeaux qui leur sont devenus étrangers. À cela, il faut ajouter l'augmentation démographique de la population (humaine et animale), le fait que des pasteurs peuls et des pêcheurs bozos se soient mis à l'agriculture, la dislocation du système politique traditionnel: tout ceci concourt à une nouvelle utilisation de l'espace et à l'occupation de terres jusque-là préservées de toute culture.

Le déplacement des champs (jachère, extension des champs, affirmations individuelles) perturbe et déprécie l'espace pastoral qui doit se contenter des terres incultes et non cultivées. Il le déprécie dans la mesure où l'extension des cultures et leur intensité modifie le couvert végétal. Il le perturbe parce qu'il entre en conflit avec le mouvement spatial de l'élevage quand les champs occupent des pistes de transhumance et enserrent dans un étau les mares, posant un sérieux problème d'abreuvement.

Il revient normalement au *Jowru* de moduler l'accueil qu'il peut accorder aux troupeaux allochtones dans le *burgu*. S'il les accueille largement, les pasteurs autochtones craignent que les réserves pastorales deviennent insuffisantes pour leurs propres troupeaux. Au contraire, le *Jowru* a quant à lui intérêt à ouvrir le *burgu* aux troupeaux allochtones. L'entrée de ces troupeaux lui permet en effet de recueillir son *conngi*, mais aussi de faire reconnaître le prestige de son *burgu* par l'attrait qu'il exerce. L'appauvrissement de certains *Jowru*, le fait qu'ils aient tendance à considérer les pâturages qu'ils sont chargés de gérer au nom de la communauté comme des propriétés privées, modifient radicalement le sens du *conngi*. A l'origine, le paiement du *conngi* devait permettre de limiter l'entrée des troupeaux allochtones dans le *burgu* et signifiait la reconnaissance, par le pasteur allochtone, de l'autorité des pasteurs autochtones du *leydi* sur les ressources de leur *burgu*. Aujourd'hui, le *conngi* est devenu une importante source de profit pour le *Jowru* et sa famille. Les allochtones, comme les 'éleveurs absentéistes', sont donc facilement admis, à la condition de verser une redevance. Ils peuvent aussi, à certains endroits, devancer les Peuls sur leur propre *burgu*: ce qui provoque de terribles conflits. Dans certains cas, la pression des troupeaux est telle et le pouvoir du *Jowru* est devenu tellement déliquescent que l'utilisation du *burgu* se fait sans autorisation préalable. Les pasteurs pratiquent alors la politique du fait accompli et entrent dans le *burgu* sans consultation.

Conclusions

La pertinence technique du modèle de l'accès et de l'itinéraire de transhumance régulé a pour conséquence de ne pas faire l'objet d'une contestation globale. Mais ses assises spatiales sont attaquées. A l'origine de ces attaques se trouvent, d'une part, la dynamique de changement des systèmes de production à l'oeuvre dans tout le Delta et, d'autre part, la faible capacité de résistance de l'organisation foncière pastorale.

Lorsque nous demandions à des *Jowru* comment ils voyaient l'évolution de leur fonction, il nous a été répondu ceci:

L'existence du *Jowru* est un avantage. Le *Jowro* n'est pas fait pour nuire. Chaque *Jowru* est comme un gardien. Si une chose n'est pas gardée, elle ne peut pas être bonne. C'est cela (la charge), de la *Diina* jusqu'à présent. Toute chose non gardée ne peut-être bonne, c'est pourquoi les *Jowru* sont faits pour garder les *leydi*. On veut le bonheur du pays. Depuis *njaahilaaku* (le règne du paganisme avant la *Diina*) jusqu'à l'arrivée de Sékou Amadou, puis des Toucouleurs, des Blancs, jusqu'à l'arrivée du Mali. A toutes ces époques, ce sont les *Jowru* qui gardent tout comme le gouvernement garde les personnes. Rien n'a changé dans le *njowraaku* (le règne des *Jowru*). C'est le temps (la nature) qui change, mais les *Jowru* gardent leur rôle de surveillant pour que les choses soient bien gérées.

Il semble désormais essentiel que l'organisation des maîtrises et de l'usage des pâturages soit reconnue officiellement comme 'mise en valeur' afin que les ressources pastorales obtiennent le même statut que les terres agricoles dans le cadre d'un nouveau Code foncier. De même, la place de l'État est à redéfinir afin qu'il joue son rôle de sécurisation des producteurs, de reconnaissance des prérogatives propres aux maîtrises locales et de justice modérée dans les oppositions paysannes.

5

The role of social networks, indefinite boundaries and political bargaining in maintaining the ecological and economic resilience of the transhumance systems of Sudano-Sahelian West Africa

Rôle des réseaux sociaux, de l'absence de limites définies et des compromis politiques dans le maintien de la capacité d'adaptation écologique et économique des systèmes de transhumance en Afrique de l'Ouest soudano-sahélienne

MATTHEW D. TURNER

ABSTRACT

Conventional common-property theory has provided the analytical under-pinnings for recent development efforts to formalize indigenous usufruct rules. The idealized goal of many rural planners seems to be to replace the political contentiousness of resource allocation within local communities with a well-ordered situation consisting of a managing group of fixed membership, allocating access to a well-defined, bounded resource through specified channels. This would undoubtedly result in significant benefits derived from increasing the ability of managers to exclude outsiders. However, in tinkering with local resource-management institutions, developers should be cognizant of the subtle difference between *excludibility* and *exclusion*. Replacing the allocative discretion of local resource managers (excludibility) with formal rules circumscribing the pool of outside candidates and specifying the acceptable means of their gaining access (legal exclusion) will result in a more ordered but rigid system. In areas such as the semi-arid zone of Africa where resource availabilities fluctuate widely, increased usufruct rigidity is likely to increase the economic and ecological vulnerability of rural producers.

Herd mobility is a critical factor affecting the ecological and economic resilience of livestock production systems in semi-arid Africa. Within a transhumance context, all herders are 'outsiders' during a portion of the yearly cycle. The formalized reality described above would significantly reduce herd mobility and herders' ability to respond to change due to a reduction in the number of key resources effectively available to each of them. Less formal tenure institutions that are typical of trans-humance systems in West Africa are politically malleable and, as such, allow more flexible response to changing ecological and economic conditions.

97

In describing two contrasting Fulani transhumance systems in West Africa, this chapter will focus on the nature of those informal institutions that provide some space for bargaining and negotiation. While their social, non-legalistic nature is more consistent with the needs of a spatially variable production system, it also makes their resource-management aspects more vulnerable to socio-economic change and political manipulation. The implications of these case studies will be discussed in the context of recent interest in the invigoration of indigenous transhumance organizations. It is argued that indigenous models of transhumance tenure are highly vulnerable to the ubiquitous threat of agricultural encroachment. For this reason, hybrid (legalistic/political) tenure models are necessary where a transhumance system based on the informal 'stranger–host' Fulani tradition is subsumed within a more legalistic, spatially bounded model governing the pasture/cropland interface. For such models to work, reforms in the representative structures involved in land-use conflict resolution may be necessary. In many areas, real representation of pastoral interests at both the local and regional levels is lacking.

RÉSUMÉ

La théorie classique de la propriété commune a servi de base analytique aux tentatives faites récemment pour donner un caractère officiel aux règles d'usufruit indigènes. Nombre de planificateurs du développement rural paraissent avoir pour objectif idéal de supprimer la répartition des ressources au sein des communautés locales au moyen de discussions politiques en chargeant un groupe de gestion à composition fixe d'attribuer de façon ordonnée l'accès à une ressource limitée et définie par des voies bien établies. Un tel système offrirait sans aucun doute des avantages non négligeables du fait que les gestionnaires seraient mieux en mesure d'exclure les gens de l'extérieur. Toutefois, les responsables du développement qui prétendent réformer les institutions de gestion des ressources locales doivent prendre en compte la différence subtile existant entre la *possibilité d'exclure* et *l'exclusion*. Le remplacement du pouvoir discrétionnaire d'attribution des gestionnaires des ressources locales (possibilité d'exclure) par des règles officielles qui limitent la liste des candidats extérieurs et l'établissement de moyens précis pour obtenir l'accès aux ressources (exclusion juridique) aboutit à mettre en place un système plus rationnel mais aussi plus rigide. Dans des régions comme la zone semi-aride d'Afrique où les disponibilités de ressources sont extrêmement variables, des régimes d'usufruit plus rigides risquent d'accentuer la vulnérabilité économique et écologique des producteurs locaux.

La mobilité des troupeaux a une influence critique sur la solidité écologique et économique des systèmes de production animale dans les zones semi-arides d'Afrique. Dans le contexte de la transhumance, tous les éleveurs sont en situation d''étranger' pendant une partie du cycle annuel. Le système le plus structuré mentionné plus haut diminuerait sensiblement la mobilité des troupeaux et l'aptitude des éleveurs à réagir aux changements en réduisant les ressources essentielles dont chacun d'eux peut effectivement disposer. Les régimes de tenure moins formels qui caractérisent les systèmes de transhumance d'Afrique occidentale sont malléables politiquement et permettent donc de répondre avec plus de souplesse à l'évolution de la situation écologique et économique.

Ce chapitre qui présente deux systèmes de transhumance Fulani en Afrique occidentale traitera en particulier de la nature des institutions informelles qui ménagent une marge de négociation et de marchandage. La nature sociale et non juridique de ces institutions répond mieux aux besoins d'un système de production mobile dans l'espace, mais elle rend aussi leur gestion des ressources plus sensible aux changements socio-économiques et aux manipulations politiques. Les incidences de ces monographies seront étudiées en liaison avec l'intérêt qu'inspire depuis peu le renforcement des organisations indigènes de transhumance. On soutient que les modèles indigènes de régime foncier de transhumance sont gravement exposés à la menace omniprésente que constitue l'avance des cultures. Pour cette raison, il est nécessaire d'établir des modèles de tenure hybrides (juridiques/politiques) permettant d'insérer un système de transhumance fondé sur la tradition informelle 'étranger-hôte' Fulani dans un modèle plus juridique limité dans l'espace qui est applicable à l'interface pâturages/terres cultivées. Pour que ces modèles donnent de bons résultats, il pourrait être nécessaire de réformer les organes représentatifs participant au règlement des litiges concernant l'utilisation des terres. Dans de nombreuses régions, les intérêts des pasteurs ne sont réellement représentés ni au niveau local, ni au niveau régional.

ENVIRONMENTAL CONCERNS HAVE very much motivated the pastoral commons debate. Unfortunately, this debate has not fully addressed the different ways in which tenure institutions may affect livestock-induced environmental stress. The productivity of African rangelands and their sensitivity to grazing vary both spatially and temporally. Environmental impact of pastoral livestock on rangeland is determined both by the regional livestock population and by the spatio-temporal distribution of that livestock population in the region. Equilibration of livestock densities to a variable vegetative resource depends on herd mobility. Both critics and proponents of a 'tragedy of the pastoral commons' have stressed tenure's effect on the demand for livestock (internally or externally owned) within a particular circumscribed pasture. Unfortunately, both sides of the debate have tended to ignore the special and often contradictory tenure characteristics that herd mobility requires.

This chapter will explore the socio-political requirements of herd mobility, their pastoral tenure implications, and the relationship of these implications to conventional common-property arguments. It will begin with a review of the strong influence of the tragedy of the commons framework on the pastoral commons debate and how its influence has become increasingly vulnerable as its ecological foundation has shifted. Growing appreciation among ecologists of herd mobility as a management parameter will be shown to modify conventional perspectives on key pastoral institutional variables.

After a review of the basic features of transhumance systems of Sudano-Sahelian West Africa, the tenure characteristics of two contrasting Fulani transhumance systems will be described. The first, that of the Diaka floodplain of the Inland Niger Delta of Mali (Maasina), has one of the most formal organizational structures in Africa. The second, that of the Fulani transhumance system in the Say Arrondissement (district) of south-western Niger, has little formal organization. Organizational differences between these two cases can be traced to:

o differences in the spatial distribution and quality of pasture resources
o the history of transhumance
o the degree of incorporation into pre-colonial state structures
o outside interventions during the colonial and post-colonial periods
o the political cohesiveness of the resident Fulani populations in these two areas.

What will be stressed in this chapter however are more informal mechanisms of control and access to pastoral resources that are common to both. Once the need for herd mobility is taken seriously, these less formal, more politically malleable features play important roles in increasing economic and ecological resilience. This chapter will conclude with a discussion of implications for efforts to reinvigorate other more severely eroded transhumance systems in the region.

Conventional views of pastoral tenure: the commons and livestock demography

Garret Hardin's *Tragedy of the commons* (Hardin, 1968) – since revised to the *Tragedy of the unmanaged commons* (Hardin, 1991) – is, despite vigorous critique from all sides (Runge, 1986; Bromley, 1989; Feeny et al., 1990; Oakerson, 1992; Peters, 1994), a persistently appealing model, especially among economists and physical scientists (Picardi & Siefert, 1976; Simpson & Sullivan, 1984; Anim & Lyne, 1994). It has proven particularly popular for analysing extensive systems of livestock husbandry since livestock and rangeland are treated unproblematic- ally as private and communal property respectively, despite a number of studies which effectively blur the boundaries of these designations (Dowling, 1975; Oxby, 1990; Ring, 1990; Hutchinson, 1992; Berry, 1993; Ranger, 1993). The strength of this model as a paradigm has effectively pulled the 'commons debate' in a direction away from the reality of most systems of pastoral management and usufruct in dryland Africa. In order to engage the tragedy narrative most effectively, critics have retained or added new assumptions to the commons model by arguing that the commons is not open access and that managers of the commons are not solely motivated by individualistic concerns (Gilles & Jamtgaard, 1982; Runge, 1986; Bromley, 1989; Feeny et al., 1990; Ostrom, E., 1990).

As a result, the tragedy model and most of its alternatives treat rights to common-property resources as spatially bounded. In addition, many of the alternative models treat these rights as uncontested, held by a definable, immut- able group of people. Such assumptions are often violated in real-world pastoral systems. Therefore, by engaging in the tragedy of the commons debate on its own terms, supporters of indigenous forms of tenure have often downplayed the role of politics and informal mechanisms in defining resource access in pastoral systems. In addition, the setting of *de facto* property boundaries by inherent biophysical limits to resource extraction is ignored, while formal, socially defined boundaries are emphasized. This has led to a fair amount of confusion in the pastoral literature, evidenced by the proclivity of many observers to equate local conflict over resources and/or a lack of fixed boundaries between common pastures as evidence of open-access situations in need of correction.

Despite academic criticism, the tragedy of the commons (TOC) framework remains influential within development circles. This is not just because of its elegant simplicity and internal logic (Roe, 1994), but also because it provides an environmental rationale for western notions of landed property (Thompson, 1991) and is consistent with economic thinking on the tenure requirements for private investment (Demsetz, 1967; Barrows & Roth, 1990; Migot-Adholla et al., 1991; Bruce, 1993). In dryland Africa, where agricultural development potential and outside investment opportunities are low, the appeal of the TOC framework derives more from the fact that it provides a 'rational', parsimonious explanation for the early range managers' views of a ubiquitously overstocked range. In development discourse, both views have worked to reinforce and support the other's claims: the TOC model provides range managers with an institutional explanation for limited social controls to regional herd growth while range management's diagnosis of widespread overgrazing has contributed the necessary ecological outcome to the TOC narrative.

Figure 5.1 places these approaches (TOC paradigm and conventional range management) within a broader, more contextualized framework of the political ecology of African pastoralism. The impact of grazing on the productivity and species composition of vegetation is felt not on a regional scale but at the level of the range site. The timing and magnitude of grazing intensity and livestock presence on this site are determined not only by the regional livestock population but by the spatio-temporal distribution of livestock across the regional land-scape. The explanatory foci of the TOC model are the incentives and dis-incentives affecting livestock accumulation by herd managers resulting from different pasture tenure regimes. Although it can be argued that pasture tenure is often one of a number of factors influencing local livestock accumulation[1], the major deficiency of the TOC paradigm is its preoccupation with the influence of tenure on regional livestock population, without considering the effect of tenure regimes on the spatio-temporal distribution of livestock.

This failing has often gone unnoticed due to the fact that range scientists working in Africa have often ignored the importance of livestock distribution as well. To the range scientist, variation of stocking rates is viewed as the major if not sole means through which livestock husbandry affects vegetation (Stoddart

---

[1] As can be seen in Figure 5.1, the TOC model oversimplifies the livestock economy, most particularly in its implicit assumptions of equating livestock owners with managers and the pastoral economy as relatively isolated from the market economy, and particular emphasis on tenure-related disincentives for livestock accumulation by pasture managers. Once these assumptions are relaxed, it can be shown that regional livestock populations are more affected by changes in the broader political economy, changes which often dominate any tenure-related incentives to stock or destock (Turner, 1993).

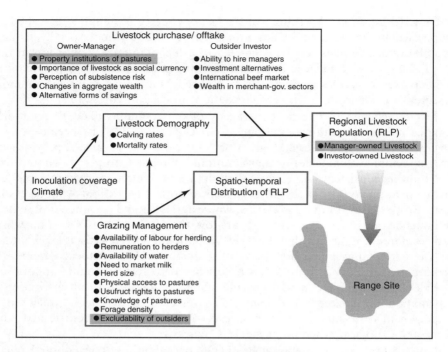

Figure 5.1 *Factors affecting the two determinants of the magnitude and seasonality of grazing pressure at the scale of a range site (with dimensions of 1 to 25ha, depending on the patchiness of vegetation)*

et al., 1975; Heady & Child, 1994). This approach reflects range management's reliance on a successional ecological model that treats grazing pressure as the major variable to which vegetation adjusts (Ellis & Swift, 1988). Unfortunately, the use of stocking rate as a management tool is difficult in Africa where rangeland is unfenced and only regional livestock statistics are available due to high herd mobility (Bartels et al., 1990; de Leeuw & Tothill, 1993). These difficulties, combined with popular notions of widespread overgrazing in dryland Africa, have led environmental analysts to focus on regional livestock population, not localized livestock density, a more meaningful but difficult parameter to obtain (see Figure 5.1). The 'tragedy' for TOC theorists and for range managers in Africa is the same: overgrazing through the unrestricted growth of regional livestock populations. In this way, the range-management approach as practised in Africa resonates with, and in so doing strengthens, the ecological basis for TOC formulations that focus on tenure's influence on incentives for livestock accumulation, and less on tenure's influence on the spatial distribution (and local densities) of livestock.

This cause-and-effect coupling of common-property theory (providing the cause) and conventional range assessments (providing the effect) has proven particularly durable to what are seen as the quibbles for nuance and local specificity on the part of anthropologists and their ilk. These two approaches combined to produce an unsentimental 'rigorous' examination of physical limits and improper incentive structures facing pastoralists, which in turn provided the scientific rationale for political/military programmes of forced destocking and the restriction of herd mobility (Ndalaga, 1982; Oxby, 1982; Peluso, 1993; Neumann, 1995). Although not the focus of this chapter, the history of 'pastoral development' provides an interesting case of the tragic real-world consequences of ivory-tower epistemology (Taylor, 1992).

The appropriateness of the conventional range-management model as practised in dryland Africa has been increasingly questioned. New conceptualizations of the dynamics of dryland rangelands (receiving less than 600 mm/year on average) argue that rangeland dynamics and livestock demography are driven less by density-dependent mechanisms (livestock–forage equilibrium) and more by fluctuations in rainfall, an abiotic external factor (Ellis & Swift, 1988; Westoby et al., 1989; Behnke et al., 1993). Therefore most changes in pasture productivity are increasingly viewed as resulting largely from rainfall fluctuations. While livestock population remains an important management parameter, the 'tragedy' of livestock population collapse is seen less as overgrazing-induced but more as climatically-induced. Increasingly, attention is focused on 'opportunistic strategies' which allow livestock to be moved quickly from areas affected by drought (Sandford, 1982; Scoones, 1994). Moreover, ecological research in the Sahel demonstrates that the timing, duration, and spatial distribution of grazing are often more important ecologically than district-level or regional stocking-rates (Turner, 1992; Hiernaux & Turner, 1996). Where overgrazing does occur, it is patchily distributed, often localized at points of unusually high and persistent livestock pressure (de Wispelaere, 1980; Haywood, 1980; Gorse & Steeds, 1987), such as near settlements where livestock mobility and dispersion are low (Niamir, 1987; Turner, 1992).

These revisions to our understanding of arid rangeland dynamics, grazing ecology and the spatial extent of grazing-induced degradation expand the pastoral management parameters of concern. No longer can regional growth in livestock numbers be automatically equated with increased ecological pressure. The spatio-temporal distribution of livestock should be given equal or greater weight than regional livestock population as a management variable. Livestock distribution needs to be adjusted in response to varying patterns of forage production and sensitivities to grazing. From a management perspective, there is declining interest in tenure institutions as incentive structures affecting the proclivity to accumulate livestock, but more as systems governing flexible access to a temporally and spatially fluctuating resource base (Niamir, 1990; Scoones, 1994).

Herd mobility and porous systems of pastoral usufruct

'Herd mobility' is an aggregate term that embodies both the timing and location of livestock movements. It has long been recognized as a major requirement of pastoral production that has significantly influenced the political and social organization of pastoral societies (Dyson-Hudson & Smith, 1978; Burnham, 1979; Dahl, 1979). Mobility plays an important role not only in reducing drought-induced risk (Swallow, 1994), but also increases livestock productivity (Cossins, 1985), and distributes grazing pressure (Behnke & Scoones, 1993). Therefore, it should be viewed as a management parameter of economic as well as ecological importance. The timing, duration, and intensity of grazing pressure experienced at a range site, as mediated by rainfall and grazing history, affect the species composition and productivity of its dryland vegetation (Turner, 1992; Hiernaux & Turner, 1996). The spatio-temporal distribution of livestock across a landscape is determined to a large extent by human-managed livestock movements.

The nature of the tenure institutions required to support herd mobility are quite different to those advocated by tragedy theory. Common-property theory points to the need for clearly defined access rights, relatively impervious to outsiders (Ostrom, E., 1990; Oakerson, 1992). On the other hand, mobility and flexible response to variable resource availabilities are facilitated by socially malleable access rights that are porous to outsiders (Baxter & Hogg, 1990; Casimir & Rao, 1992; Unruh, 1995). Resource access in the former is viewed as being best governed by formal legal rules, economic arrangements, or lineage membership, while the later allows access through *ad hoc* social relations as well. The economic benefits to livestock herders of a more porous system of usufruct stems not only from a wider array of grazing options in situations of resource scarcity, but also from the fact that they are able to 'trade' a range of assets (cash, grain, labour, livestock, future pasture access, social prestige, political favours), which are often not readily convertible, to obtain access to the pastoral resource (Berry, 1989). More formalized, legalistic systems of usufruct often have fewer channels of access, increasing economic vulnerability with shifts in resource availabilities.

What about the ecological costs and benefits? More porous systems of usufruct do allow adjustments in the distribution of livestock to changing forage and water availabilities. The classic comparison of open to enclosed rangeland is a case in point (Ostrom, E., 1990). Do such adjustments lead to the overstocking of certain areas? The answer depends upon the degree of pastoral mobility, access to pasture information, and the political and economic motivations of the usufruct owning/managing group and those of the outsiders. Take the case of a pure subsistence pastoral economy where a herder's dominant motivation can be assumed to benefit his livestock. In such a situation, the individual decisions of livestock herders will lead to an even distribution of grazing pressure across the foddered and watered landscape *if* herd mobility is not restricted. Wide distribution of livestock herds 'naturally' occurs not only because herders search out ungrazed areas accessible to water but often more importantly because of herders' avoidance of management difficulties under crowded conditions including:

o greater exposure to disease
o higher demands on herders to avoid mixing of livestock

104

o greater chance of losing livestock
o the attractiveness of livestock concentrations to both livestock rustlers and government officials.

All these factors limit the attractiveness of otherwise biologically attractive range sites.

Real situations deviate to some degree from that described above. First, there may be constraints to herd mobility. A reduction in herd mobility due to labour scarcity, greater management control by sedentary peoples, or land use and legal barriers to herd movements can produce situations of 'artificial' scarcity where overgrazing can occur. In addition, outsiders who are attracted to an area for reasons other than its pastoral resources (e.g. access to milk markets) may be less responsive to declining health and nutrition in their animals, leading to situations of over-demand, to the detriment of livestock and local range. Usufruct managers may choose to provide access to too many outside herds for private gain or in response to external demands.

In summary, gaining access to pastoral resources through informal social networks is very much associated with the maintenance of high herd-mobility within pastoral production systems. Often, resource use is restrained not by formal rules and boundaries but by social and ecological characteristics of pastoral production systems (Sobania, 1990). In such cases, tenure experts should think twice before attempting to legislate restraint, since by doing so they may make the system more rigid and inflexible. However, such systems are by nature vulnerable to new barriers to herd mobility and changes in the motivations and resource endowments of pastoral producers and usufruct managers. Cases exist that so diverge from the simple pastoral model described above as to make more formalized impermeable forms of usufruct necessary.

Transhumance in Sudano-Sahelian West Africa

The major bioclimatic zones in West Africa are determined in large part by annual mean rainfall and length of growing season with both declining dramatically from south to north (Le Houérou, 1989). Forage quantity declines, while forage quality increases from south to north (Penning de Vries & Djitèye, 1982). Transhumance as practised by the Fulani can be defined as the seasonal movement of domestic livestock along (usually) this gradient on a more or less annual basis (Map 5.1). One-way distances range from 40 to well over 500 kilometres. Multiple productive reasons exist for transhumance movements including:

o access to initial growth from earlier rains to the south (in north (N)–south (S) direction)
o escape from cultivation pressure during the rainy season (variable but generally S–N)
o escape from floodplain dry-season pastures during the flood season (variable direction)

105

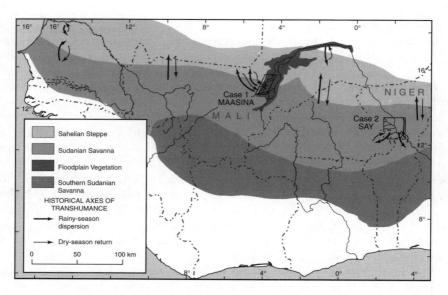

Map 5.1 *Major vegetation zones of West Africa, historical axes of Fulani transhumance, and location of two case studies*
(Source White, 1983)

o greater access to sparser but higher quality forage to the north during the rainy season (S–N), and

o escape from tsetse challenge in the Sudanian and Guinean zones (generally S–N) (Stenning, 1960; Gallais, 1975; Beauvilain, 1977; Diallo, 1978; Traoré, 1978; Benoit, 1979; Bonfiglioli, 1982; Breman & de Wit, 1983; Gallais, 1984; Bassett, 1986; Grayzel, 1990).

Actual transhumance cycles (seasonality, direction, and distance) are constrained by spatial patterns of cultivation pressure; the existence of biophysical barriers to movement such as large rivers; military threats posed by government agents, insurgents, and bandits; and the effective access of herders to pasture and water resources.

Within these transhumance systems, herd movements can be classified into two categories:

o daily/diurnal grazing movements to and from a central point where non-weaned calves are allowed to rejoin their mothers at milking

o travel movements where herds and their managers move from one encampment (central) point to another.

There is considerable variation across herders, households, villages, clans, ethnic groups, and regions in the degree of mobility observed within both of these categories (White, 1984; Turner, 1992; Moulin, 1993). The point-centred nature of these transhumance systems derives from a basic requirement of milch

106

pastoralism – the need to separate calves and mothers to obtain milk. This requirement has reinforced a natural tendency among most herders to navigate from point to point, with more detailed two-dimensional knowledge of the landscape reserved for those areas within a grazing radius (< 7km) of encampment points (Thorton, 1980).

While on transhumance, individual family herds will often travel between encampment points in groups. Moving together increases security for members and also facilitates the sharing of knowledge, labour, milk, and social contacts along the route. Not surprisingly, one often finds a mix of herders by age and livestock wealth within these transhumance groups. The size and membership of these groups is fluid. While members usually come from the same home territory, herds of close agnates often travel in different groups formed through marriage and friendship ties (Dupire, 1970; Lewis, 1981). Group composition not only changes from year to year but also within a particular transhumance season, with larger groups often splintering into smaller groups upon reaching target pastures.

In the extant, more formally organized transhumance systems of the Fulani, leadership positions exist for these transhumance groups (usually constituting 5 to 10 family herds). Examples of such indigenous positions include: the *garso* in western Niger (Laya, 1984; Beidi, 1993); the *amiiru daanδé* in the Maasina (Lewis, 1981; Ba & Daget, 1984); the *ardoβe* among the Wodaaβe (Bonfiglioli, 1988); and the *gaalel* in Futo Toro (Schmitz, 1986). In situations where transhumance is organized within or in close interaction with a State political structure, a second-tier political position, representing a number of transhumance leaders to the state bureaucracy, is common such as: the *rugga* (*ruggaajo*) in western Niger (Laya, 1984; Beidi, 1993); the *amiiru nai* in the Maasina (Lewis, 1981; Ba & Daget, 1984); the *laamiδo* among the Wodaaβe (Bonfiglioli, 1988). Unlike the Fulani political positions associated with the home territory/village (where women are located)[2], both first-tier and second-tier transhumance positions are non-hereditary leadership positions, elected or appointed from within transhumance groups. In addition, there is greater social control over the transhumance leaders' exercise of authority. In many cases, these leaders (especially first-tier leaders) are viewed as defenders/preservers of proper Fulani social conduct – similar, but in a more circumscribed fashion, to the role of the *mawδo laawol pulaaku* (Laya, 1984). Although the Fulani code of conduct (*laawol pulaaku* or *pulaaku*) is variable and malleable to social context, in a transhumance context, the characteristics of leadership most consistent with *pulaaku* would include: hospitality, social reserve, fairness, political deliberateness, and consistent concern for welfare of livestock. Actions that seriously deviate from these characteristics are grounds for the leader's dismissal.

---

[2] Such political organization is often two-tiered as in the case of transhumance political organization. Examples of first-level positions (village chief) include the *jooro* in Liptako (Kintz, 1985), the *jom wuro* in Futa Toro (Schmitz, 1986) and the *amiiru ngenndi* in the Maasina (Ba & Daget, 1984). Examples of second-level territorial political positions include the *laamiδo* in Liptako; the *arδo* in the Maasina (Ba & Daget, 1984) and the *jom leyδi* in Futa Toro (Schmitz, 1986).

Tenure institutions associated with Fulani transhumance systems vary tremendously across Sudano-Sahelian West Africa. The major reason behind this variation is the different histories of Fulani occupation and political control relative to other groups. In those areas where pre-colonial Fulani states existed (e.g. Futa Toro, Futa Jallon, Maasina, Sokoto, etc.), regional political control by Fulani élites provided a political and military environment conducive to the development and maintenance of usufruct institutions associated with transhumance during the pre-colonial and early colonial periods. By contrast, those areas that were either weakly controlled by broader regional authority or controlled by an agriculturally-oriented political elite, tend to have much weaker forms of transhumance organization and tenure institutions.

The usufruct conventions associated with even the more highly organized transhumance systems do not fit western models of secure, well-defined tenure systems. Major differences are as follows.

o Access rights are socially malleable. Either temporary usufruct rights or more permanent rights through membership in managing groups can be gained by outsiders through a myriad of different channels (as discussed above). Resources such as cash, livestock, political favours, pasture access, and labour can all be traded for pasture.

o Usufruct systems typically display an unbounded, point-centred spatial pattern (nucleated) with rangeland access governed by grazing radii around tenured points (ponds, wells, encampments) rather than bounded rangeland territories. The *de facto* boundaries around these points will expand and contract in response to the local availability of fodder.

o Usufruct rights to rangeland, even when made explicit, are often tied, not to land, but to the forage resources produced by the combination of land and variable rainfall. Pastoral Fulani are often best viewed as owners of grass not of land. Grazing rights are often only seasonally enforced, when the forage in question has nutritive value and/or when forage of comparable quality is in short supply (demand or supply-determined).

o Usufruct conventions more often than not specify the timing of access to what is frequently an ephemeral resource rather than assign exclusive access to particular individuals. Timing conventions usually consist of orders of precedence among different user groups with or without additional specifications as to the minimal time interval between the arrival times of successive groups and their maximum time of stay. Such systems can govern more graduated forms of access (to variable portions of the vegetation resource) without becoming over-extended in their enforcement requirements.

o Beyond the timing conventions described above, the quantity of fodder that can be consumed by outside herds is not specified. However, observers have long noted a sense of obligation among transhumance herders to leave some grass for the next (Monod, 1975, p. 39). With or without the timing conventions described in the point above, strong distinctions are often made between 'travel encampments' and 'destination encampments' by herders with respect to the *proper* length of stay for a transhumance herd. This herding norm is strongest in the case of travel encampments found along transhumance corridors, where stays exceeding 2 to 3 nights are often viewed as improper even when forage quality and quantity is known to be worse further along the transhumance trek (Bellot, 1980, p. 134).

These basic characteristics of the more organized Fulani transhumance systems of West Africa deviate from the TOC-influenced conventional view of a secure, clearly defined tenure system. Rights over pastoral resources are divisible within poorly defined spatial boundaries controlled by a socially malleable group of people. Despite this, access to pastures is not open to all. Instead, household resources (including political and social capital) are expended to gain access, with expenditures generally increasing with resource scarcity. Rules of access are not fixed but are often politically malleable. As has been argued above, such systems are better able to facilitate flexible but enforceable access to variable resources. However, static comparisons between western and indigenous forms of pastoral tenure are insufficient, especially in socially and ecologically volatile areas such as the Sahel. In fact, the flexibility of indigenous systems may actually make their ecological and economic benefits vulnerable to two general trends observed in Sahelian pastoralism.

o Changes in the productive logic that underlie herd and usufruct management. Cases where the incentives facing usufruct holder(s) or outside herd manager(s) diverge from that of the biological requirements of livestock in their care may lead to situations of overly aggregated distributions of livestock. For example, large shifts in livestock ownership from 'herding peoples' to outside investors have led to a growing separation between the day-to-day managers and the beneficiaries of livestock products as well as a shift in investment emphasis away from livestock as productive assets and towards livestock as stores of wealth (Bonfiglioli, 1985; Little, 1985b; White, 1990; Turner, 1992).
o Livestock mobility is affected by a whole range of factors beyond pastoral tenure. Changes in agricultural land-use, pastoral labour availability, war and political instability, and the need for marketing livestock products will all have important effects on herd mobility (Bassett, 1986; Bonfiglioli, 1990). Such changes are influenced by the broader regional economy. Therefore, a flexible tenure system should be viewed as a necessary but insufficient condition for maintaining/stimulating herd mobility.

In order to provide the basis for a deeper exploration of the strengths and vulnerabilities of indigenous forms of Fulani transhumance tenure, two cases of Fulani transhumance will be presented in this chapter. The first involves a return to the Maasina region of the Inland Niger Delta of Mali (see Map 5.1) already presented by Pascal Legrosse in Chapter 4 of this volume. A return to the Maasina is necessary due to the apparent consistency of the tenure system there with the dictates of the commons paradigm – namely, specific groups governing access through a formalized, economically based system (grazing taxes) to spatially bounded pastures. This case has very much shaped analysts' perceptions of the nature of viable pastoral tenure institutions and, as such, must be revisited, paying careful attention to the role of informal mechanisms in *de facto* pastoral usufruct.

Case 1: The Maasina region of the Inland Niger Delta of Mali

The indigenous system of pastoral usufruct of the Inland Niger Delta (Delta) has long been used as the example from Africa of a sophisticated territorial form of pastoral tenure (Cissé, 1982; Riddell, 1982; Schmitz, 1986; Gilles, 1988). In the rush to identify an indigenous system with features that satisfy TOC-derived institutional requirements, analysts have tended to consider only the formalized skeleton of pastoral usufruct in the Delta. This has led to misleading descriptions of the systems of pastoral usufruct in the Delta, along with overly optimistic evaluations of the applicability of Delta institutions elsewhere. In this way, these descriptions have affected development analysts' perceptions of how transhumance systems can and should be organized in West Africa. In his more sophisticated description of pastoral usufruct in the Maasina, Pascal Legrosse (see Chapter 4) has provided a strong foundation by which to evaluate the strengths and weaknesses of the Delta model. Relying heavily on this excellent chapter, the aim here is to stress those less formalized features of the Delta system that, while often ignored by analysts, are common in other Fulani transhumance systems as well. Their repeated occurrence across West Africa is indicative of the strong economic and ecological imperatives that these features address.

Access to floodplain pastures of the Maasina Legrosse describes the division of the floodplain into pastoral territories or *leyδe*. These territories are historically derived from the pre-existing territorial structure of the Maasina *Ardoβe*. During their rule, the *Ardoβe* had provided floodplain land to subordinates for farming, which led eventually to the bounding of their spheres of influence into *leyδe*. In his attempts to regularize the administration of his Delta subjects, Cheikou Amadou adopted the territorial model of the *Ardoβe* to force Delta households to settle into *leyδe* composed of the pre-existing *leyδe* of the *Ardoβe* and new *leyδe* carved out of the remaining floodplain. Therefore, depiction of the *leyδe* system as an indigenously developed form of *pastoral* tenure (Gilles, 1988) is somewhat misleading since its agriculturally rooted territorial form was imposed upon and openly resisted by Fulani pastoralists.

Closer inspection of the workings of pasture usufruct in the Maasina reveals the 'unnaturalness' of the territoriality of the *leyδe* structure to Fulani transhumance systems. *Leyδe* in the Maasina are often home to different Fulani lineages (descendants of the pastoral Fulani subordinated by the *Ardoβe*) with or without a fictitious common ancestor. These lineages control different encampment points (*bille*) and paths (*gumpti*) within the same *leyδe* and make independent decisions as to the outsiders allowed to follow them through the *leyδe* and the timing of movements between most *bille* (including the entrance date into the *leyδe*). There are generally no boundaries between adjoining *bille*

110

controlled by different lineages although there may be agreements as to intervals of time separating first arrivals. Therefore, while more highly organized, the pasture usufruct system of the Maasina, upon closer inspection, resembles the unbounded, point-centred form typically found in transhumance systems of the region.

The formalized system of payments for gaining access to pastures (*tolo*), the second feature which has made the Maasina system an attractive model to western analysts, is also revealed upon closer inspection to over-simplify the reality of usufruct access on the floodplain. Outsiders have managed, and continue, to gain access to floodplain pastures through a myriad of ways other than paying the *tolo* grazing tax. For example, of the approximately sixty outside herds admitted to a Maasina sub-*leyδe* in 1989, only 22 paid a grazing tax. This is despite the fact that at the time this clan was in conflict with another over pasture and was therefore under severe pressure to raise cash (through collection of *tolo*) to support its claims with the government (Turner, 1992). Perennial (although revocable) access can be obtained by outsiders through marriage, fostering of children, and reciprocal agreements between *jooro'en* lineages. Over the past twenty years, cash-poor Maasina clans in need of cash to support their control over pastures with the government, have actually begun selling perennial access to their pastures to certain outsiders able to make large lump-sum payments of cash. Such marketing of kinship-based access is understand-ably a sensitive issue in the Maasina, with clans who make such sales insisting that these newcomers remain outsiders despite their having been accepted into the clan's corporate herd (*eggirgol*).

More typical however, are the *ad-hoc*, single-year grants of access to outsider 'friends'. Members of pasture-controlling lineages will often petition other clan members to allow free entry to a friend. More often than not, outside herders become friends by providing favours to the clan member such as: political favours; acting as hosts in northern pastures; sharing of herding tasks in northern pastures; livestock loans; or gifts (cash, animals). Such favours may have occurred in a single year many years ago, over multiple years, in the same year (labour or cash gifts), or even in future years. For example, labour-deficient Maasina herders will often pool their herd with that of an outsider during the rainy-season. The outsider will often attempt to convert his rainy-season labour into dry-season pasture by requesting free access into the Maasina herder's *leyδi*. Pasture access is but one of the various assets (political influence, livestock, labour, pasture) needed to be a successful herdsman. In many pastoral and agropastoral areas, cash markets are undeveloped and therefore open 'barter' of these assets through the social contract of a 'friend' provides herdsmen with greater abilities to gain access to productive assets in short supply, thereby reducing the vulnerability of the transhumance system.

Such a system, while conflict ridden and prone to political manipulation, does not degenerate into an open-access situation. There are strong social norms as to what is an appropriate number of sponsorships. A clan member, who may have a number of friends approach him, must decide which friend to sponsor. Unless powerful within the clan, sponsorship of more than one friend would not be

111

looked upon favourably by other clan members. In fact, it is in the interests of the political élite of the clan (personified by the clan head or *jooro*) to maximize the fraction of outside entries who pay *tolo* since these cash revenues are controlled at the level of the clan. Benefits accrued through friendship sponsorships are generally less divisible and accessible to clan élites.

Some analysts would react negatively to the revelation that pasture access is often granted more through political processes than the Maasina's formalized tenure structure would suggest. Economists in particular may argue that the allocation of pasture could be easily manipulated to the benefit of inefficient but politically powerful producers. Political influence is, however, only one of the assets being 'traded' within these social networks. In fact, due to the survival imperative facing most actors in this system, there is necessarily a strong materiality to these 'trades'. While hard-pressed to argue for the inherent fairness of such a system, it is arguable that given the imperfect nature of cash-based markets in most rural areas of Africa, rural producers controlling variable unmonetized asset bundles, have always depended (and will continue to depend) heavily on such social networks in their subsistence struggles. From a resource-management perspective, social networks and *ad hoc* social contracts should be viewed as critical in allowing flexible response to changing availabilities of physical resources.

Limitations and weaknesses of the Maasina model There are significant weaknesses associated with the Maasina model of transhumance tenure. The Maasina model has proven vulnerable to external or internal changes that have in some cases worked against pastoral production and resource management (Gallais & Boudet, 1980). For example, the colonial and post-colonial states, through their assertion of a parallel state-controlled system of usufruct, have contributed to the erosion of the system developed during the *Dina* (Moorehead, 1991; Vedeld, 1994). Given its regional nature, any transhumance system must have state support to operate effectively. One of the major barriers to existing or new forms of transhumance organization is the establishment of alternative tenure conventions by the state which, either through direct contradiction or through the establishment of overlapping authority, undermine the tenure prerogatives of transhumance.

Another major threat to the proper workings of transhumance in the Maasina and West Africa as a whole is agricultural encroachment. Most pre-colonial systems of transhumance such as that of the Maasina are only concerned with usufruct rights to vegetation, water and salt licks. Therefore, they were internally focused in their organization, with little explicit consideration of other forms of land use. Control over agriculture occurred either through the 'pastoral master– agricultural slave' relationship as on the Maasina floodplain or through the threat of violence and crop damage more generally. Since the second half of the colonial period, the effectiveness of these forms of control has declined significantly. Pastoral tenure on the Maasina floodplains and elsewhere now more accurately refers to ownership of grass found on uncultivated land.

Attempts on the Maasina floodplain (i.e. 'Opération bornage' in the Cercle de

Tenenkou) and elsewhere, to prohibit cropping within certain radii from encampment points or transhumance paths, have generally been failures. Failure results in part from the fact that the administration of these programmes is given to local government authorities who are more prone to being influenced by local agricultural interests. Given the greater economic benefits derived from cropping a parcel of land rather than leaving it for grazing or livestock transit, local interests are dominated by agricultural interests. The resulting parcelization of transhumance corridors and associated pastures has greatly increased the physical and political barriers to herd mobility, thereby straining transhumance organization.

Case 2: The Say Arrondissement of South-west Niger

The Say Arrondissement of Niger (Map 5.1) spans the 500 and 800mm long-term isohyetes with natural vegetation varying from Sahelian scrub in the north to mixed Sudanian woodland in the south. In contrast to the Maasina, this area during the pre-colonial period was sparsely populated and, with the exception of the town of Say during the mid-nineteenth century, a political hinterland. Transhumance has developed relatively recently and without state support. Physical geography, including the Niger River (to the north and east) and tsetse fly infestation to the south[3] have, along with political insecurity presented barriers to movements out of the zone. Moreover, local pastures were adequate up until the mid-twentieth century due to low population density and cultivation pressures. As a result, the Say area has not been associated historically with long-distance transhumance.

History of transhumance   Prior to the 1970s, the pattern of herd mobility in the area has historically been very diverse (Dupire, 1972), varying between three poles:

o truly sedentary management (herds staying within 5km of village year-round)
o local circuits between key pastoral resources (salt licks, watering points, pastures) found within a 20km radius of village
o more pronounced north-to-south transhumance movements during the early rainy season (20 to 50km), remaining there for a period ranging from one month to the whole rainy season (3 to 4 months).

Even today, dominant strategies vary within and between Fulani clans and villages. This variability reflects not only the diversity of different circumstances which herd managers are under but also the narrow benefit–cost calculus of pursuing transhumance in this area and the lack of a pre-colonial power stipulating uniform herd management, as in the case of the Maasina.

---

[3] During the early colonial period, tsetse fly challenge was reported to have limited the prevalence of livestock husbandry and permanent settlement in the area encompassing what is now Parc 'W' and the Tamou Reserve (see Map 5.2) – described as a no man's land in 1929 (Urvoy, 1929).

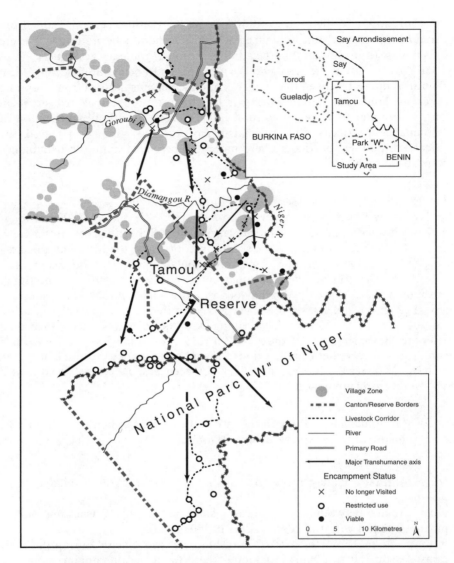

Map 5.2 *Major transhumance routes and camps in the Say Arrondissement of south-western Niger*

The impetus for longer transhumance towards the south was two-fold. First, contrary to many situations in the Sudano-Sahelian zone, it was to the south, not the north, where uncultivated pastures existed. Along the eastern edge of the Arrondissement, the whole area lying south of the Goroubi River remained only sparsely settled – it was to these areas (including what is now the Tamou Reserve and parts of the National Parc W) that herders in the immediate Say area brought their livestock during the rainy season. Secondly, the longer of these movements (40 to 50km) allow the herd to benefit from the earlier sprouting of grasses to the south resulting from earlier rains at the onset of the rainy season[4].

Those managers pursuing longer distance transhumance did so after leaving a milk herd in their home territory. Southerly transhumance movements were conducted in group herds (referred to locally as *eggirði*) along transhumance paths (*burti* or *gurti*) linking encampment points (*bille*, singular: *winde*) located near surface water points[5]. Pastures within the grazing radius of a particular *winde* could be grazed by local sedentary, circuit, and milk herds during the rainy season and/or mid- to late dry season as well as by outsider transhumance herds moving through for short periods at the beginning, middle, and end of the rainy season (Bellot, 1980). Historically, access to encampment points already used by local livestock or located near agricultural settlements was gained by outsiders in an *ad hoc* fashion through the development of friendship ties and marriage alliances with host (resident) groups (Wilson, 1984). Informants of the Say and Tamou cantons describe conventions giving particular *eggirði* (led by their *garso*) rights of first passage (with or without a specified delay) to particular *bille* that are or are not associated with a resident Fulani. These 'first passage' rights were most likely enjoyed by those outside transhumance groups who were the first to start visiting these encampment points.

Land-use squeeze: 1970–1996 A number of changes have occurred during the colonial and post-colonial periods that have increasingly strained local pastoral resources. The first has been the expansion of cultivated area especially in the Torodi, Say, and Tamou cantons. Pushed by the droughts of 1931, 1973 and 1984, immigrant farmers from Tera and Ouallam looked to the Say arrondissement for arable land (Institute de Recherche en Sciences Humaines, 1977, p. 54). They, along with merchants and government officials in Niamey, were able to circumvent village-based authorities and gain access to land by working through the canton chiefs or the sub-prefect. Some of the areas that have been seen as having the highest agricultural potential by immigrants were Fulani encampment points, due to their higher fertility resulting from long-term manure deposition.

Key wetlands in the area, which in the past provided valuable dry-season forage, have increasingly been converted to rice or counter-season vegetable production (Bellot, 1980). This has eliminated the dry-season pastures of some resident herders as well as increased the influx of pastoralists into the area. Bellot (1980) describes the conversion of 'borgou' pastures near Niamey in the late 1960s, to farming schemes controlled by government officials, as a major reason for the large influx of Bitinkooji Fulani to the Torodi and Tamou cantons.

---

[4] Even the fulfulde term for the direction south, *hoore huɗo* meaning literally the 'head of the grass', refers to the earlier sprouting of grass to the south (Laya, 1984, p. 10).
[5] In contrast, northerly herd movements were not coordinated on return from the southern pastures. This is due not only to the variable return times of individual herds but also to the practical difficulties of regrouping after dispersal across southern pastures.

As a result of the 1972–73 drought, many pastoralists from the Tera region, like their agricultural neighbours, moved south and settled in the Say Arrondisse-ment. Influx by these outside pastoralists not only increased the resident livestock population but also put strain on the Fulani ethic of hospitality with respect to pastoral resources (see the discussion above). In addition, local livestock popula-tion grew as a result of the growth in per capita livestock ownership among long-term residents who were able to accumulate livestock sold cheap by northerners during recurrent droughts. Livestock Service estimates of livestock population in the Say Arrondissement increased over the 1967–77 period by approximately 3 times for cattle; 2 times for sheep; and 4 times for goats (Bellot, 1980, p. 80); a period of time across which livestock populations either were maintained or declined in the Sahelian region as a whole due to the drought of 1972–73.

These trends led, during the 1970s, to a situation of growing scarcity of local pastures and growing numbers of resident livestock. As a result, resident herders who had been experimenting since the late 1940s with transhumance to more distant southern pastures[6], increasingly adopted such strategies despite continued trypanosomiasis challenge in southern pastures (Bellot, 1980, pp. 145–148). By the mid-1970s, pastoral strategies had shifted decidedly toward a greater reliance on southern pastures in Burkina Faso and Benin, with distances covered (50 to 70km south) significantly longer than in the past (Bellot, 1980, p. 128). In addition, evidence suggests a widening of the time period in which herds were away from home pastures, with some herders leaving as early as February and not returning until after harvest in October (Bellot, 1980, p. 128). In the Say Canton, where cultivation pressure is the highest, herder–agriculturalist problems have evolved from struggles over pastures to struggles over pastoralists' attempts to move animals in and out of the area.

During the 1960s and 70s, livestock managers in Say increasingly *adopted* transhumance as a livestock production strategy in response to cropping pressure. Since that time however, there has been another change that has complicated this trend. In response to both increased livestock pressure and heightened international pressure to protect nature reserves, the Service des Eaux et Forets, the government service in charge of administration of the National Parc W (see Map 5.2), increasingly took measures to block movements of livestock through the park to Benin and Burkina Faso[7]. Similar trends,

---

[6] One of the two present *rugga*s of Say, Rugga Manga Hoore, is reportedly one of the first residents of the Say Canton to practice long-distance transhumance to the south (pastures in Benin and Burkina Faso). He reports that he learned of the pastures there by first following the Udaaβe, sheep herders long known for their long-distance transhumance movements (Dupire, 1972). He began in around 1946, taking only half of his herd. Such experimentation found that if the cattle did not die of trypanosomiasis they would be more healthy and productive than those left behind in home pastures.

[7] Livestock and farming are prohibited within Parc 'W' while farming alone (except for villages with prior rights) is prohibited within the Tamou Reserve. According to herders in the area, policing vigilance and the nature of enforcement of the livestock prohibition within park boundaries varies between the three national services (in Niger, Benin, and Burkina Faso) from year to year. Despite this variability, livestock prohibitions have increasingly tightened during the post-colonial period. The mid-1970s and late-1980s are most often mentioned as periods during which the livestock prohibition in gazetted areas in Niger were progressively tightened.

116

heightened by a bias against foreign herders, have occurred in both gazetted and non-gazetted areas in the destination pasture areas of Benin and Burkina Faso. These trends have significantly increased the risk associated with longer trans-humance to the south for herders from the Say area. As a result, the benefits of transhumance have declined, resulting in a re-diversification of herd-manage-ment strategies (Gavian, 1993, p. 48). The narrowed benefit–cost calculus asso-ciated with transhumance decisions is today much more onerous than that faced prior to the 1970s. Say residents must now decide between two less satisfactory choices: (1) remaining in the Say area to graze their animals on poor pastures found on non-arable lands, roadsides, or in narrow fallows during the cropping season, or (2) conducting transhumance to the south with risks of trypanoso-miasis and draconian enforcements by forest guards. Faced with such options, it is not surprising that a significant number of herds and herding families have permanently left the area to the south.

Organization of transhumance Increasing land-use pressure and government restrictions on herd mobility are barriers to effective transhumance occurring in much of the southern Sahel and Sudanian zones of West Africa. While more severely affected than many, the Say area provides an interesting case of the response of an indigenous transhumance organization to such pressures. As described above, transhumance in the Say region has never been highly organized. This is due to the fact that unlike the Maasina, the historic cost–benefit calculus was, up until the 1970s, never decidedly in favour of transhumance, nor was there a higher-level political power providing both the stability and support for transhumance. In addition, the land-use pressures described above worked to erode the organization of herd movements that did exist in the region. Agricultural pressure increasingly hampered the movement of large corporate herds (*eggirði*) through the area. Encampment points became increasingly more circumscribed and numerous in heavily cultivated areas in the north (see Map 5.2). As a result, encampments where *eggirði* formed increasingly shifted southward during the 1960s and 70s. With increased enforcement of the livestock ban in Parc W, herders soon found that only by moving through the park in smaller herds could they avoid detection. As a result of this land-use squeeze from the north and south, *eggirði* movements remain possible only within a narrow strip of land lying north of the park and are no longer practised.

A two-tiered political organization exists among Fulani pastoralists in the region, built around the positions of *garso* and the *rugga*. Fanned by the politics associated with defining 'tradition', there exists a fair amount of uncertainty and controversy concerned with the historical importance of these positions in the Say region. As reported by Bellot (1980, p. 42), the chief of the Say Canton has persistently maintained that these are not customary positions among the Fulani of Say, the implication being that the local Fulani adopted these positions so as to defend their interests to State authorities, who had recognized these as positions of pastoral authority elsewhere. Others have argued that these positions predate Independence and that their power and legitimacy declined during the 1960s as a result of their not being recognized by the Diori regime (Wilson, W., 1984, p. 174).

117

Herders can name the succession of holders of some these positions at least as far back as the 1950s. What is clear is that, starting in the late 1970s, there has been a proliferation of holders of these positions, most notably of the *garso* position. This was the result of a strategy by Fulani pastoralists in the Say Arrondissement to counter the local village-based political power of the chief and the newly instituted position of *Maasamari* (chief of youth). As we will see below, the nature of the position was most likely reoriented as well during this period from that of a transhumance leader to that of an encampment manager.

Both the *rugga* and *garso* positions are life-long, elected positions. There are usually one or two *rugga* in a canton, each overseeing many *warsooβe* (plural of *garso*). Because of this, the *rugga* and *garso* positions are often seen as comparable in responsibility to canton and village chiefs respectively (Laya, 1984, p. 47). In fact, in those cantons of Niger where there is no *rugga*, the Niger government has pushed local herders to elect individuals to fill the position. Despite this superficial congruence, a *rugga*'s territorial authority may very well not coincide with canton boundaries. While the customary role of the *rugga* involves other responsibilities such as transhumance leader and veterinarian (Laya, 1984), increasingly in the Say Arrondissement, the *rugga*'s activities are dominated by the responsibilities of being an intermediary between herder associations and the canton- and prefect-level administrations.

As described by Laya (1984, p. 47), the customary position of *garso* is seen as the chief of the transhumance group, with his responsibilities split between resolving disputes (within and between his group and outsiders); utilizing his greater knowledge of pastoral geography to provide guidance about pasture and watering locations; and overseeing decisions as to the timing and destinations of herd movements. The *warsooβe* of the Say and Tamou cantons today act less as leaders of particular groups of herders and more as managers of particular encampment points open to different herders. Especially in the Tamou Canton where settlement is relatively recent, *warsooβe* are typically respected Fulani elders who maintain almost a permanent presence (farming and maintaining milk herd) near the encampment point(s) under their control. This change can be traced to land-use pressure; the proliferation of *garso* positions; the disruption of Fulani clan organization; the splintering of corporate transhumance herds (*eggirδi*); and large drought-induced fluxes of herders into and out of the area. The presence of resident Fulani families at a location depends on the viability of the local pastoral area to support livestock. In this sense, the responsibility to group and to encampment are closely interlinked. However, the 'death' of an encampment point will not result in the group moving together to settle elsewhere – more likely the group will disband and so, with the 'death' of the encampment comes the loss by the encampment manager of the title of *garso* as he often finds it necessary to move to the encampment point managed by another *garso*.

Interviews of sixteen of the approximately twenty-five *warsooβe* within the study area (see Map 5.2) reveals that their self-described responsibilities today are less directed to transhumance leadership and more toward the management of encampment point(s) under their supervision[8]. With the 'death of the *eggirgol*', it is the younger males who are actually going on transhumance that make most of the transhumance decisions, with the *garso*'s role being simply to act as counsel prior to their departure each year[9]. Encampment management involves three types of responsibilities: protecting encampment pastures from field encroachment and negotiating crop-damage settlements, settling disputes among resident Fulani and acting as host to outside transhumance herders. Upon arrival in an area, outsiders are expected to notify the local *garso* of their presence. Except in cases where the outsider has diseased livestock within his herd, the *garso* rarely refuses an outsider. These rare refusals are directed to transgressors of *lawol pulaaku* or trouble-makers. In certain circumstances the *garso* will suggest locations suitable for the outsider's camp. The ethic of hospitality, a major component of *lawol pulaaku,* does not allow a *garso* to refuse an outsider for reason of local pasture shortage. However, overcrowding does create problems for the *garso* and his uneasiness in such circumstances, while never directly stated, will always be clear to the outsider. To the outsider, the garso is seen as a local negotiator and information source. Therefore, the establishment of cooperative relationships with the *warsooße* along his trek is seen as a way in which to increase his security against various, largely human predations (theft, field owners, forestry guards).

A major preoccupation of most *warsooße* in the study area is that of protecting pastoral areas from agricultural encroachment. The more place-based authority of the *garso* today is, compared to a transhumance leadership position, a much more effective model to counter this threat. The *garso* must maintain a working relationship with local village chiefs. Being able to pick and time confrontations with agricultural and State interests is an important political skill required for an effective *garso* (or *rugga*). A *garso* who is overly combative will quickly deplete his political capital. In situations where confrontations move to the canton or sub-prefect levels, the local *garso* may call in his *rugga* for assistance, for it is the *rugga* who is more likely to have developed working relationships with the canton chiefs and the sub-prefect. Having the option to bring the conflict to the attention of the *rugga* gives local *warsooße* more leverage in local disputes.

---

[8] Locations of encampments (*bille*) are mapped in relation to village zones, major axes of north–south transhumance, and national park/reserve boundaries. Three categories of encampments are distinguished: those that have been abandoned and no longer used; those that are still used but on a restricted basis by smaller herd groupings due to either field encroachment and/or forest guard surveillance; and those that are still viable for larger herd groupings (*eggirði*). Village zones represent the 'infield' perimeter of permanent villages in the area. This perimeter is approximated by a circle of area equal to (2 hectares/person) (1988 censused population of village) with the circle's minimum and maximum radius fixed at 0.5 and 5km respectively. Due to the large amount of cropping by outsiders, these village zones should not be taken as the cropped area. For example, most of the area along the main north–south *gurtol* between the Goroubi and Diamangou Rivers is cropped.

[9] Limited *de facto* control by elders of transhumance decisions is actually very common among transhumance Fulani. Even in the Maasina, unless the senior male actually goes on transhumance, his instructions as to destinations are often not heeded by juniors. This is not due simply to the rebelliousness of youth but to the fact that many decisions are best made *en route*.

Programmes to reinvigorate transhumance institutions: lessons from the two case studies

Transhumance as a production strategy is very much threatened in the Sudano-Sahelian zone of West Africa. This is not due to a reduced need for inter-regional movements of livestock. Despite a resurgence in the popularity of the mixed-farming model among western analysts (Landais & Lhoste, 1990; McIntire *et al.*, 1992; Winrock International, 1992; Bourn & Wint, 1994), the growth of population densities in the Sudanian zone will, unless nutrient inputs are provided from outside, actually increase the need for transhumance movements to maintain livestock and cropland productivity (van Keulen & Breman, 1990; Lercollais & Faye, 1994; Turner, 1995). Transhumance is threatened by:

o a significant shift in livestock ownership from traditional pastoralists to sedentary farmers and investors (Bonfiglioli, 1985; Grayzel, 1990; Turner, 1993)
o agricultural encroachment onto important transhumance corridors
o the diversion of labour within agropastoral households from rainy-season herding to agricultural production
o large inter-regional migrations of people and livestock
o violence and insecurity in many pastoral zones.

Because of these changes, transhumance organization, even if existing in earlier times, is often very much weakened. This has resulted in a corresponding deterioration of tenure institutions related to transhumance.

What insights do the Maasina and Say cases provide for programmes directed at the reinvigoration of transhumance tenure institutions in the region? A closer inspection of the way in which pasture access is actually governed in these areas reveals common institutional forms. Even in the case of the Maasina, transhumance tenure is much more similar to institutions found in other Sudano-Sahelian transhumance systems than is typically acknowledged. The maintenance of informal institutions that can be reshaped by face-to-face negotiation and bargaining in both these cases suggests that such institutions are shaped by strong cultural, ecological or economic imperatives. Development programmes that attempt to reinvigorate or introduce tenure conventions into existing transhumance systems must be cognizant of the high probability of failure of opposing these imperatives[10]. For example, spatial organization of Fulani transhumance tenure is inherently point-centred and territorially unbounded. More bounded forms to protect transhumance points from agricultural encroachment are probably most feasible in key resource areas and if

---

[10] It should be noted that despite their populist rhetoric, most 'gestion de terroir' projects in Sudano-Sahelian West Africa do not incorporate the interests of non-resident transhumance herders in their implementation. Instead, ignorant outsiders are used by local agricultural interests to tighten their grip on local resources (e.g. counter-season gardening) by legitimating exclusion of non-residents from pastoral resources to which they may have historic usufruct rights.

introduced in a hybrid form where a territorial agropastoral tenure system encompasses a point-centred pastoral usufruct system.

As has been argued above, ecological and economic imperatives strongly reinforce socially malleable resource access as mediated by *ad hoc* social contracts ('friendships') and social networks. Development programmes that attempt to formalize systems of pastoral usufruct either legalistically or through formalized economic arrangements, not only often presume a government's capacity for enforcement (Ostrom, 1990), but may threaten to make usufruct access more rigid ('rigid' meaning here a narrowing and formalization of the available channels of access to a pastoral resource). The definition of exclusionary powers should reduce as little as possible the potential pool of outside users. A reduction in the available ways in which pasture access can be obtained will increase the economic vulnerability of transhumance herders. Specification of usufruct rights in and of itself does not lead to such rigidity. Rather, it is the way in which usufruct rights are specified that is most important.

In many areas transhumance has been so disrupted that it can no longer be viewed as a socially organized activity. Those individual herders who continue to move through populated areas to reach northern pastures are highly vulnerable to predation by local populations and government agents. Attempts to increase the security of these herd movements could build upon the indigenous host–stranger model as described in the Say case study. Settled Fulani, found near most agricultural villages and transhumance encampment points in the southern Sahel and Sudanian zones would act as local 'hosts' to transhumance herders. This has spontaneously occurred in many areas. At present, local 'hosts' benefit financially from their position by acting as mediators in farmer–herder disputes, providing livestock-marketing services to outside herders, and negotiating manure contracts between local farmers and herders.

In heavily cultivated zones, outside intervention may be required to formally grant long-settled pastoralists authority to administer access to certain key encampment points and corridors. The benefits gained through such authority would be countered by the assumption of more political responsibilities, namely the protection of key pastoral resources from agricultural encroachment and the collection of compensation for crop damage caused by sponsored strangers. Settled pastoralists are most often the only members of local communities who have an interest in protecting pastoral resources from agricultural encroachment. Moreover, their appeals to the local government official (canton chief) are usually futile. It is only at higher levels within the administration that the importance of maintaining transhumance paths through the area is increasingly recognized. Therefore, for the stranger–host model to be successful, local hosts must have a representative who could access the administration at the district level (arrondissement). Such a representative (a role played by the *rugga* in Say) would filter grievances prior to presenting them to the district administrator.

In many areas, potential 'hosts' are politically weak, considered to be the guests of the founding agricultural lineage of the area. Their access to land is vulnerable to repeal in cases where they oppose agricultural interests. This is a fundamental weakness of this model. It may be necessary in highly contested areas to take measures to increase the security of the agricultural tenure of 'hosts'.

With respect to sanctions against agricultural encroachment, typical approaches in which cropping is prohibited within a certain distance of pastoral points have not been successful. The time required for local pastoralists to shepherd their grievances through the different layers of political authority is too long – often exceeding the length of the growing season. An alternative would be to add a specification that all fields within protected zones are not eligible for crop-damage compensation. It could be argued that this may fan conflict and lead, through delay, to low compensation of crop damage occurring outside protected zones. However, by assigning financial responsibility to the host for the collection of compensation for crops damaged by sponsored herds, the problem of under-compensation due to pastoralist flight would be lessened. The host is in a much better position to trace the herder and/or his family; to revoke future access to local pastoral resources if the herder should refuse; and to distribute the financial burden should he fail in his collection attempts (due to his control over local pastures).

Conclusions

Tenure institutions play important but limited roles in the operation of contemporary pastoral systems in a number of different ways. It has been argued here that the 'tragedy of the commons' paradigm, which focuses on individual disincentives to reduce grazing pressure on deteriorating common rangeland, has so dominated common property debates as to exclude from consideration other ways in which tenure institutions affect the ecological and economic sustainability of pastoral systems. One notable exclusion is the way in which tenure systems promote or inhibit herd mobility, an ecological and economic parameter of increasingly recognized importance. Tenure regime characteristics which foster herd mobility often run counter to those promoted by TOC and common property analysts alike. Even the system of indigenous pastoral usufruct in the Maasina, long exemplified within the common property literature as unusually sophisticated and organized, displays characteristics of spatial unboundedness, porosity, impermanence, and continual social/political re-negotiation. The maintenance of these characteristics within the historically unique transhumance system of the Maasina reflects their importance in maintaining flexibility and resilience to highly variable ecological and political economic circumstances.

The introduction of overly formalized tenure institutions runs the risk of rigidifying transhumance systems, thereby increasing ecological and economic vulnerabilities. When and if the political will develops within national governments to support transhumance, the task for applied social research is to creatively promote tenure systems that retain as much flexibility of herd movements within the more formalized territorial structures as is necessary to protect against agricultural encroachment. Such solutions will most likely develop from

122

indigenous models. The 'stranger–host' tradition is an obvious candidate in the case of Fulani transhumance. However, as the discussion above makes clear, such models will not be successful if approached from pure rural populist or devolution perspectives. Political and economic support of transhumance against agricultural encroachment tends to weaken as you move from regional to local levels. In the many areas where agricultural interests dominate local power structures, significant barriers to herd mobility exist and will continue to exist at least on a seasonal basis. Without active support from central governments, further elaboration and development of these indigenous pastoral institutions will not be possible.

6

Customary law and ways of life in transition among the Nuer of south Sudan

Droit coutumier et modes de vie en cours d'évolution chez les Nuer du sud du Soudan

WAL DUANY

ABSTRACT

This chapter focuses on how the Nuer achieve patterns of sustainable development in the modern world. The Nuer are a Nilotic people who live in the Upper Nile region. The area is subjected to extreme variability of seasonal flooding and drought. The variety of grazing lands provide pastoralists with grazing throughout the dry season, but this requires constant movement to take advantage of the irregular availability of water and the life-cycle of different grasses.

The intrusion of the British Empire through its exercise of hegemony over the Egyptian-Sudanese Protectorate resulted in the imposition of subject status upon the Nuer. The British imperial commitment to indirect rule, presuming to rely upon traditional institutions, had a corrupting effect on the Nuer system of order. After the British, the Sudan emerged as a centralized state. State institutions are centralized, restrictive, and predatory. Despite this, the Nuer precolonial structures with adaptations and transformations have remained viable. Autonomy and federation provide a way for the Nuer as a transhumant people to meet modern requirements of order.

RÉSUMÉ

Ce chapitre traite des types de développement durable que les Nuer adoptent dans le monde moderne. Cette population nilotique vit dans la région du Haut Nil qui est exposée à des inondations et des sécheresses saisonnières violentes. La grande variété des pâturages permet aux pasteurs de faire paître leur troupeau pendant toute la saison sèche, mais à condition de se déplacer constamment pour tirer parti des disponibilités irrégulières d'eau et des cycles biologiques différents des plantes.

L'hégémonie exercée par l'Empire britannique sur le protectorat soudano-égyptien a eu pour effet de transformer les Nuer en sujets. L'administration indirecte pratiquée par l'Empire britannique qui reposait en théorie sur les institutions traditionnelles a eu pour effet de corrompre l'ordre établi chez les Nuer. Après le départ des Britanniques, le Soudan est devenu un État centralisé dont les institutions sont autoritaires et avides. Malgré tout, les structures précoloniales des Nuer sont restées viables au prix d'adaptations et de transformations. L'autonomie et la fédération permettent aux Nuer, population pratiquant la transhumance, d'assurer l'ordre nécessaire dans le monde moderne.

Introduction

THE FOCAL PROBLEM of this chapter is: How do the Nuer achieve patterns of sustainable development in the modern world? In order to begin to answer this challenge, we need first to understand the Nuer as a stateless society. Stateless societies tend to be deeply ingrained in tradition. They work well as long as the central organizing principles of traditional ways can be maintained. The members of Nuer society need to develop an awareness of new patterns of order so that they can learn how to modify their organizational structures in light of the changing conditions of the modern world. Problems may arise in accommodating change because essential and deeply engrained traditions may be destroyed by such change.

Nuer society has the potential for the exercise of self-governing capabilities in relation to empires and states. The processes of governance are built through the dynamics of conflict and conflict-resolution. Contestations are common at various levels of Nuer society. This poses a problem in a state that concentrates state powers in one centre of supreme authority.

In addressing the question of how the Nuer are developing a sustainable way of life in transition, this chapter will, first, describe the traditional system in reference to natural resources among the Nuer. Second, it will consider the imperial intrusions – initially, by the British Empire, and then, by the Sudan as a Nation-State. Third, it will attempt to describe how, in the presence of such circumstances, an indigenous transhumant culture demonstrates its viability in the contemporary world.

The Nuer and their place in the world

The Nuer are a Nilotic People. They live in the flood region which covers most of what is now administratively called Upper Nile and Bahr el-Ghazal in the Sudan (see Map 6.1). The area is subjected to extreme variability of seasonal flooding and drought. This situation limits its use for cultivation in the *tot* (rainy) season. A mixed local economy is thus what one would expect under these conditions. Indeed, one finds in the region a heavy emphasis on cattle husbandry.

The variety of grazing lands found in the Nuerland provide pastoralists with grazing throughout the *mai* (dry) season, but this requires constant movement to take advantage of the irregular availability of water and the life cycles of different grasses. The most valued grasses are those found in the seasonally river-flooded grasslands, the *toich*, which are exposed late in the dry season as the flood waters recede. All communities, therefore, try to ensure access to some form of *toich*, wherever their wet season villages may be.

Natural resource systems in traditional Nuer society
The Nuer perceive sandy ridges or highland, *toich* or grazing grounds, hunting fields, fishing reserves, rivers, pools, *khors* and other bodies of water, as natural resources that are shared by the community of users. The list of shared resources and facilities is both long and diverse. Each of these resource can have a fixed

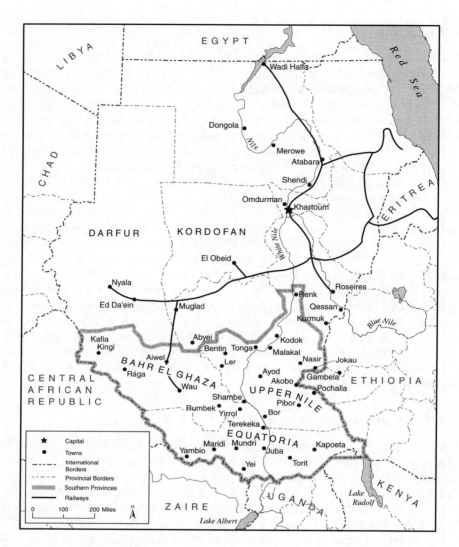

Map 6.1 *The Sudan*
(Source: *Duany, 1992*)

location such as grazing grounds or fishing reserves. The commons can be renewable or not, such as grazing grounds and water pools. The nature of these natural resources is such that individuals cannot easily divide them into discrete pieces in order to avoid sharing the use of such resources with others in the community. In addition, resource units extracted by one person cannot be extracted by any others (Ostrom, E., 1990).

Resource systems are best thought of as stock variables that are capable, under favourable conditions, of producing a maximum quantity of a flow without harming the stock or the resource system itself. Among the Nuer, the resource

126

systems include grazing grounds, fishing reserves, streams, lakes, and other bodies of water. Resource units are what individuals use from resource systems. Resource units are typified, for example, by the tons of fodder consumed by animals from a grazing ground, or by the quantity of aquatic resources used each year. The difference between the resource as a stock, and the harvest of use-units as a flow, is useful in connection with renewable resources, where it is possible to define a replenishment rate. As long as the average rate of withdrawal does not exceed the average rate of replenishment, a renewable resource is sustained over time.

Cattle and common-pool resources among the Nuer   Nuer are predominantly cattle people but grow some cereals. Cattle are their dearest possession and the Nuer will risk their lives to defend their cattle or to increase their herds. Most of their social activities concern cattle, especially during the dry season. Their attitudes towards their neighbours are influenced by their respective interest in cattle and the desire to expand their herds and control pastures.

Among the Nuer, cattle are used for many socio-economic purposes. They are used as compensation for life and limb. The union of marriage is brought about by payment of cattle, and every phase of that process is marked by their transfer or slaughter for ritual purposes. The success of bridewealth exchange and blood-wealth negotiations depends entirely on the wide-scale acceptance of cattle as a medium of exchange and for judicial reckoning.

Nuerland was, and still is, owned by lineages and clans in communal systems and the management of lands, water and fish reserves was entrusted to indigenous leaders, while the right of use was held by the whole communities.

Property rights and making the commons work   Nuer cosmology gives prominence to the individual's right to own property. This tenet is considered the basis for the principle of being in the right in a case. *Dungda*, my thing, *dungdu*, your thing, *dungdan*, our thing, and *dungdien*, their thing, are concepts organizing the way persons act toward one another in an acephalous system of order. Harold Demsetz gives a working definition of property rights:

> Property rights are an instrument of society and derive their significance from the fact that they help a man form expectations which he can reasonably hold in his dealings with others. These expectations find expression in the laws, customs, and mores of a society. An owner of property rights possesses the consent of fellowmen to allow him to act in particular ways. An owner expects the community to prevent others from interfering with his actions, provided that these actions are not prohibited in the specifications of his rights (1967: 347).

Demsetz is discussing personal property rights, but this definition can be expanded to include group property rights as well. Among the Nuer, property rights give the holders of such rights certain presuppositions as to the actions of others. They also give others presuppositions as to the actions of the right holders. These presuppositions help persons devise strategies when they are interacting with respect to the resources over which the rights are defined. The specific form of the rights does proscribe certain actions and enhance other actions.

It is through the system of property rights that a mechanism of rightdoing and wrongdoing can be determined by both the law users and the law enforcers. It is the privilege of the right holders to restrict others from claiming benefit to a resource, over which they themselves have clearly defined property rights. Being in the right depends upon the recognition of the right possessed by another.

Land tenure, grazing rights, and rights for water, are held by those who are recognized as descendants of the original occupiers of the area. This is particularly true of the Nuer east of the Nile, for such rights are held by descendants of the original colonizers (Evans-Pritchard, 1940; Jal 1987; Johnson, 1994). An understanding of land tenure requires an analysis of the lineal structure of Nuer society. Such an analysis is briefly given in the section below entitled 'Patterns of Village Life'. Land tenure here means the holding of, and rights to, the ridges and sandy outcrops which the Nuer occupy during the rainy season (Howell, 1954b). Nuer build their permanent villages on these highlands. Households have their cattle byres and cultivate specific pieces of land and carry out various forms of land use for pastures and fishing. Certain communities have definite rights to distinct areas by virtue of traditional conquest which is clearly recognized by all Nuer. This right is enforceable as a rule by the Nuer. There are, however, different property rights depending on the type of resource, such as *toich* pools and lagoons, fishing reserves, and rights to drinking water. Because the Nuer society is segmented into household, lineages, clans, and wider society, the rights of access to natural resources tend to overlap. The village, for example, owns the land around the village, pools, and lagoons; lineages own the *toich*, the use of water of the River Nile, the River Sobat, and major lakes jointly used by the Nuer people as a whole.

Individuals can also hold some rights in resources as a private property. They hold rights to use, and to transfer to heirs, resources such as farmland, livestock, wells, and pools. The technology that renders these resources productive, including byres, granaries, ploughs, and hand tools are also held as private property. The general rules in regard to land are that all cultivated land, well water, and pools surrounded by cultivated land have a proprietor, and that claims to these resources can be publicly affirmed, disputed, and enforced.

Rules for access and boundary conflict The Nuer pastoral economy is much more complex than simply maximizing the sustainable yield of cattle on a certain grazing ground. The most important reason for this complexity is the variable and uncertain rainfall in the region. Not only must access to the grazing land be controlled, but the risks of uncertainty of rainfall and drought also must be managed if there is to be long term, sustainable use of the common-pool resource.

Access to *toich* in most cases depends on some sort of established relationship among the users of the grazing ground. Such relationships may be organized through marriage alliance and age-set. Marriage alliances and age-set arrange-

128

ments facilitate access to resource sharing between clans and lineages in Nuer society. As John R. Commons (1959: 21) wrote, 'an institution is collective action in control, liberation, and expansion of individual action.' This generates trust and confidence between disparate people so that they can carry out mutually beneficial relationships among themselves. Covenant-making and reaching mutual agreements between groups to share use of a common-pool resource is another way to deal with foreigners who want to benefit from the use of a common-pool resource.

When cattle are few and grazing lands are adequate, no serious problems exist, and boundaries among regional segments from one grazing ground to another are not closely guarded. In conditions of scarcity, such boundaries become essential and their violation may then cause conflicts that often result in violence and bloodshed. Sometimes unusual shortages, both of grass and water, exist and people who normally allow the intrusion of outsiders less favourably provided than themselves, turn upon them and demand their withdrawal. Epidemics of some contagious disease such as rinderpest or contagious bovine pleuro-pneumonia, also make exclusion of others necessary. Fear of the spread of disease in general compels the owners to guard their boundaries more closely. Bringing contaminated cattle into an area recognized as the exclusive right of other segments, where strangers are only allowed on sufferance, is a particularly serious infringement of grazing rights and leads to hostilities, unless strangers promptly withdraw. Although fights have not been infrequent, and boundary disputes are a common cause of blood feuds, disputes of this nature are often settled peacefully by the council of elders, age-set leaders or custodians.

Patterns of village life

A Nuer village is organized on the basis of segmentary lineage systems in which there is a *diel* (dominant) clan to furnish a *mar* (kinship) framework on which political aggregations are built. The smallest political and territorial body is made up of a *cieng* (village). Each village is composed of homesteads consisting of a *luak* (byre) and *duel* (houses). Each homestead may contain a simple family group or a polygynous household. The household is often referred to by the Nuer as *gol*, a word that means 'hearth'. The *cieng* is further divided into *dhor* (hamlets), which consist of a group of homesteads and the gardens and grazing land surrounding it. A hamlet is generally occupied by close kinsmen, often brothers, and their households. This kind of group is considered a joint family.

Joint families or lineages provide the conceptual framework of the village structure. A village comprises a local community, linked by common residence and by a network of kinship and affinal ties. Members cooperate in many activities including provision for collective defence and regulating the use of common-property resources. Other activities carried out by households in the village include: cultivation, marriages, and establishment of age-set system.

The most important activities centred in village life include cultivation, marriages, and the rite of *gar* to establish age-set. The need for cereal food to supplement their insufficient milk diet prevents the Nuer from being entirely transhumant. Sorghum, maize, beans and pumpkin production are, therefore,

129

important for survival. Research in the region has shown that the greater part of the village's diet comes from the grain crops they grow[1].

Lawful marriage is a village activity for a number of reasons. First, marriage requires the presence of many people, especially the bride and the groom's families and kin. The participation of such a large group of persons is necessary to develop ways of insuring commitments to the long-term arrangements essential for productive undertakings. Village as sedentary life facilitates the participation of the villagers in the processes of marriage negotiations. The residence of each member of the extended families involved in any given marriage is known by everyone. They must be notified and be present at a certain date for the negotiations and distribution of bridewealth cattle. Because kin do not necessarily camp together during the dry season, it is difficult, if not impossible, to conduct marriages with everybody involved moving with cattle camps.

Second, conducting marriage contracts involves discussions of issues regarding building marriage alliances and common problems likely to confront the individuals entering into such an alliance. People negotiating marriage agreements need food, accommodation, and drinks. It is the village that has the resources to meet such expenses for marriage. There are more houses available for accommodation in villages than in cattle camp. Because of these reasons and others, marriages usually take place in the village between mid-July and mid-December when there is plenty of food.

The age-set initiation is a key activity taking place in Nuer villages. The formation of a new age-set system is accomplished through the ritual of *gar* (cuts) or initiation. The gar consists of cutting six marks across the forehead. The operation is performed by a man known as *gaar* (marker) who is skilled at the task. *Gaar* uses a small sharp knife for the cicatrization.

Initiation into age-sets has two major objectives. The first is to create each age-set and to train young men in the rules of conduct governing the relationships between initiates and non-initiates on the one hand, and between the members of different age-sets on the other. The second is to make initiates unafraid of the spear or blood and to accustom them to bearing pain when defending herds of cattle, villages and cattle camps. Third is to observe other communal rules such as those rules on common-pool resource use and grazing land access, and the formation of a new set of decision-makers.

The Nuer traditional judicial system

The Nuer judicial system is composed of different levels and groups involved in conflict resolution. Among the Nuer as an acephalous order, there is no central mechanism for conflict resolution. Thus, each decision unit mediates conflict within its jurisdiction. This suggests conflicts are resolved in different ways and in different contexts. The judicial system in Nuer villages, hamlets, and clans, depends upon the action of autonomous individuals who pursue their

---

[1] The Equatorial Nile Project 1954: map E 12.

interests in ways that grant reciprocal respect for the autonomy of others. Disputants argue their case before elders and witnesses. Decisions are made based upon the principle of who is in the right in the case. Such decisions of the elders can be appealed against by going to the next level of segmentation in a lineage system. However, there are identifiable ways that can be considered as a common judicial system which operates among the people and across the land. These include the processes of mediation and contestation in the process of conflict resolution.

Conflict resolution among the Nuer was, and is still, through mediation. Mediation in the Nuer context is an adaptive and responsive process; in fact, it is altogether a rather loose process that captures considerable diversity. The forms of mediation reflect the Nuer political structure in which it occurred and the nature of the conflict.

The openness of the Nuer society in which contestation is the norm makes development of different forms of cooperation possible. Getting to the negotiating table itself is extremely important, since it is only through interaction, dialogue, and debate that understanding can develop. It is within such a process that the parties to a conflict are able to delineate and compare the actual differences in their positions, differences that in many cases do not prove to be so very different.

In order to promote their common interest in maintaining a village, residents established standards of conduct. These standards are enforceable as rules of law by and for individuals who voluntarily covenanted with one another to form governing structures. Officials (elders, lineage leaders, custodians, age-set leaders, etc.) entrusted with rule enforcement are expected to discharge their responsibilities according to the concept of legal equality.

All villages and cattle camps maintain a regular police-like force. This *ngueni* (police force) is a component of the traditional *ream cieng* (home guards). The home guards are established to maintain lawful relationships within a village or a cattle camp and to protect against external aggression. Home guards or militia are trained by their parents, peers, and age-group members to understand lawful relationships. They are also trained in how to use weapons by their elders. The decision taken by both the assembly of villagers or camp residents authorize the council to use the police force in order to maintain order in the jurisdiction involved.

Pastoral institutions   The Nuer *luak* (cattle byre) is the basic pastoral institution that is associated with the family. Each family must have a *luak* to keep cattle during the rainy season for protection from the swarms of mosquitoes and other insects that breed in standing water. The *luak* is where decisions are taken regarding movement to grazing grounds, land and water rights, and acquisition and distribution of cattle in marriage. The Nuer family is also known as the *luak* (household). A number of *luak* congregate into a *dhor* or hamlet. *Luak* as a pastoral institution accommodates the *riek kuoth* and *gol*. The *riek kuoth* institution is a sacred *nyiot poll* (special tree) with *ngapni* (hooks) where special spears, small sacred gifts, and other items are kept. Libation in the

131

form of milk, beer or water is poured under the *riek poll*. The action of pouring libation is an act of thanksgiving to God for the welfare of the *luak* or household. It is also asking for forgiveness of the wrongs committed. *Riek* institution symbolizes, to the Nuer, *kot liec* where the Nuer were supposed to have emerged. It is an important symbol of common religion and culture. It is, therefore, a national symbol.

Gol is another pastoral institution of general significance accommodated in the *luak*. *Gol* means a joint family, whose symbol is a hearth – a circular structure made of mud and placed in the centre of the *luak*. Dried cattle dung is burned in it to protect animals from mosquitoes and insects. Thus, *gol* means both the hearth and the joint family. *Gol* is also located in the *geau* (windscreen) in the cattle camp during the dry season. The structure of *gol* is made of grass to last for the duration of the dry season. *Gol* and *riek* constitute the centre of the household of the cattle culture of the Nuer people.

Wec jiom is a temporary cattle camp made of kin from the same village or hamlet. Young men and girls move cattle to nearby grazing grounds after the old grass has been burned and green grass is coming up. These cattle camps are usually established around mid-November through December each year after the pools in the village have dried up and intermediate grazing is exhausted. The cattle can graze on marsh plants that abound in depressions. Each section of the village may have different *wec jiom* depending on the grazing grounds, availability of water and size of the village. Smaller villages may camp together. People who camp in *wec jiom* are mostly kin and members of the same village.

Wec jiom has developed as an accommodation of some generational conflict. Young people would like to take the cattle to cattle camp as soon as the rain stops and leave village maintenance work to older people, even though grazing grounds and water are available within the village. Building and repairs generally take place early in the dry season when there is plenty of straw for thatching and enough millet to provide beer for those who assist in the work. Young people are required also to help in these activities. A second crop of *paan* (sorghum) still has to be harvested. Cattle are usually retained nearby to eat the sorghum stocks after harvest. *Wec jiom* as an institution meets the desire of the young for being away from the village, and also the desire of the elders to keep the young near enough to help in village work.

Wec mai is dry-season settlement. It is usually a mixture of people of different villages of diverse lineages or clans. Sometimes members of different clans may camp in one cattle camp. In most cases, large camps are placed along the rivers such as the Sobat, Gile, or Pibor. Lakes such as *Dual dap*, pools and depressions which are deep and large containing water that lasts through summer months, are also used as camp site.

Grazing patterns in Nuerland vary according to the physical and biological conditions of each locality. Some areas are near the grazing grounds, such as the grazing patterns of the Gaawar, and do not have to move far from their village, and others like that of the Lou Nuer have to move almost every season.

Variations of grazing and natural endowments call for a high degree of flexibility and adaptation among users of natural resources.

Principles behind mobility   The dry-season movements from permanent Nuer villages to cattle camp sites, where water and grazing are expected to last the summer months, is an annual event. However, the extent of such movements varies from one region to another. Such movements may be understood by focusing on the principles behind the movements of clans such as Lou, Eastern Jikany, and the Zeraf valley communities. These movements are shown in Maps 6.2, 6.3, and 6.4, showing movements of different communities during the dry season. The principles that underly the dry-season movement include: the need for mobility, how rights overlap, how rights vary depending on the season, and how rights are negotiated.

The needs of mobility in each season are designed to make the best use of the natural resources. The herds from villages are moved from one grazing field to another, although existing pastures are not exhausted at each move, for three reasons: first, to find water for humans and animals; second, to prevent over-grazing and the consequent spoiling of pastures; and third, the drive for food. The dry season is a time of relative scarcity as compared to the wet season. Securing food makes the mobility of a household a necessity.

Rights of where to move from village to cattle camp depend on the season.

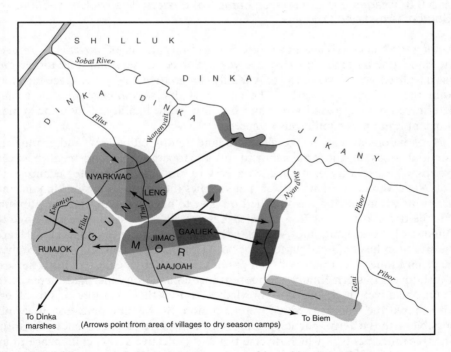

Map 6.2 *Villages and dry season camps of Lou Nuer sections* (Source: *Evans-Pritchard, 1940: 56*)

133

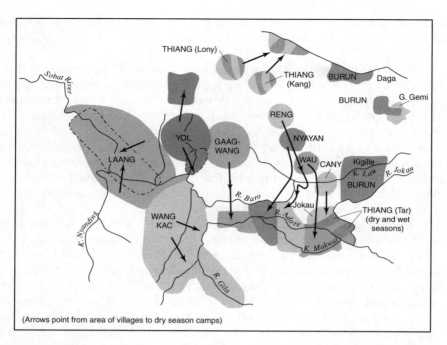

(Arrows point from area of villages to dry season camps)

Map 6.3 *Villages and dry season camps of Eastern Jikany Nuer sections*
(Source: *Armstrong, 1940*)

Among the Nuer, no one follows exactly the same patterns of movement nor uses the same grazing areas in successive years. Subject to seasonal limitations, one has a wide choice of where to go. A family can go to virtually any place far away from their village, or nearby. The choice of dry season pasture is not only determined by its proximity to water but also by availability of food for human consumption and negotiations of access.

Negotiations in reference to access to land among individuals and groups are essential steps in the establishment of a cooperative arrangement between groups. The necessity of negotiated access to land-use applies, for example, to Lou Nuer who do not have a river that supplies them with water for humans and their animals in the dry season. The Lou have to negotiate with their neighbours (for the dry season most Lou move to Eastern Jikany and to a lesser degree to Dinkaland, Murleland, and Anuakland) for access to use grazing fields. Hence the situation has enforced patterns of both transhumant and sedentary use over space and time. A corollary of such patterns was interdependence. Both the less transhumant Eastern Jikany and more transhumant Lou Nuer had therefore to acquire relatively high levels of knowledge *vis-à-vis* their relationships to others who shared the resources which were common. Such communication was only possible with the active participation of those involved. The existence of institutional arrangements in which diverse types of collective actors can engage in an open contestation and find ways of accommodating complementary interests has been a blessing to the Nuer.

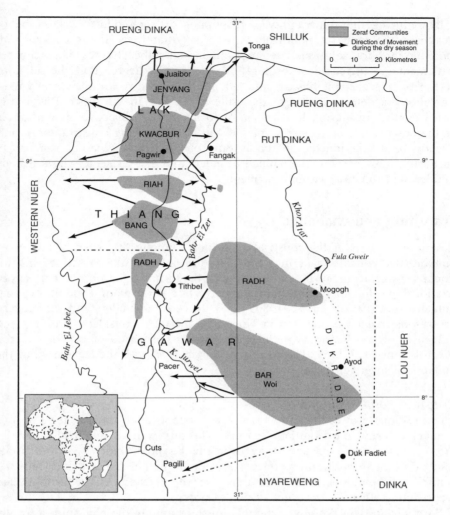

Map 6.4 *Dry season transhumance of the Zeraf Nuer communities*
(Source: *Lewis, 1940*)

Land-use conflicts between segments are solved by *kuar muon* (earth custodian) and *wud ghok* (cattle custodian). The conflict-resolution efforts of these religiously motivated third parties do more than merely offer a negotiating mechanism, a method of communication, or any other such purely procedural assistance. By introducing the authority of religion into the negotiating equation, they enable the parties to concede lands, assets, or claims to that authority itself rather than to their antagonists. Concessions previously regarded as intolerable evidence of a lack of fortitude, become politically acceptable when they are presented as acts of deference to religious teachings and religious personnel.

135

By reducing the vulnerability of disputants on each side to accusation of weakness, the range of politically feasible negotiating positions is expanded, more options for solutions become available, and the chances of reaching a settlement are increased accordingly. Hence, to the degree that the religious authority is evoked, it has an internal political validity for the disputants – warranting a deference that neither side may show to the other. Religiously based conflict resolution has been able to ameliorate objective circumstances of the conflict and not merely operate within unchanging constraints. As has already been mentioned, religion is only one of the sources of third-party authority; most notably, the council of elders and age-set leaders have usually fulfilled that function within a segment.

Transition and challenge

A serious analysis of the constitution of order among the Nuer, or any other human society, has to take into consideration not only the prescribed institutional arrangements but the daily practices, the formal and informal patterns of relationships, that are fashioned by a society's internal dynamics and the external influences on it (Ostrom, V., 1980, 1987). In order to understand the modern transitions that have occurred in Nuer life, one must consider the events and influences of two historical periods. The first is the period of British imperial control from 1899–1955. The second begins in 1956 with the emergence of the Sudan as a sovereign state.

British imperial challenge

The traditional way of life among the Nuer probably confronted its most serious challenge during the twentieth century. The intrusion of the British Empire through its exercise of hegemony over the Egyptian-Sudanese Protectorate resulted in the imposition of subject status upon the Nuer. The British imperial commitment to indirect rule, presuming to rely upon traditional institutions, had a corrupting effect on the Nuer system of order.

Formalizing the rules governing relationships among the Nuer into a formal code of law, separating those rules from the processes associated with conflict resolution, and imposing a system of chief's courts subject to British concepts of penal law associated with terms of imprisonment, were inconsistent with the conception of governance among the Nuer. The later commitment of British colonial officials to a better understanding of the Nuer way of life led to the anthropological investigations of Edward Evans-Pritchard, commissioned by the Colonial Office. Modifications were made in the native courts that allowed for more viable accommodations to the Nuer way of life.

British Empire and indirect rule The colonial government was interested in building a system of rule-ordered relationships in Nuerland which were grounded in British conceptions of government. This effort resulted in the creation of institutions alien to the Nuer. In the Native Court Ordinance of

1931, the protectorate government decreed that the only legal way of administering justice in the (Sudanese) territories was by indirect rule. Resorting to the traditional system of justice was regarded as a violation of the proclamation of the Native Court Ordinance (Howell, 1954a). It was through their native courts that Britain administered Nuerland.

Warrants issued to chiefs appointed by a governor on the recommendation of the district officer authorized the chiefs to sit in British courts for the purpose of judging cases. The warrant also authorized each of the chiefs to exercise executive powers that were territorial in scope. The assumption of exclusive executive and judicial powers ran counter to the Nuer tradition; no tradition recognized one man or a group of men as paramount rulers over other persons or communities.

The basic rule changes introduced by the Imperial Government during the 1930s and the 1940s included the following:

o the elimination of contestation and argumentation in the process of conflict resolution, especially in the settlement of feuds
o the introduction of capital punishment, terms of imprisonment, and collective cattle fines as deterrents
o the elimination of the right of vengeance of the relatives of the slain
o the redefinition of several forms of homicide that, for the Nuer, warranted compensation.

The colonial administration also did not uphold the Nuer belief that the type of weapon used in a slaying (that is, whether club, fishing spear, fighting spear or gun) should affect the sentence of the court.

The courts of the chiefs tried persons for deviating from accepted conduct and imposed fines ranging from *ruath* (bull-calf) to a pregnant cow, or even more. They also had the power to assign the wrongdoer to a term of imprisonment, if the British district officer agreed to this (Howell, 1954a: 61–67, 230–237). Premeditated killing was beyond the jurisdiction of the chief's court. Persons charged with this offence were tried by a magistrate using a British statute known as the Sudan Penal Code (Howell, 1954a: 66). Thus, capital punishment was imposed on Nuer by an alien law. When a person was sentenced to death by a Magistrates' court and executed, there was no payment of compensation by his kin to the kin of the victim.

The introduction of the colonial prison system, a hated penal innovation, undermined the Nuer forms of punishment for serious offences. In Nuer society, persons found by the council of elders to have committed a fault were obliged to repair the damage by payment of compensation or suffer banishment. In addition to imposing terms of imprisonment, the native courts also administered corporal punishment of which public flogging was the most detested. Both men and women were flogged, a practice previously unheard of in Nuer society.

The most immediate effect of the establishment of the native court system was, however, its impact on the traditional institutions for adjudicating disputes. Prior to the colonial period, the assembly of the people, that is, the council of elders, age-set leaders, custodians, prophets, the supporters of the disputants,

137

and the public following the court proceedings, all played important roles in dispute resolution and the administration of justice. The establishment of the native courts drastically curtailed the ability of these participatory groupings to play their traditional juridical roles. Instead of complaints being sent to elders, age-set leaders, custodians, prophets, they were now sent to the courts or to the district commissioner.

Dependency and corruption   The effect of the establishment of the native court system was to bring about dependency and corruption. By empowering the warrant chief with both executive and judicial powers, the government overturned that portion of the indigenous political system of the Nuer that the British considered to be lacking in accountability. Formalization of the rules governing relationships among the Nuer into a formal code of law was perceived as an essential characteristic of a rational legal order. The Nuer had no written code of law.

The concentration of government authority in the hands of chiefs throughout Nuerland brought about an exclusive reliance on the courts established by the protectorate government for adjudicating disputes. The courts curtailed the participation of the council of elders, age-set leaders, custodians, and prophets in dispute resolution. All cases were brought to the courts or to the district officer instead of to community leaders. The court system fostered a new passion for litigation that proved injurious to the Nuer way of life. Certainly, the rapidly growing number of cases brought to the courts reflected a heightened degree of social and political disintegration.

With the increase in the number of cases, the chiefs could not give sufficient attention to every individual problem. Nor were the chiefs disposed to settle the cases. Most of them were ready to take advantage of opportunities for personal benefit including the expectation of gratuities (bribes) rather than to simply serve their own people. The result was that many innocent persons were convicted (Bum, 1989). In many cases, the corruption of the chiefs was the cause of the miscarriage of justice (Bum, 1989; Ekechi, 1989).

Although the large number of cases were burdensome to the colonial officials who oversaw the work of the chiefs, the increased number of cases was important for the financing of government in the localities. The courts provided revenue for local administration. The revenue produced largely by court fines and fees rose in proportion to the number of cases disposed of in the court. The need to increase revenue levels also led to the appointment of court clerks to record court proceedings and to keep account of court fee receipts. Another purpose of clerks was to aid the District Officer in overseeing the work of chiefs who were illiterate. There had to be a record of cases in order to hold chiefs responsible and accountable. Appointment of the 'missionary boys' as they were known by the Nuer enhanced the popularity of the office of the chief. The clerks, most of whom had only four years of formal education and whose salaries were relatively low, tended to enrich themselves from the court proceeds and from bribes taken from litigants. Examples can be given of these clerks pocketing court fees and fines. As cases at Diror, Walgak, and Pathai Courts revealed, clerks often failed to enter the correct amount of revenue collected into cashbooks (Bum, 1989; Banak, 1989). In some cases, no entries were made at all.

Missionaries and western education   Every empire has religious and educational components. The Arabs and the Arab empires relied upon the mullahs, dervish orders, and places of prayer to carry out these functions. Likewise, the British, French, Dutch, Spanish, and Portuguese all turned to churches and monastic orders. The missionaries were crucial participants in British rule of the South Sudan. One cannot logically separate the British policy of indirect rule, which included an intention to respect indigenous institutions, and the work of the missionaries.

Responsibility for literate education and other services was assigned to missionaries by the protectorate government with some government support (Sanderson & Sanderson, 1981). The type of education the missionaries were expected to offer in South Sudan was 'to fit the ordinary individual to fill a useful part in his environment with happiness to himself. . . .' (Lugard, 1930). In keeping with the principles of indirect rule, this required a conscious effort to retain 'native institutions' while at the same time teaching the African 'to adapt himself and his institutions to changing ideas and conditions' (Lugard, 1930). From this principle it logically followed that a type of education that attracted an African away from his social unit was wrong. These principles were not always easy for missionaries to implement (Sevier, 1975: 192–193).

Developing education among the pastoral Nuer in the late nineteenth and early twentieth centuries was no small problem. Education was made a part of the religion as among biblical Jews. In order to adapt to climatic conditions, schooling was conducted during the rainy season so that students had the choice of carrying their readings – the Book – with them to the cattle camp or to leaving it in the village. Missionaries provided Nuer with the opportunity to gain access to a literary tradition somewhat similar to Nuer covenantal theology. Arabic literary traditions would have likely been introduced as an alternative system. Missionaries taught the Nuer the English language, Bible, science and technology with reference to medical, agriculture, and veterinary sciences.

Anthropological inquiries: a search for stability
The Nuer's unceasing struggle to maintain representation in their local courts led to important changes in the Chieftainship. Extensive negotiations and adjustment periods were required and tolerated during which diverse, conflicting, and ineffective solutions to the problem of governance survived. Nuer institutions had developed over time a character that discouraged arbitrary structural changes; sometimes Nuer changed their environments rather than adapt to them (Lugard, 1930; Evans-Pritchard, 1940). Administrative insensitivity was reduced as a result of the recognition and admission by the British of the inappropriateness of the system of executive chiefs for Nuerland.

The search for a new social order resulted in a return, with widely accepted modifications, to the customary way of life. In the 1940s and 1950s the colonial administration made a sustained effort to understand how the Nuer governed themselves. The primary objective of this effort was to determine how best to adapt British rule to Nuer patterns of governance. Reports by Edward Evans-Pritchard, the American Presbyterian Church, Catholic Church missionaries, and British protectorate administrators provided insight into Nuer institutions and the general pre-colonial history of some segments of regional communities. These reports contain the first recorded versions of the oral traditions of most Nuer segments (Johnson, 1980; Jal, 1987).

Using the information collected by these various groups, Nuer elders, the chiefs, and British administrative officers restructured the courts in order to better reflect the Nuer concept of justice. Thus, in place of the small number of territorial courts that had forced unrelated and autonomous segments under one chief, the government created additional courts so that each original group had its own. This action restored equality and the rule of many chiefs replaced rule by a few.

The Sudan as a centralized state

The withdrawal of the British Empire led to the emergence of the Sudan as a nation-state. The European concept of a nation state refers to a group of people who share a common language, literature, and cultural tradition. It is this concept of the state that has become increasingly important in the organization of Sudanese society, and other Third World societies, following the collapse of the British, Dutch, French, German, Japanese, and Portuguese empires in the period immediately following World War II.

Fundamental to the conception of the state is the belief that the peoples of the world are (or should be) organized as nation-states in a global family of nations. The state is defined as a monopoly of the exercise of authoritative relationships and the legitimate use of coercive power in a society; it is associated with a command theory of sovereignty. Sovereignty is authority to govern. In the Sudan, this meant the centralization of the prerogatives of rule in the hands of the government. Territorial administrations were established to facilitate central government control and property rights dominated by state ownership of the public domain to the exclusion of communal and private property rights. But it was evident that the Nuer had not depended on a state. The basis of their economic well-being rested on the community rather than the state. The Nuer economy centres on relations among human beings rather than the relation between a Nuer and the State.

The Sudan as a nation-state superimposed its centralized structures upon the array of social structures and processes that constitute Nuer civil society and upon other peoples. State institutions are centralized, restrictive, and dominate economic activities. State powers are concentrated in the executive branch of government, especially in the cabinet. The state claims ownership of land, including *toich*, and all natural resources while it has no effective institutional mechanisms for providing sustainable management.

The origin of these problems can be traced to the condominium administration of 1899–1956.

The Sudan as a nation state, therefore, faced the challenge of creating a new social framework within which Sudanese institutions would work themselves out, reaching new accommodations to promote more productive human relationships and a viable social order. The Sudan has not, however, been fully capable of fashioning new social arrangements to address its problems. The state's institutional order and how it defines problems and possible solutions, are themselves a substantial source of the tension and conflict.

Tension and conflict have moved down to localities. The matter of legal boundaries associated first with the British Empire and then with the Sudan to delimit jurisdictions has done great damage. They have disrupted customary reciprocal arrangements that had internalized rights to the grazing *toich* that had been a regional common-pool resource. Unless these rights were specifically mentioned and awarded, one clan or lineage could not graze its animal with that of another. The conflict between the Jikany and the Lou Nuer is indicative of such tensions arising between more transhumant (Lou) and more sedentary (Jikany) elements of Nuer population which have erupted into destructive warfare.

Autonomy and federation as a way of coping with empire and the autocratic state

The most important problem in the Sudan has to do with the nature of relationships between the state and the Nuer and other peoples in South Sudan. The general expectation was that the new government of the independent Sudan would be democratic, which meant the people would meet, discuss, and decide on ways to solve their problems, and that government actions would reflect these decisions. Such expectations were not realized in practice after independence was achieved. Instead, all crucial decisions were taken by a bureaucracy ill-trained but entrenched in authoritarian habits and practices. The major portion of the state's meagre resources went into building governmental coercive capabilities, increased centralization, and larger, more inefficient, bureaucracies and military capabilities which, in turn, produced greater predation and oppression.

Given conflict and civil war in Nuerland, the appeal of revolutionary movements, accessibility to modern weapons, explosives, and opportunities for pillage and plunder, the indigenous transhumant culture felt a challenge to demonstrate its viability in the contemporary world. How the Nuer are able to maintain their autonomy and adapt to the changing environment hinges upon the value placed on autonomy and freedom of association. This takes into account the internal dynamics of Nuer social processes, the nature of local institutions, and considerations of tradition, custom, religion, and other opportunities and constraints.

How the Nuer organized their relationships with one another to enable their indigenous development to continue despite alien disruptions is considered here first. The pre-colonial structures with adaptations and transformations have remained viable. As shown earlier in this chapter, a different configuration of institutional arrangements which support a productive and resilient system of order has evolved in Nuerland. Second, to be described is how the Nuer related to Nilotic and other peoples, and what pre-colonial forms of public participation might still be useful in the evolution of post-colonial forms of democracy relevant to current problems in the Sudan and other African countries.

Autonomy and federation among the Nuer

The Nuer as an acephalous society have a close kinship to federal societies, if the term 'federal' is construed as pertaining to a 'covenantal society' (Elazar, 1980: 3–30; Ostrom, V., 1991: 53–68, 252–53). The Nuer have a strong covenantal tradition but without the 'formal' political apparatus that is commonly associated with two or more overlapping governments in a federal system of government. There are, however, important features of Nuer society that pertain to the ideas of covenant, breach of covenant, and the re-establishment of the bonds of covenantal relationships that are similar to those of all covenantal societies.

Autonomy and federation have been key elements in the constitution of order among the Nuer. Clans, lineages, and sub-lineages, conscious of their separate identities, have all desired and maintained self-governing capabilities yet achieved vital objectives by combining their efforts and domains into a confederation of Nuer, both distinct from, and interacting as a unit with, other cultural groups.

The Nuer confederation is made up of eleven federations based on kinship and local affiliation. The confederation occupies a continuous landmass from east to west containing water and land resources. A confederate council of elders is called in to take leadership whenever the need for joint action arises. The habits of self-governance, the spirit of compromise and self-reliance, and the strong moral basis of Nuer religious beliefs are sources of social capital available to each member of the confederation. Each of the federations has known boundaries, a police and militia, its own law and order, and a common language.

The objectives that the Nuer achieved through federation, combined with a division of powers were a common defence of their way of life against the influence of external military forces and external social influence. Federation protected individual and communal rights in grazing lands, fishing reserves, and water resources. Military and economic concerns, however, have usually not been sufficient in themselves to create Nuer federation. There had to be an awareness of a prior covenant of 'peoplehood' with a single God, a bond and a hope for the preservation of equality, autonomy, and equal justice among persons. In William H. Riker's words, 'some deeper emotion than mere geographic contiguity with cultural diversity' must be present (1964: 35).

The conception of federation among the Nuer derives from their belief in the equality of persons. Theirs is not a caste society. The presumption of equality is grounded in the belief that *nei dial* (all Nuer) are descended from a common pair of ancestors. Equality is related to the Nuer concept of accountability. As people decide to hold others responsible for their actions, as they allow the same principle to extend universally, they create a particular kind of moral environment for each other. They expect each other to be held equally accountable. Thus, their principles of governance are based on self-governance and not on command and obedience.

142

Nuer covenantal arrangements give individuals the opportunity to take advantage of both a large regional organization and small independent autonomous units. Political units such as villages and/or cattle camps vary in size. The small size of a political unit increases its vulnerability to aggression by more powerful opponents. On the one hand the ecosystem and the management of the natural resources require units that can benefit from ecological niches and variability. On the other hand, small size means greater vulnerability. Federation is one way that the Nuer resolve this size dilemma. It is a means of securing autonomy and self-preservation. Different levels of organization can combine to thwart aggression and internal corruption. Federation is also used as the means of building the Nuer commonwealth and of enhancing collective action.

The use of natural resources shared by two or more federations is regulated by the peoples involved. The Nuer as self-organizing and self-governing people had instituted arrangements for a cycle of conferences to deal with constitutional matters. The elders and the chiefs across clans met every five years at Fangak in the Central Nuer Region to see that basic constitutional arrangements kept pace with their changing environment and to address new challenges to the Nuer way of life. Since the Sudan became independent, such conferences have been discouraged by the representatives of the central government in Nuerland. They were viewed as arrangements usurping governmental functions. The last conference of elders, chiefs, and officials was in 1963, more than 30 years ago.

However, there were and still are multiple agents with limited authority to resolve conflicts arising between the groups using a resource held in common.

Federation with the other Nilotic peoples   The Nilotic peoples have different resource endowments. These differences are due to variety in altitudes, rainfall patterns, flood conditions, land types and vegetation cover. These variations provide the reasons for cooperation, population movements, and the alternations of grazing and cultivation patterns. In the upper Nile region, adversity is to be expected from time to time.

The social dynamics working within and among the peoples in the region are accordingly concentrated around the need to minimize the risks of a harsh environment. Harmonious achievement of common interests in trading, grazing lands, use of water sources, fish resources, and other productive endeavours, require development of patterns of interdependencies. This has led to the institutionalization of diverse communities of relationships.

In order to mitigate the effects of adversity, the peoples created voluntary arrangements within a number of regional economies and federations that helped to link them together. Markets, trading centres in villages and cattle camps and towns, are the most formal nodes in the voluntary network of exchanges that operate within particular societies and bind together peoples from differing regional economies and climatic conditions. The emergence of the Nuer–Dinka federation was initiated by association of the cattle owners. The exact date of the Nuer–Dinka federation's emergence is not known. Douglas Johnson (1980) suggested that Guek Ngundeang, a Nuer prophet, established the right of compensation by cattle. This meant if a Nuer or a Dinka were killed, the relatives of

143

the deceased would have the right to claim cattle to compensate the kin of the dead. This rule was instituted in the mid-1800s. The Nuer–Dinka cooperative arrangements might have already begun earlier and Guek Ngundeang only formalized it. In 1839, an Egyptian flotilla succeeded in penetrating the vegetation-choked channels of the Upper Nile and gaining access to the Bahr el Jebel, which was explored as far south as Bor and Aliab Dinka territory (Gray, 1961). The report of the voyage indicated that the Nuer by 1939 already occupied both banks of the Bahr el Jebel which were formerly Dinklands (Johnson, 1980).

During the emergence of Nuer federations with Dinka, herders associated themselves together in cattle camps, while *kuaar* (chiefs) of adjacent villages associated themselves in an effort to solve common problems between their peoples. One of the problems was cattle raids in the *toich*. The federation was a device to normalize relationships and to make individual members accountable for their actions. Federation as a method of problem-solving was applied to Anuak and Murle in the region. Such arrangements were not made with the Shuluk because competition over the use of common-pool resource was not as acute compared with others. It was also likely due to the Shuluk not having as many cattle as other communities bordering the Nuer.

Rich networks of voluntary associations within and between communities frequently involve ties of reciprocal obligations and social contracts. The loaning or bonding of livestock, the sharing of labour, and ultimately the arrangement of marriages, forge sets of more profound and lasting links between individuals and within communities. These functional and voluntary associations were complemented by rules that take account of communities of relationships that were, and still are, multinational in character.

The Jikany–Lou Nuer conflict and the Akobo conference The Jikany–Lou Nuer conflict can be used as a case study to illustrate how autonomy and federation have worked and are still working in coping with changing conditions among the Nuer. The outburst of warfare between the Jikany and the Lou Nuer has made Nuer way of life vulnerable to disintegration and destruction. Patterns of behaviour manifested in this conflict have violated the accepted norms among the Nuer. It is not accepted as customary for Nuer to raid other Nuer, to kill women, children, and elderly persons. These categories of persons are perceived to be non-combatants. The burning of homes, destruction of property, and killing of persons who have run into someone's home for refuge or sanctuary, were prohibited by Nuer customary law. The Jikany and the Lou Nuer war, however, included Nuer raiding cattle from one another, the loss of sanctuary, the burning of homes, and the killing of children, women, and elderly persons. With the use of weapons of modern warfare, it was not possible to know from where the fatal bullets come. Personal responsibility was lost, moral precepts collapsed and lawful order disintegrated into near anarchy (see Map 6.5 showing the location of the Jikany and the Lou Nuer).

The causes for the conflict were the following: rights to fishing grounds; access to water; and rights to grazing lands. In addition to the above reasons, the Lou settlers in the Sobat Basin had requested authority from Jikany through their

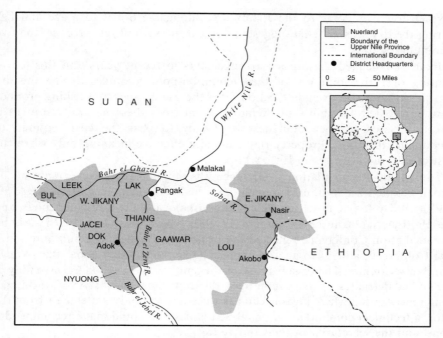

Map 6.5 *Nuerland within Upper Nile Province, 1954*
(Source: *Johnson, 1994: xxii*)

chiefs in 1991 to create a head chief position for themselves that would be accountable to the Waat administration instead of the Jikany administration at Nasir. The Jikany response was the creation of sub-chiefs or executive chiefs for Lou migrants but within the existing system of Jikany chiefs under the Nasir administration. The other reason was that John Garang, in 1989, in an attempt to curb the movement and expansion of Lou southward, issued a decree annexing all lands south of Sobat River to the Bor Administration. This decree was resisted by the Ngok Dinka, Lou, and Jikany as this divided their peoples.

The Nuer considered lands, water and fish reserves as common properties. However, the Nuer are divided into units (clans) of governance. Rights to a portion of Nuerland are divided among these units. Rights to resources are further divided among lineages. Each lineage has its own cultivation land, pastures, water supplies, and fishing reserves. Members of lineages alone had rights to exploit the resources.

The Jikany and the Lou represent the two largest Nuer groups on the east side of the Nile. The Lou traditionally have occupied the belt between Kongngor and Akobo, including the Ayod, Waat, and Yuai areas. The Jikany are located along the Sobat River, including the Ulang, Nasir, Chotbora, Jekou, and Maiwut areas. These communities extend across the Pibor and Akobo Rivers into Ethiopia. The Lou have had their rainy season grazing areas and permanent villages in the area where there is no river. But from December through May they have customarily come to the Jikany area along the Sobat to graze and fish during the dry season. This suggests a joint ownership of resources by the two communities.

Jikany challenged these prior arrangements in recent years and this led to fighting over the joint use of these common-pool resources during the dry season. Attempts to exclude the Lou from the use of *toich* and fishing grounds were costly. The feasibility of excluding them from these natural resources is directly related to the willingness of Jikany to bear the cost required to accomplish that task. Property rights acquire effective meaning only when the cost of denying access is worth bearing.

The Akobo Reconciliation Conference was initiated by the leadership of the Sudan People's Liberation Movement and the Sudan People's Liberation Army (SPLM/A-United) led by Riek Machar, the Jikany and the Lou Nuer chiefs, and the Presbyterian Church of Sudan (PCOS). The constitution of the Akobo Reconciliation Conference reflected the principles of federation and autonomy. The Conference sought to acquire more reliable information about the Jikany–Lou Nuer Conflict. The organization of the conference included the assembly of the invited delegates, *ad hoc* committees, the traditional court style of working, a conference setting where anyone can ask questions or make statements from the floor, a technical committee that analysed issues and could make recommendations, and the secretariat of the Conference.

There were 18 delegations of mediators (Duany, 1994). Members of the delegations constituted the assembly representing 11 different federations within the Nuer confederation and specialized groups such as the South Sudan Women's Association, church groups as a religious association, and the Nuer in diaspora. There also were mediators from different Dinka federations such as the peoples from Ngok, Twic, Dungjol and other peoples from different parts of the South Sudan to help in the process of conflict resolution. Chiefs from the Liech (Bentiu) federation functioned as the court who would hear the case and manage the process of reconciliation in the Nuer way of doing things. Malual Wun Kuoth, an elder drawing on his 44 years of experience as a chief, was nominated to preside at the Conference. Malual was recommended by the organizing committee to the Conference and confirmed by the Jikany–Lou Conference as the Chairman of the Conference. He was assisted by Head Chief Yot About of Dongjol Dinka and Paramount Chief George Along of Latuka region of Eastern Equatoria.

The choice of Liech or chiefs from Liech Federation to act as the court was based on the Nuer tradition which maintains that when there is a conflict an impartial third party would be called to help solve the problems of the parties in conflict. The chiefs of Liech Region were considered by the Jikany and the Lou Nuer as impartial. Thus, the role of Liech chiefs was a transient responsibility.

The issues the delegates to the conference deliberated on included:

o should both sides to the conflict compensate each other for their losses or should the losses be cancelled?

o was the administration of Cieng Dhuor Bang and Cieng Dongjak to be under
  Lou or Jikany administration?
o the rights to grazing land, and
o the right to fishing.

An open process was used to call forth testimony, allow for free expression of grievances, and provide wide input to finding solutions to the conflict. The design of the Conference institution at all different levels of activities reflected the principles of autonomy and federation. The Nuer believe they are equal to one another and such belief is grounded upon their religious presuppositions. The presumption of equality before their Creator gives individuals equal standing in the establishment of order. The development of a group is primarily viewed as a process of mutual agreement among individuals. This implies that the Nuer constitution is an artefact made by people to achieve important tasks of value for themselves.

Contestation helped the mediators at the Jikany–Lou Reconciliation Conference to learn from one another what somebody else had to say about the problem they were facing. Open deliberation was appreciated because it permitted the Nuer to come to terms with conflict and conflict resolution. This is how they learn to live together in a self-governing society. The disputants argued their side of the disputes to the other members of the Reconciliation Conference.

The Jikany and Lou Nuer agreed to share the use of the grazing land as a common-pool resource. This *toich* has been used as common-pool resources by the two communities for centuries. The agreement was in fact re-affirming the traditional pattern of use. Agreement over the grazing grounds came about as the consequence of learning from the open discussions on the floor of the conference. Open deliberation provided information and clarified issues. Disputation and contestation were accepted as part of the 'nature' of things at the Conference. The most important indication that both sides to the conflict perceived their agreement as mutually satisfactory was the acceptance for sealing of the covenant with blood. The Nuer will not accept the action of sealing an agreement with blood unless they are completely satisfied with the agreement.

The spirit of reconciliation signified by the agreement is known in Nuer as *nguot*. The word *nguot* means 'to cut'. It means here 'to cut a covenant'. People negotiated and compromised to reach an agreement, but the formalizing element of the *nguot* commitment was essential to establishing the relationship among the parties. The Jikany and the Lou Nuer made a *nguot* commitment verbally to declare the nature of the agreement that was in the making. Even when the Peace Agreement was committed to writing, symbolic action was required to accompany the verbal as well as the written Agreement. These actions included the offering of a sacrifice by killing oxen in order to seal the covenantal commitment with blood and to participate in a feast in celebration of reconciliation. The dismembered animal(s) represent the curse that the *nguot* maker calls down on himself if he should violate the commitment he has made. This would be the punishment that will be prescribed by the covenant mediator designated by God or His representative.

147

To implement conference resolutions, including the agreement on grazing land, the Jikany–Lou Reconciliation Conference established a special court, a police force-like institution, and a military unit to enforce decisions of the conference. Copies of the Peace Agreement and the creation of institutions for implementing and monitoring of the conference resolutions were deposited with every federation of the confederation and other peoples who participated in the Conference. The police force, the church groups, the women's associations and ordinary citizens were the monitoring institutions. Resolving disputes and challenges to the accord were the processes to be exercised by the court.

Conclusions

Federation is a viable option or solution to the conflict in the Sudan. Advocacy of self-governance or autonomous administration of the Nuer is not a repudiation of all centralized institutions. Experience has shown in many places that centralized institutions have helped to protect the rights of local minorities and achieve the aspirations of a society (Sawyer, 1992). In the Sudan, majority desires supersede minority interest, and matters concerning fundamental values recede under pressures for immediate gain. Therefore, centralized institutions do have an important role to play in the processes of governance and should be considered as among the array of essential institutions in the governance of the Sudan. The challenge is to create configurations of institutional relationships that foster self-governance in such a way that centralized institutions can function without transforming the structure of governance into the monopoly of a few.

The creation of the macro-constitutional superstructure (Ostrom, E., 1989) needs to involve a role for the people, especially those in the South Sudan who never had the opportunity to express their preference on separation, federalism, or unitary centralization in relation to the North. It is the Council of Ministers in Khartoum that invents and reinvents government for various communities and not the people of the locality who make their government. Basic to a theory of effective choice is the view that communities in pursuit of their freedom, safety, and happiness, against their own background and culture, and knowledgeable about both the opportunities and limitations prevailing in the context of their local circumstances, can themselves enter into covenants to form a community of communities. The practice of self-governance in these circumstances is an essential foundation for democracy and human rights.

148

7

Conflict management and mobility among pastoralists in Karamoja, Uganda

Gestion des conflits et mobilité chez les pasteurs du Karamoja, Uganda

MARYAM NIAMIR-FULLER

ABSTRACT

Changes in mobility patterns, constraints imposed by a reduction in territory, decreasing herd sizes, and increasing population, have all resulted in a lowering of the standard of living of the average Karimajong. Herders are no longer able to exploit the variable natural resources properly, leading to decreasing livestock productivity. Increasing poverty, proliferation of guns, and erosion of the power of elders, have led to an increase in insecurity, banditry and other conflicts. Resolution of the conflicts in Karamoja must benefit from both preventive and curative mechanisms. Ways must be found to reduce or eliminate the causes of conflict: to legitimize mobility of livestock in accordance with the non-equilibrium nature of the environment, to establish formal and informal mechanisms for negotiated access to grazing resources, to ensure equitable distribution of development benefits, and to allow effective participation of the Karimajong and their leaders in the development process. Preventive mechanisms must be based on traditional systems, and should include early warning systems. Conflict-resolution mechanisms are best based on existing and traditional mechanisms. However, innovative structures and formal processes must also be found not only to resolve conflicts, but also to sustain the gains into the future.

RÉSUMÉ

Les modifications de la mobilité, les contraintes imposées par la réduction du territoire et des troupeaux et l'accroissement de la population ont provoqué l'abaissement du niveau de vie moyen des Karimajongs. Les pasteurs ne peuvent plus exploiter convenablement les ressources naturelles variables, de sorte que la productivité du bétail diminue. L'aggravation de la pauvreté, la multiplication des armes à feu et l'affaiblissement du pouvoir des anciens se sont traduits par une augmentation de l'insécurité, du banditisme et des autres conflits. Il est nécessaire de prévoir des mécanismes préventifs et curatifs pour régler les litiges au Karamoja. Il faut trouver les moyens de réduire ou de supprimer les causes de litige, de légitimer la mobilité des troupeaux en harmonie avec l'absence d'équilibre de l'environnement, de mettre en place des mécanismes formels ou informels d'accès négocié aux pâturages, d'assurer une répartition équitable des fruits du développement et de faire participer effectivement les Karimajongs et leurs dirigeants au processus de développement.

Les mécanismes de prévention doivent être fondés sur les systèmes traditionnels et comporter des dispositifs d'alerte rapide. Les mécanismes de règlement des litiges doivent de préférence être basés sur les instruments existants et traditionnels, mais il faut aussi trouver des structures et des processus novateurs pour régler les conflits et aussi pérenniser les résultats obtenus.

Introduction

KARAMOJA HAS LONG been considered a 'hot spot' of internal conflicts. Early anthropologists documented a history of shifting tribal warfare and alliances. Early colonial policy was to 'pacify' the roving pastoralists. Uganda's troubles in the 1970s and 1980s aggravated a fluid situation of raids, conflicts, and revenge killings. Recently, under the regime of President Museveni, innovative conflict-resolution mechanisms are being experimented with. The internal and external stresses causing the conflicts are as strong as ever, and solutions need to be found quickly to prevent the situation from deteriorating further.

This Chapter looks closely at the dynamics of conflicts over land, mobility and people's access rights to the resources. It first describes the physical and social setting, both in the past and today. It then traces the causes of present-day conflicts by analysing the changes occurring in leadership structures, customary and modern institutions, grazing rights and natural-resource tenure. It then describes new conflict-resolution mechanisms, whether *ad hoc* or formalized, that have emerged spontaneously in the past five years. It concludes with an analysis of these mechanisms, and provides design principles, and perhaps guidelines, for successful conflict-management mechanisms in Karamoja and elsewhere.

Karamoja's resource base

Karamoja is located in the north-eastern corner of Uganda, bordering Sudan and Kenya (see Map 7.1). It has two main seasons: one long and dry, and the other short and wet. The long-term average annual rainfall (1939–1995) is 620mm, but it can vary between 420mm and 1260mm. Map 7.2 shows the average rainfall isohyets. Rainfall decreases from west to east, and from south to north. This spatial variation is a major factor in the land-use patterns of the different tribes using this area.

The distribution of rainfall in time is another determinant of the ecological potential of this land. Three major droughts preceded this century: the first recorded in traditional folklore occurred between 1706 and 1733, the second about 1800, and the third from 1876 to 1900 (Pazzaglia, 1982), during which the estimated loss of cattle was between 70 and 90 per cent. Between 1924 and 1952, almost every third year resulted in crop failure due to droughts (Dyson-Hudson, 1966: 76–77). The 1960s were relatively good, but since the 1970s crop failure

Map 7.1 *Uganda*
(Source: *UNCDF, 1991*. Integrated development in Kotido District, Karamoja. *UNCDF, New York*)

has returned to its three-year cycle. In addition, rainfall is not always distributed adequately during the wet seasons, leading to false starts and flooded harvests.

The uncertainty of the rainfall is the primary reason why the Karimajong do not depend solely on cultivation for their livelihood, and the reason why they consider the land better suited to livestock production. In contrast to the three-year cycle of crop failures, major droughts leading to starvation of at least 20 per cent of livestock occurred only once every decade between 1927 and 1995[1]. The most severe was probably that of 1957–58, reported to have decimated a total of 200 000 cattle from a herd of 700 000 (Rattray & Byrne, 1963).

The coefficient of variation (CV) of rainfall is very high in Karamoja. Long term CVs are 30 and 35 per cent depending on the recording station[2]. Some

---

[1] In 1927–29, 1932–33, 1943–45, 1957–58, 1964–65, 1973–74, 1984–85, and 1994.
[2] Average annual rainfall for Kotido Town, between 1948 and 1995 was 648mm, with a coefficient of variation (CV) of 31 per cent. The average for Kangole (Moroto) in the same years was 618mm

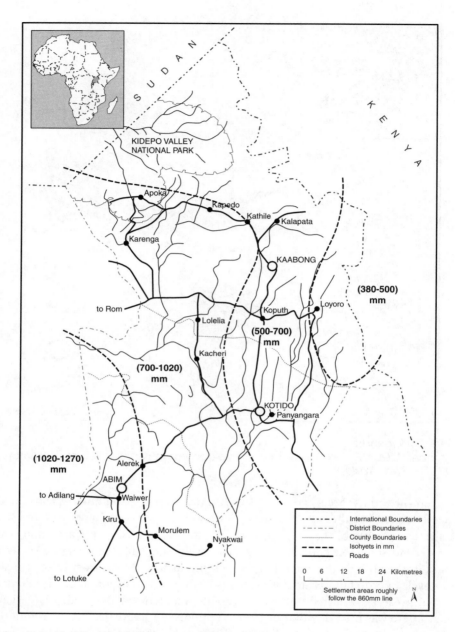

Map 7.2 *Kotido District mean annual rainfall*
(Source: *Niamir-Fuller in UNCDF, 1991*)

with a CV of 34 per cent. Calculated from the following: 1939–1957 rainfall statistics from four stations in both Kotido and Moroto Districts (Dyson-Hudson, 1966; data derived from the Annual Reports of the colonial administration and the Meteorological Department); 1961–1995 data from four stations, recorded by the Meteorological Department and the Church of Uganda.

researchers have proposed that arid lands with CV greater than 30 per cent be considered 'non-equilibrium' ecological systems (Ellis & Swift, 1988). According to this view, arid ecosystems never achieve equilibrium because of the high degree of variability. The ecosystem is constantly changing from one level or state to another. Thus, it is difficult to predict vegetation responses to stresses, such as overgrazing, fire, drought, etc. Furthermore, sustainable use of the resources must be flexible enough to follow the constant climatic changes.

The Karamoja region covers 24 000 square kilometres (Map 7.3). A steep escarpment, commonly called the Turkana Escarpment, defines the eastern border with Kenya. The central plain is the most striking feature of the region, with abruptly emerging mountains, some of which are volcanic (Mamdani et al., 1992). The ecology of Karamoja can be described as a three-stage drainage pattern (PARTICIP, 1992): the steeper eastern slopes drain into the central plains, where the rivers have deposited alluvial soils and sandy beds in between clayey soils. These, in turn, drain into the western flatlands and seasonal swamps.

The Karimajong distinguish at least 16 types of water points, depending on their source (river, rain), location (plains, mountains, slopes, swamps), permanency, and method of capturing it. This indigenous technical knowledge (ITK) forms a 'grid' over known stretches of territory (Dyson-Hudson, 1966), and is another factor in determining a herder's movement plans over the year and between years.

Given the variations in rainfall from one corner of Karamoja to the other, and variations from year to year resulting in non-equilibrium conditions, the vegetation of Karamoja can be quite varied. However, three general landscape forms can be distinguished for descriptive purposes:

o the eastern slopes covered with fairly dense deciduous trees and shrubs (mostly Acacias and some *Combretum* spp.), and a good cover of mostly annual and some perennial grasses
o the central plains with fewer trees and shrubs and a few perennial grasses, and
o the western flatlands with few trees and shrubs, mostly hydromorphic trees (*Acacia nilotica, Acacia seyal*) and perennial grasses (Map 7.4).

Nested within these general vegetation groups are micro-niches of specialized vegetation, adapted to the variations in soils, topography, drainage and micro-climate (shade, rocks, etc.). The Karimajong have learnt that mobility of live-stock can use the micro-niches to their fullest. Estimates of the carrying capacity average out high-quality micro-niches, and result in figures below the actual potential. Tracking[3], dispersion, and mobility are variables with which the pastoralists can stock the rangeland at higher rates than that suggested by an average carrying-capacity figure.

---

[3] Tracking is the process or method by which the herder balances the need of his livestock with the productivity of the pasture, within a continuous feedback loop based on daily monitoring (Niamir, 1997).

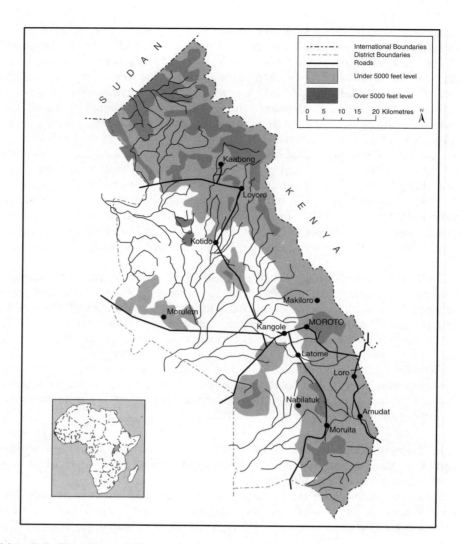

Map 7.3 *Elevation of Karamoja*
(Source: *Rattray & Byrne, 1963*)

The Karimajong have never relied solely on the resources of Karamoja proper, but have always taken advantage of natural resources outside. Examples are the Acholi and Teso lands to the west, and neighbouring Kenya to the east. In recent years, this mobility has been curtailed through administrative decisions and political insecurity. In addition, grazing lands are increasingly coming under cultivation, because of the relative security of crops since they are not the object of raids. Land degradation and resource scarcity have been direct results of these trends, as more and more people and livestock congregate on less and less land.

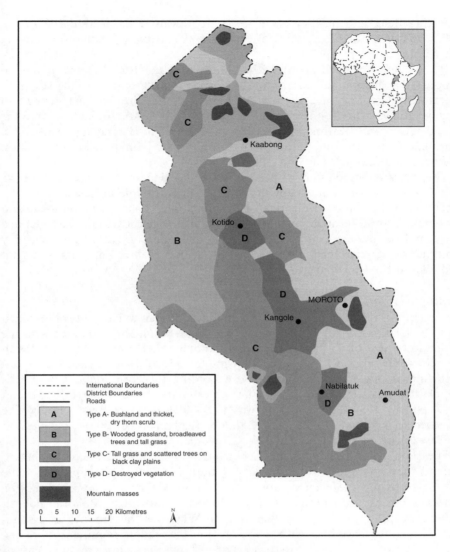

Map 7.4 *General ecological map of Karamoja*
(Source: *Trapnell & Wilson, reported in Rattray & Byrne, 1963*)

The Karimajong 'cluster'

Prior to the year 1500, the area now known as Karamoja was inhabited by cultivators. These were Cushite-speaking peoples such as the Tepes, Mening, Nyakwae, Nyangia, Labwor and the Ik, and Kalenjin peoples such as the Pokot, Sebei, and a bushmanoid group called the Woropom. Between 1520 and 1550 the Somali in the Horn of Africa put pressure on the Galla, pushing the latter toward the Sidamo Mountains. The Galla, in turn, met the Nilotes of the Plains,

155

defeating them and pushing them westward toward Lake Rudolf (Pazzaglia, 1982). It was around the shores of this Lake that tribal differentiation among the Nilotes began (Map 7.5).

By the late sixteenth century, the shores of Lake Rudolf were unable to support the large influx of pastoralists, and many groups moved west into Karamoja. The Teso moved closer to Lake Victoria, the Turkana decided to stay in Kenya, the Toposa pushed onwards into southern Sudan, and the Jie, Dodoth and Karimajong settled in Karamoja (Pazzaglia, 1982). The Karimajong proper split into various groups based on previous clan lines: Matheniko, Bokora, Pian and Kadam (see Map 7.6).

The 'Karimajong cluster' of peoples are today defined as those pastoralists and agropastoralists that occupy the Province of Karamoja in Eastern Uganda, whether they be Nilotes, Kalenjins, or Bushmen. They share common social and economic features through years of assimilation: an age-set system, the belief in a deity, and the pre-eminent social and economic value of cattle. However, they continue to perceive themselves as separate tribes, each claiming its own land and backed by its own armed forces (Map 7.7).

Population and production systems

Compared with the rest of Uganda, Karamoja continues to be a thinly populated land. Population density is 13.5 people per square kilometre compared with the national average of 85 people per square kilometre[4]. However, the population growth rate is relatively high, even allowing for difficulties of measurement and reporting among mobile populations. According to census data, Karamoja's population was 26 000 in 1919. Further estimates were made in 1931 and 1948, and by 1959 the population was 172 000 (Rattray & Byrne, 1963). The 1969 census recorded 270 300 people, that of 1980 recorded 350 000 people, and the last census of 1991 recorded 370 423 people[5].

The production system of the Karimajong varies with the ecology, but in general they are agropastoralists, raising both livestock and crops. For some groups crops are more important (e.g. Labwor, Mening, Nyangia) while for others livestock are paramount (Jie, Dodoth, Matheniko, Bokora, Pokot). Others are primarily hunter-gatherers (Ik). Whatever the production system, the Karimajong have one feature in common: the feeding requirements of the livestock pull the herd-owner to the west or east, but the security requirements of the family pull it to the centre, where the bulk of the settlements and farms are. Herd-owners have reconciled these conflicting needs by splitting the herd into a main herd and a milk herd (Dyson-Hudson, 1966). The head of household has to ensure that the settlements have enough milking cows to feed the settled population, but not to leave so many as to attract raids.

The permanent settlements (*ere*) are in the crop-growing areas (usually higher ground, sandy or clayey soils) where most women, children and older people reside. The men, older boys and girls (and sometimes recently married couples)

---

[4] Statistics Department, Ministry of Finance and Economic Planning, 1996.
[5] Statistics Department, Ministry of Finance and Economic Planning, 1992.

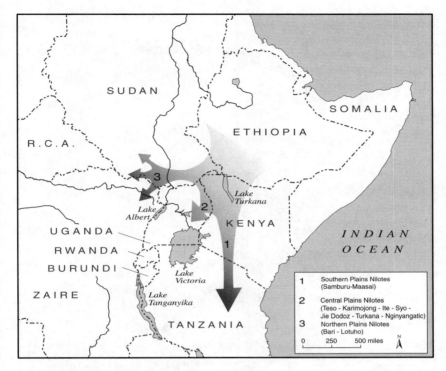

Map 7.5 *Historical migrations of the Plain Nilotes*
(Source: *Novelli, 1988*)

are almost always on the move with the bulk of the livestock. When milking cows finish their lactation period, they and their grown calves are sent to back to the 'main' herd, and replaced with new milking cows.

Traditional agriculture is carried out using simple hand-held tools. Ox-power has only recently been introduced by missionaries and projects, and is not yet practised by all. A few traditional technologies for soil conservation are still prevalent. For example, the Dodoth practise trash-bunding on their cropped land, the Tepes terrace their mountain slopes, and the Labwor practise strip cropping (Rattray & Byrne, 1963).

Livestock ownership and distribution

The Karimajong ideal is to possess a large enough herd to support an extended family of three generations who, in turn, provide the labour for the herd and crop production. They do not consider wealth in a speculative manner as do capitalistic entrepreneurs, but as a form of economic security over the long term. The frequent droughts create a 'boom and bust' situation, forcing the owner to raise a large enough herd so that a core group of surviving breeders can reconstitute it after a drought.

157

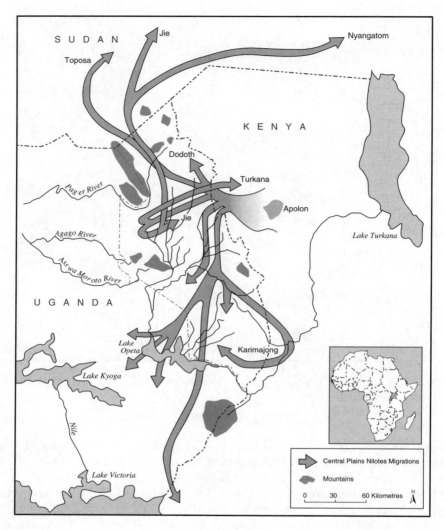

Map 7.6 *Migrations of the Central Plain Nilotes*
(Source: *Novelli, 1988*)

Livestock records are even more unreliable than those of the human population. However, available data suggest that livestock have decreased from 603 000 head of cattle in 1959, to 321 845 head in 1982 (Annual Reports of the Veterinary Department). By contrast, the population of sheep and goats in the same period increased from 376 391 to 554 165 head. This is probably due to the effect of constant raids across the border being aimed primarily at cattle and not at sheep and goats.

In the 1990s cattle populations are estimated to have increased, while sheep and goats have remained stationary. Although no formal census has been

158

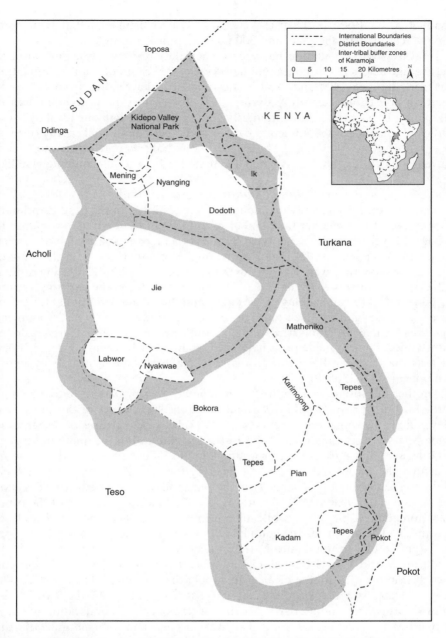

Map 7.7 *Ethnic groups and inter-tribal buffer zones of Karamoja*
(Source: *adapted from Gubert, 1988; UNCDF, 1991*)

conducted, by 1995 the District Veterinary Officer estimated that there were 536 000 cattle, 230 000 goats and 300 000 sheep in Karamoja.

Total livestock population has yet to reach the pre-1959 level. Furthermore, the average herd size per household has decreased because of human population increase. Between 1900 and 1950 it is estimated to have fallen from 100 to 50 head per household (Rattray & Byrne, 1963). Today the ratio is no more than 28 (assuming 10 people per household). By contrast, the Karimajong ideal of a self-sufficient, viable household economy is 6 cattle per capita (Baker, 1974), or at least 60 head per household. The average household income at the moment is currently below the level considered adequate for long-term economic viability.

Socio-economic and political significance of raiding

Raiding is a traditional activity among all plains Nilotes. In the past, its primary purpose was to regenerate the number of animals in the herd as a response to drought. It was therefore sanctioned by society and given spiritual and material support by all (Ocan, 1992). The Karimajong were not criminals in the sense of robbery or rape; they were warriors on a noble mission with the blessing of their tribal spirits (Rattray & Byrne, 1963). It was unthinkable in the past to carry out a raid without the full backing of the elders and the divine. Traditional leaders of successful raids gained much prestige from their prowess, and apart from being the choice catch for unmarried women, could become potential organizers of future raids. Some notable raids have been recorded in Karimajong oral history, such as that between the Jie and Bokora at Nakoret Amoni, near Panyangora (Lamphear, 1976: 205–8).

Another purpose of raiding for the individual warrior, was to obtain enough cattle for bride price (Novelli, 1988), although the significance of this purpose is played down by some researchers (Ocan, 1992) because increases or decreases in bride prices are seemingly unrelated to the increase in raiding and banditry.

Guns were first introduced to Karamoja in the 1880s by ivory traders. The Jie benefited the most, and became the leading force in Karamoja at that time, under the leadership of Loriang. This phenomenon saw the emergence of raiding for private gain (Ocan, 1992). But it was also the first time that one of the tribes was potentially able to consolidate a centralized state structure for all of Karamoja. The advent of the colonial administration and military might effectively put an end to this centralization process.

During World War II the colonial administration instituted forced sales and used Karamoja as a source of meat to supply the war effort (Ocan, 1992). The modern State, which replaced the colonials, is also perceived by the Karimajong to be a formidable raiding party, because of its policies of confiscation of cattle as punitive measures against raids. The State has lost some of its moral authority in Karamoja, because of its lack of understanding of the Karimajong mentality, and the impact of its policies.

The leading cause of cattle loss, however, must still be attributed to raids by the Karimajong and their neighbours, (such as Turkana, Pokot, and Didinga). Although guns have been around since the 1880s, violence and banditry did not explode until the 1970s under President Amin and later President Obote

160

(1980s). Amin as an officer in the Colonial King's African Rifles pursued the Karimajong and Turkana ruthlessly. As President in 1973, he ordered the mass killing of hundreds of people in southern Karamoja (Ocan, 1992: 16). In 1979 after the collapse of the Amin regime, which coincided with a major drought, the Karimajong broke into military armouries and equipped themselves. The same occurred after Okello's regime in 1986. A new pattern of heavily armed raiding had begun to emerge.

Ethnic conflicts are now sharper than ever before because of privately motivated raids (Belshaw et al., 1996). Not all raids are preceded by ritual and community consensus. Many raids are organized on an *ad hoc* basis by individual warriors. It is no longer a question of only stealing livestock, as it once was, for now even household goods are taken. Banditry on the roads, aimed mainly at passenger vehicles and buses, was of major detriment to the local economy and led to suspension of most development aid in the late 1980s and early 1990s.

The Karimajong have lived with and adapted to conflict for a long time; conflict touches many areas of their lives, including their houses. All houses are clustered together in a defensive ring, with livestock parked for the night in the middle. The outer ring is protected by a strong stockade of stout trunks and thorns. Gates through the stockade are only half the height of a grown man, requiring all visitors and intruders to crawl while entering.

One precaution against raids, and, indeed, against other calamities such as drought, disease, etc., is to distribute one's animals among several herds. Distribution helps to spread the risk, in the hope that only part of the herd will be lost. Another precaution against raiding is to leave the bare minimum of animals at the settlement to feed the homestead family, so as not to attract the attention of raiders. Despite their imposing stockades, homesteads are not as well defended as the mobile camps because very few warriors stay at the home-steads permanently. By far the most effective precaution against raiding is to have, and to show, as many guns and able-bodied men as possible.

Another defence against raiding is to vacate remote areas and settle close to population concentrations. Maps of the 1960s show that the southern part of Jieland, bordering on Bokora, was fully settled, but these lands have now been abandoned for Kotido Town and the area to its north, leading to greater land pressure and degradation. Gulliver (1970) reports that the Jie would use the eastern zones (bordering Turkana) as wet season pastures; oral reports confirm this. However, by the 1990s and with renewed conflicts with the Turkana, these areas were also abandoned. The Dodoth have found a different way of adapting to raids. They prefer to disperse into the crevices and valleys of the mountains to the north, venturing into the plains only for the cultivation of rapid, short-season crops. These land-use changes have been the primary cause of land degradation in Karamoja. There is now overgrazing on marginal hillsides, in the case of the Dodoth, Ik and Nyangia, and over-concentration of settlements (and therefore overgrazing and deforestation around settlements) in the case of the Jie, Bokora and Matheniko.

161

Local institutions

The term 'local institutions' in Karamoja describes both customary and modern/ government institutions. The term is used in its wider sense and includes all structures, whether individual or communal, formal or informal. These institutions have all changed considerably over the years.

Traditional institutions

Apart from the sole attempt by the Jie to forge a centralized authority in the 1880s, the Karimajong remain essentially as fragmented tribes and sub-tribes. Traditional political authority is based on the age-set system (Dyson Hudson, 1966). All Karimajong adult males progress through a series of age- and generation-sets in their lifetime. Five age-sets comprise a generation-set, of which there are two: elders and juniors. The elders are responsible for the community's welfare and are authorized to direct its activities. The juniors are subordinate and are the instruments of policy, such as conducting raids. The elder generation (for example, among the Matheniko, the 'Mountain' generation) is a corporate body with ritual sanctions, which can intercede with God, and is empowered to arbitrate disputes. It makes decisions through consensus on behalf of the community, such as moving cattle, scheduling traditional ceremonies, and negotiating access to resources with other tribes.

Among the elders there is no established hierarchy, but power is bestowed through consensus by virtue of their age, the number of cattle they possess, and their prowess and successes in battle. Political priorities are fluid, and depend on fortune and skill, not on inheritance. When the number of elders in the Mountain generation becomes too few to run society, they will have to hand over power to the juniors (at the moment the 'Gazelle' generation), who upon becoming elders, will initiate their sons (the 'Rats') to become the junior generation.

While men take all the decisions regarding political issues, war, alliances, and cattle movements, women are the decision-makers in matters of daily life. Everything regarding the family, children and village activities are women's responsibilities. Recent efforts to introduce mobile education adapted to pastoral needs among the Karimajong was only able to proceed once the women, in a major gathering, agreed to allow their children to attend and to adopt modern education. They were then assisted by the elders in a ceremony called 'Unearthing the Pen'[6].

In addition to the age-set, the Karimajong recognize a clan system based on heredity. A clan is headed not by one elder, but by as many as are accepted into that generation from that clan. Clan members are scattered among different locations and tribes. The same clan name is to be found among the Jie and Toposa, as among Matheniko and Bokora (Pazzaglia, 1982).

Clans are, in turn, divided into sections, called *ekitela,* which are territorially based and associated with the settled agricultural areas. The *ekitela* are

---

[6] 'Unearthing the Pen' refers to a ritual cancellation of a curse that was placed on 'the pen' after the Colonial Administration's punitive and forceful campaigns to educate the Karimajong.

institutions primarily for organizing religious and social events, but also for taking production decisions. In a conflict among persons of different clans or sections, one usually sides with the people of one's own section rather than those of the clan, because territory and land use are better determinants of everyday cooperation and loyalty (Novelli, 1988).

Each section has many '*kraal* leaders'. These are younger members of the elder generation, or older members of the junior age set. *Kraal* leaders are in charge of the main herds of the section, and take responsibility for all decisions on mobility, camp location, scouting, watering, labour allocation, dairy cow distribution, etc. Several herders work under the *kraal* leader, depending on the size of the herd. Selection of the *kraal* leader is done through consensus, and depends on his prowess, intelligence and social factors.

The *aruanit* are highly respected *kraal* leaders who have a special position in all public debates. The Jie, for example, have 21 *aruanits*, and the Dodoth 30. Each year *kraal* leaders will decide which *aruanit* to join for the dry-season movements, or whether to strike out alone. Although movements and yearly allegiances of the *kraal* leaders are fluid and opportunistic, each *aruanit* has his geographically specific movement pattern; the starting point is always the settlement area of his section, going through specific water points towards a specific dry-season grazing area. The *aruanit* will not waver from this pattern, except to change the speed of movement or to decide how far into Acholi and Teso land to venture.

Local government

Beginning in 1916, the British Administration imposed a hierarchical system of chiefs based on the Buganda model from Southern Uganda. The chiefs and sub-chiefs were given various administrative responsibilities, among them to enforce the dress code, compulsory education, and to collect taxes. Dyson-Hudson (1966) remarks that imposition of such a new form of organization met with failure since it never obtained the approval of the elders.

Until World War II, the Colonial Administration did not consider that Karamoja could contribute to the economy of Uganda, and its policy was based on containment, pacification, and sedentarization of pastoralists. By 1945 'development' of pastoralists replaced the old view, and for the first time a Veterinary Officer was stationed in Karamoja (Rattray & Byrne, 1963).

The early post-colonial period saw the continuation of colonial administrative structures and policies. But the political events of the 1970s and 1980s led to the breakdown of law and order. Government administrative organs barely functioned, rural development was forgotten, and militarism took over.

Local institutions today

Local government today is based on a hierarchical system of districts (of which there are two in Karamoja: Moroto and Kotido), divided into counties, sub-counties, and parishes. A local government representative is present at all district and some county levels. The government of President Museveni has initiated a system of local 'Revolutionary Councils' (RC) at each administrative level,

163

which are supposed to elect members from the local population. In practice, council members are usually young, educated and/or urban folk (traders, teachers), who have relatively little contact with the traditional leadership.

The mass media have reported extensively on Uganda's wars, insurgencies, cattle raids and other security problems. Concerns about political violence and insurgency are more prevalent in the northern region of Uganda (i.e. Kitgum, Lira and Gulu) than in Karamoja at the moment. However, the northern conflicts do have added impact on Karamoja, in the form of an active market in guns and ammunition, and refugees arriving into Karamoja. The war in Sudan has affected the north and Karamoja, increasing the conditions favourable to successful raiding. There are also residual effects from the troubles of the 1970s and 1980s: decaying infrastructure, lack of skilled men and women, poverty, looting, idleness and redundancy (Barton & Wamai, 1994).

As banditry becomes more prevalent than community-sanctioned sorties, elders see their powers gradually being ignored. A contributing factor has been that soldiers of war, under Amin and Obote, brought their ideas of independence back to the tribe (Novelli, 1988). This and other factors led to an erosion of the authority of the elders, and decreasing capacity to control the activities of the young warriors. Gradually, a few of the *kraal* leaders, gaining more successes than others through banditry and raids, have become the so-called 'warlords' of today. These shadowy figures are known to be behind most of the organized banditry in Karamoja.

The national government structure after Independence was never able to come to grips with the situation in Karamoja. Apart from a well-armed military presence, the administration's capability was weak, with few, and poorly motivated personnel. With the coming to power of the National Resistance Movement, the 'problem' of Karamoja became a national priority. After 1986 the National Resistance Army of President Museveni gradually restored order in the land and the internal supply of guns diminished. Today, guns still filter through from southern Sudan, and after the overthrow of Siade Barre in 1990, from Somalia via Kenya and Ethiopia. The collapse of Mengistu's regime in 1991 saw additional guns and ammunition come into the market.

Early development efforts by the current government focused on the equitable distribution of dry-season water points ('valley tanks'). A Ministry of State for Karamoja was created, and provided with an implementation mechanism called the Karamoja Development Agency (KDA). This institution was to coordinate all development activity in the region. These early efforts have not had much success. The causes of failure have been identified as insecurity, little or no community participation, and poor assessment and analysis of the basic problems facing the Karimajong (KPIU, 1996).

Another emerging institution is democratic representation in parliament. The recent parliamentary elections (1996) have brought four very active and young Karimajong to Kampala, whose expressed mission is peaceful resolution of conflicts, and rural development in Karamoja.

Beginning in 1994, with the appointment of a new Minister of State for Karamoja, a Special Presidential Adviser for Security Issues, and a Brigade

Commander, all of whom are Karimajong, a greater convergence between the Karimajong reality and government policy is being achieved. Yet, leadership among the Karimajong has now diversified and political power has been dispersed. Traditional leadership has disintegrated (elders, auranits, kraal leaders, warlords) and modern leadership is trying to establish a niche of its own (revolutionary civil leaders, local administration, parliamentarians, intellectuals). There are too many leaders now (most of whom fortunately wish to see an end to the conflicts) but there is too little consensus on how it should be done.

Land tenure

Customary tenure
Customary land tenure in Karamoja is organized around membership in tribes. Each tribe has its own clearly defined 'home territory' where the settlements are located. The home territory also includes all wet-season and some of the early dry-season grazing areas. The boundaries of the home territory are more or less fixed. Successful raids can sometimes expand the boundary, but the new territory is only used for grazing, not for settlement or cultivation.

Tribal home territories are separated from each other by buffer zones, where no settlement or cultivation occur. Buffer zones become usable for grazing if friendly relations are negotiated between neighbours (Novelli, 1988). Today's buffer zones are shown in Map 7.7, and include the more or less 10km buffer of non-cultivation seen on satellite imagery between Acholi and Karamoja.

The traditional system recognized a set of formal and informal rules for enforcing customary tenure. Within each home territory, tribal members are free to use any part of the land; however, *kraal* leaders and *aruanits* have habitual routes based on the location of settlement areas and ownership of water sources. Open, natural sources of water are free to all tribal members. Outsiders must seek permission from section elders to use both the water and land around it. Improved water sources (hand-dug wells, and *'ngaperon'* or hand-dug small catchment basins) are the property of the individual or section that invested in them.

Impacts of the colonial administration on land tenure
The early policy of the colonial administration was to stabilize pastoralists within strict boundaries. They defined these boundaries according to where they found the people at one given time (Gulliver, 1970: 9). As a result, the pastoralists were restricted to only one of their seasonal migration areas.

Administrative boundaries are irrelevant to herders concerned with utilizing available grass and water at need, since it is mutual pressure among competing groups that eventually sets up effective boundaries (Dyson-Hudson, 1966). However, the boundaries were patrolled by the colonial administration with the zest and vigilance normally reserved for international borders, because Karamoja was still considered to be an 'occupied territory' needing 'pacification' (Mamdani et al., 1992: 29). These efforts were met with resistance and defiance from the various groups. The Karimajong continued to compete for natural resources in the traditional way, and to raid each other's livestock.

Karamoja's border with Kenya was first delineated in 1940 during the colonial era. In so doing, the Karimajong formally lost about 15 per cent of their southern grazing lands to Kenya (Dyson-Hudson, 1966: 783). This transfer affected four of the Karimajong tribes, but primarily the Pian. Most of the land was given to the Pokot of Kenya, thus sparking off a major conflict that still lasts to this day. The Karimajong also lost their *de facto* rights to use the fertile dry-season grazing land in Teso (Rattray & Byrne, 1963: 8; Cisterino, 1979: 79), and only do so these days when they have to, and in stealth.

Another major factor was the *cordon sanitaire* that was established on the border with Sudan in 1919 to contain Sudanese rebels. Settlements near the border were translocated to more southerly areas. Wild herbivores spread into these abandoned areas, bringing with them tsetse flies into Karamoja for the first time. By 1948 the fly had occupied all the grazing grounds up to Labwor[7], in spite of a 32km bush/tree clearing around Kotido to halt its spread. This pushed the Jie and Dodoth out of many of their dry-season pastures into Labwor, Mening, Ik and Bokora lands.

Another impact on land use was the establishment of forest and hunting reserves, and national parks. By 1958, 25 per cent of Karamoja was gazetted by the colonial administration (Mamdani et al., 1992: 47). The gazetting was done under the assumption that pastoralism was a major threat to the eco-system, and communities had to be kept out of these areas. Only now are some environmental proponents recognizing that mobile pastoralism is not only compatible land use, but that pastoral communities rely on the resources within these gazetted lands. For example, a recently approved regional Global Environment Facility project is willing to consider people's participation in the conservation of wildlife areas. A survey done in 1995 of herd movements among the Dodoth and Jie, showed that some herds continue to use the lands within the forest reserves and even the Kidepo National Park on a seasonal basis for their main herds (Karamoja Integrated Development Project, Phase 1), but most prefer to avoid these areas for fear of rangers and police.

A total of 40 per cent of the original lands of the Karimajong (not counting rights to Teso) have been taken away during the colonial period. The Karimajong now are restricted to only a small portion of the remainder, due to a combination of: tsetse flies, the need to avoid the Turkana and Pokot, as well as conflicts with the Teso, and increasing insecurity between the Karimajong. The mobility of the Karimajong is no longer determined by non-equilibrium ecological forces and political alliances, but also by increasing insecurity and conflicts between tribes. This change has resulted in land degradation – for the moment only around settlement areas – and in lower productivity of animals, because of the inability to properly 'track' feed resources.

---

[7] Tsetse Control Department Reports, 1954.

166

Land tenure today

Administrative boundaries of today in general coincide with the home territories of the major tribes (Map 7.8), but they do not include all their annual grazing areas, which can sometimes cover several districts or provinces. In addition, minority tribes, such as Mening, Nyangia and Ik, are subsumed under the main tribes, potentially causing conflicts. The Labwor have more land than they can use today (part of which is in Jie land) because they no longer herd livestock as before.

It is only in recent times that the notion of administrative boundaries has begun to make an impact, and only because of the adoption of democratic rule in Uganda. Now, parliamentarians are elected from each District and County, and gradually the majority of Karimajong are beginning to understand the political significance of such boundaries, if not their significance for land utilization.

One factor that has affected land-use patterns between the tribes was the construction by the colonial administration of at least 108 dams and water-retaining structures, mostly in the central plains and the western flatlands. The rationale was to increase the land area available for dry-season grazing within Karamoja so as to avoid the need to use Acholi and Teso lands. The structures were put in place and seen as public goods with, therefore, open access to all. This could have resulted in a 'tragedy of the commons', but not everyone changed their grazing movements, because of inadequate forage around the water points. The post-colonial government was not able to maintain the structures because of financial constraints, and many fell into disrepair.

Anectodal data suggest that in the early 1900s, the central and eastern zones were covered with vast grasslands, and had very few shrub and bush thickets. The present-day bush encroachment is often considered to be due to overgrazing by the pastoralists, proponents invoking both the 'cattle complex' and the 'tragedy of the commons' theories (e.g. Wilson, 1962; Bredon, 1963; Baker, 1967). Others believe that the water-retaining structures built by the colonial administration resulted in overgrazing in the eastern slopes, and were therefore the cause of bush encroachment (Mamdani et al., 1992). It is highly unlikely that the dams had such a profound effect, since they barely operated for ten years on average, and by the early 1960s only nine were operational (Baker, 1967). In addition, many of the water structures were not fully utilized because of inadequate quality forage around them. What is more likely is that the redrawing of the international boundary between Kenya and Uganda heightened the conflicts between the Turkana/Pokot and the Karimajong, leading to the establishment of a *de facto* 'no man's land' in the eastern slopes that continues to be respected to this day. Under-utilization of these arid and semi-arid lands, and the accompanying reduction in bush fires normally set by herders, are more likely causes of bush encroachment (Sabiiti, 1990).

Another external influence of major importance was the designation, in the 1970s, of Lolelia as a resettlement scheme. The idea originated with the colonial administration in the 1950s but was not acted upon until much later (Rattray & Byrne, 1963). Lolelia is an area with a distinct micro-ecosystem: good soils,

167

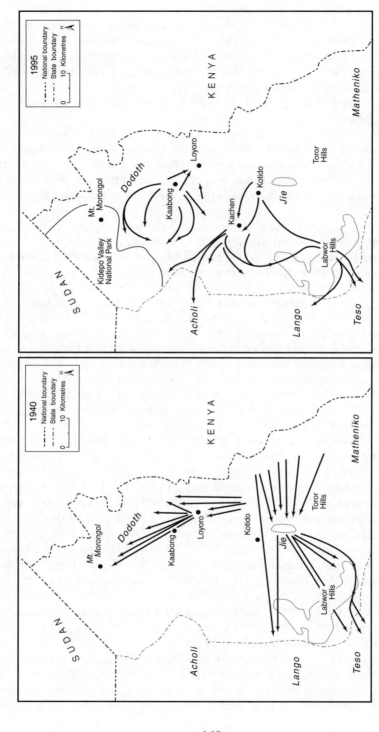

Map 7.8 Dry-season grazing patterns in northern Karamoja, 1940 and 1995
(Sources: 1940 data from Randall reported in Dyson-Hudson, 1966; 1995 data from Niamir-Fuller in UNCDF, 1991)

shallow aquifer, sheltered by a ring of mountains, and irrigated by a major drainage system that has water all year round. It was considered to be an excellent site for crop cultivation, and in 1976 a number of Jie households were relocated, provided with material and financial assistance, and encouraged to sedentarize.

The scheme was quickly a failure. By the early 1980s no one was left settled in the area. The reasons were simple: Lolelia is a major dry-season congregation point for both Jie and Dodoth herders – at times peaceful, at others quite conflictual. It is in a buffer zone between the two tribes and therefore property rights can be contested. In addition, a series of Turkana raids on the settlements sparked off a mass dispersal of the households. This was a clear case where appropriation by a few of a commonly held resource was unacceptable to the majority.

The idea of a settled community based on irrigated agriculture is still alive among administration personnel. It may only be a matter of time before a private entrepreneur decides to expropriate the land to establish a large farm, unless steps are taken to secure the rights of the community to this and other valuable resources.

The New Constitution of Uganda (1995) has attempted to clarify the land-tenure laws in Uganda under Article 237. It clarifies the processes of private ownership and acquisition of land, and clearly empowers the local government to control those lands that are for the common good of the citizen. The constitution recognizes four land-tenure systems, all related to private ownership of land. It does not explicitly recognize communal ownership of land but it can be assumed under 'customary tenure'. It is silent on the issue of mobility of animals on common land, and the need for negotiable 'inclusive'[8], not exclusive, rights to grazing resources.

Grazing rights and rules

The Karimajong herder and *kraal* leader takes daily decisions on where to guide his animals based on monitoring and evaluation of the condition of resources; in other words, a continuous process of 'tracking' the resource and adjusting live-stock behaviour to changes in the resource (Behnke & Scoones, 1993; Niamir, 1997). Camps are moved fairly often, not only to avoid exhausting the range-land, but also for security. Scouts are sent ahead to evaluate future sites for the camp. Once in a new camp, the herders decide on where to graze each day, based on the availability and condition of the patches, or micro-niches of the resources (Niamir, 1997). The little information that exists on the ITK of grazing in Karamoja, points to quite a sophisticated knowledge of rangeland dynamics and fairly complex grazing rotation systems (see Box 7.1). The ITK of the

---

[8] 'Inclusive rights' to land are based on the recognition of a tiered priority system of property rights. For example, customary owners have first priority, neighbours have second priority, and occasional users (e.g. during droughts) have third priority (Niamir, 1997).

> The Jie and Turkana very frequently move camp to avoid overgrazing and to 'track' resources. They are known to make 5 to 10 major camps each dry season, and more temporary camps around each major one (Gulliver, 1970: 44). The Karimajong recognize that certain areas should be left ungrazed in the wet season so as to leave enough fodder for the dry. An example in point is the western flatlands, which could be occupied all year round, but are not; and anyone who tries to do so is faced with intense public disapproval. The Pokot practise wet season deferment of areas with termite-resistant grass, to allow good fodder for the dry season (Ostberg, 1987: 48).

Box 7.1 Examples of Karimajong pastoralists' indigenous technical knowledge

Karimajong has not been studied adequately and needs much greater attention, especially by local and national researchers.

In times of drought or resource pressure, there is increased mobility and dispersion, a rule that is common to most pastoralists (Niamir, 1987). However, among the Karimajong this rule is tempered by the need for security. The greater the insecurity, the more concentrated the groupings around the *aruanits*. In the dry season, movements are less frequent but of longer distance. In bad years, the main herds may move several hundred kilometres into Acholiland.

It has long been accepted by researchers and development workers that the Karimajong herders do not follow any schedule nor pattern of movement, but will opportunistically follow rainfall wherever security and natural conditions allow (e.g. Dyson-Hudson, 1966; Gulliver, 1970). However, recent investigations show that there is more consistency in movement from year to year than was hitherto believed. *Kraal* leaders try to go to the same grazing area every time if possible; *aruanits* and their followers, too, have definite grazing patterns; and tribes stay within their home territory as much as possible. Although movements have to be flexible in order to respond to a non-equilibrium ecosystem, this does not mean that there is no visible pattern or consistency when conditions allow.

The Karimajong gain access to grazing land through negotiation on a yearly basis. An alliance between two tribes, once it is obtained, is a transient, but *de-facto* recognition of the right of tribal members to use each other's land without specific permission. Alliances, however, can be fragile, and a herder exercises his rights with caution. The 'rural radio' (word of mouth) is indeed an effective means of information dissemination in Karamoja, with which the herder keeps abreast of all raids and alliances.

The Jie reportedly pay grazing fees to the Teso each dry season (one bull for every 'large' herd). This information remains to be verified, since it has not been recorded for the other tribes, but if true, then it represents a more secure mechanism for acquiring temporary grazing rights.

Periodic raids distribute livestock from one rangeland area to another, and back again, resulting in a 'rotation' of livestock between different tribes (Table 7.1). Per capita livestock for all of Karamoja in the 1950s was 4 cattle

Table 7.1. Evolution of per capita cattle distribution among the four major tribes in Karamoja (Uganda)

Period	Jie	Dodoth	Matheniko	Bokora
1980[1]	0.16	0.16	1.5	1.5
1985[2]	3.5	0.03	2.07	0.93
1989[3]	2.1	2.1	3.5	3.5
1995[4]	2.8	1.06	1.2	2.1

1 = Euroconsult, 1983; 2 = Gubert, 1988; 3 = District Veterinary Officer (DVO) estimates on livestock & 1991 population census; 4 = Statistics Department and DVO reports.

and 4 sheep and goats. By 1980 this had fallen to, on average, 0.83 cattle per capita. The successful raids by the Matheniko on the Jie in the late 1970s resulted in higher per capita livestock among the Matheniko/Bokora than the Jie/Dodoth. This was reversed in the mid-1980s when the Jie raided both the Bokora and the Dodoth. In the late 1980s the Karimajong replenished their herds by raiding outside Karamoja (Turkana, Pokot). The Dodoth then bounced back slightly with successful raids against the Jie in the early 1990s. At the same time, the Bokora raided the Matheniko.

The pattern of raiding is not one-on-one revenge, but opportunistic raids against any neighbour. Whoever has plenty of animals should expect to be raided by anyone. The effect of raiding resembles rest–rotation grazing schemes, since parts of Karamoja will be clearly understocked for a few years, before being stocked again. Thus, it has the indirect effect of preventing continuous overstocking on rangelands.

Livestock mobility has changed considerably since the 1940s. At that time, during the colonial administration, movements were determined by territorial rights, proximity to settlements, distribution of tsetse flies, inter-tribal allegiances, and non-equilibrium variations in the ecosystem. Most of these factors are still important, but other factors also influence the mobility: the presence of the Kidepo National Park, Turkana movements, recent raids by the Sudanese, and banditry and insecurity.

There are no traditional procedures for changing the rules of negotiation for access to grazing land, because the mechanisms have always been flexible. However, insecurity and banditry have changed the context and there is now a need to formally recognize some of these mechanisms, and to reinforce them through procedural rules for negotiation, mediation, etc.

Conflicts and conflict-management mechanisms

The Karimajong classify non-Karimajong peoples in three categories (Novelli, 1988: 98):

o '*Friends* are those who help the Karimajong attain his pastoral goals; the occasional allies who allow him to graze his herd in their territory, or who join him in raids against other groups . . .

o *Enemies* are those who oppose the attainment of those goals . . .

o *All others* [are] those who neither help nor hinder, are looked on with in-difference which borders on contempt, because they have no livestock.'

Enemies and *all others* are to be exploited at will, while *friends* are for building mechanisms with in order to avoid or minimize conflicts.

The causes of conflict in Karamoja are diverse. The main source is the non-equilibrium ecological system which creates booms and busts in livestock population and people's livelihoods. This, in turn, leads to the need to revive herds, with is either done slowly (natural reproduction) or quickly (raids on those with relatively more livestock). New sources of conflicts have now emerged. Banditry for personal gain, and conflicts stemming from land-resource scarcity can now be added to the list.

Since the 1920s grazing rights and access to pastures in Karamoja have drastically changed. Pressure from the Turkana and Pokot on the eastern and southern frontiers, and from tsetse flies in the north, and restrictions on using western Acholi and Teso lands, combined with increasing population, have increased inter-sectional conflicts (Rattray & Byrne, 1963). Resource pressure has also increased land degradation in the central plains, leading to more stress.

Another potential source of conflict is the recently identified prospect for mineral exploitation in north-eastern Karamoja. A South African-linked company has already obtained government approval to explore the site for gold and other minerals. According to the land tenure laws of Uganda, all mineral rights are the Government's to bestow, and royalties, if any, accrue to the national coffers. The site is currently used by the Dodoth, Ik and Turkana as grazing land. It is not certain what impact the mineral exploitation, if it arrives, will have, not only on the health of the rangeland but also on access rights to those resources. Currently there is an internal debate on whether royalties and other benefits should also accrue directly to the local residents in a 'benefit-sharing' system (KTA, 1996).

In the 1980s and 1990s, the trends toward decentralization, deregulation, and popular participation have offered important insights into the analysis of conflicts over natural resources all over the world. There has been a move away from formal judicial processes towards local-level, more participatory processes, such as conciliation and mediation, or what is known as 'alternative dispute resolution (ADR)' processes (Cousins, 1996). In Karamoja too, there is a movement toward mediation for peace talks and consensus-building.

Typology of conflicts

Conflicts can be classified as 'management problems', 'disputes' and 'conflicts' (Burton & Dukes, 1990: 8). This classification provides a typology that helps to identify the appropriate processes and procedures for resolving the problem.

o 'Management problems' are those that arise from differences and arguments over choices of alternatives among persons having the same goals and

172

interests; these require *problem solving*, improved communication and improved personal interaction.

o 'Disputes' refer to issues that require *settlement* processes, such as judicial procedures, negotiation, and bargaining.

o 'Conflicts' are reserved for more fundamental differences, such as cultural conflicts, and competition over natural resources. They require *resolution* processes, i.e. an in-depth understanding of the cultural, economic and political causes of the conflicts, and often the assistance of a third party as mediator.

This typology is not very easy to apply to Karamoja, as some problems can be both disputes and conflicts, such as raiding. Other problems, such as conflicts over land resources, not only require mediation but also improved communication and management, as well as negotiation and other 'settlement' processes. Perhaps a better typology is to look at the 'phases' of a conflict (Rothman, 1995; Rupesinghe, 1995), on a case-by-case basis, and to find appropriate responses to each phase. For example, the early stages of a conflict (*formation*, and *escalation*) would require early warning and crisis-intervention mechanisms. Conflict *resolution* itself would require empowerment, mediation, negotiation, and problem-solving processes. The final stage, or conflict *transformation*, may require new institutions and mechanisms to sustain the resolution (Rupesinghe, 1995).

The typology of conflict resolution and prevention mechanisms can borrow from the literature on institutions and institution building. Mechanisms can be formal or informal. Formal ones can be constitutional or transient – in other words, permanent or temporary. Some mechanisms can appear to be transient in that the actual agreements or rules may change, but the fundamental institutional structure of the mechanism remains permanent. Furthermore, the mechanisms can be classified according to whether their objective is to prevent or to resolve conflicts (Table 7.2).

Traditional mechanisms for conflict prevention

Informal sanctions, reciprocity, spontaneous adaptations, and negotiated alliances are all mechanisms by which the Karimajong seek to prevent conflicts. Informal sanctions by the community, through the threat of social ostracism, ridicule, satire, and other overt or covert demonstrations, are flexible and powerful means by which the community prevents the formation and/or escalation of conflicts. In Karamoja today, informal sanctions continue to function, even if formal rules have broken down, but they too will be ignored if the basic fabric of the community is destroyed.

Another mechanism for preventing conflicts is the social convention surrounding *reciprocity* of favours. Reciprocity is the backbone of the community spirit among the Karimajong. Giving and receiving permeates their life. Receiving entails the creation of a new bond requiring gratitude or a reciprocal exchange, leading to a web of supports (Novelli, 1988). *Friends* are for giving and receiving, while *enemies* and *all others* are for exploiting.

Conflicts are also prevented by *adapting* to the stress leading to the conflict.

Table 7.2. Typology of conflict-prevention and resolution mechanisms in Karamoja (Uganda)

	Mechanisms	Typology[1]	Phase of conflict	Main actors
Conflict prevention	Informal sanctions, and reciprocity	Informal	Formation	Public
	Spontaneous adaptations	Transient	Formation	Public
	Alliances	Transient	Formation and escalation	Elders
	Education	Constitutional	Formation	NGOs
	Regional development	Constitutional	Formation	Government
	Land-use planning	Constitutional	Formation	Government
Conflict resolution	Informal sanctions, and reciprocity	Informal	Formation	Public
	Tribal mediators	Transient	Escalation and resolution	Elders, *aruanits*, parliamentarians
	Akiriket	Constitutional	Resolution	Elders
	Tribunals	Constitutional	Resolution	Elders
	British-based judiciary	Constitutional	Resolution	Government
	Vigilantes	Transient	Transformation	*Kraal* leaders and warlords
	Peace talks	Transient	Resolution	Local government, parliament, elders
	Elders' councils[2]	Transient? constitutional	Resolution?	Elders

Notes: 1 Mechanisms can be categorized as informal and formal. The latter can be transient or constitutional.
      2 Elders' councils are not yet fully operational, therefore their nature, scope and responsibilities are not well defined.

Some notable examples can be cited. The people of Labwor, after a series of raids and incursions, completely gave up livestock by the 1980s. They were able to do so since the Labwor Mountains and valley ecosystems were of higher agricultural potential and the people could subsist almost entirely on crops. It is only recently that the more affluent people of Labwor have started to re-invest their surplus into cattle, but even so they have not revived their tradition of herding, and would rather entrust their cattle to the Jie to be taken care of. The relationship between the Jie and Labwor has swung around from one of overt conflict, to negotiated alliances and trust.

Another form of spontaneous adaptation has been by the Ik. They have concentrated more on hunting-gathering, and cropping, although they continue to raise a few livestock. Being caught between the Dodoth and the Turkana has never been easy for them, but they have learned to play off one side against the other. Thus, most conflicts are resolved on an *ad hoc* basis, through skill, political savvy and luck. However, this does not protect them from occasional severe raids and attacks, misunderstandings and reprisals (Ayoo, 1995).

Yet another way the Karimajong seek security from conflicts is through *alliances* with the greatest number of people possible. This is made through the exchange of gifts and reciprocated favours. Alliances guarantee survival (Novelli, 1988). This can operate at the level of individual herders, women and even children, as well as at the level of sections and tribes.

Such alliances and enmities are fluid over time. Necessity and opportunity dictate the composition at any point in time. Today, the power relationships, and who raids whom, can be characterized as in Figure 7.1. For now, the Jie are friendly with the Matheniko, Labwor and the Ik. They continue to raid the Dodoth and the Bokora and the Turkana. However, the Turkana are friendly with the Matheniko but not with the Jie. Potential power-brokers and mediators could be found within this fluid system. For example, the Matheniko could potentially mediate between the Jie and Turkana, but this role has not yet been exploited.

Alliances are established through the offices of the tribal elders. Little is known about this traditional process – how often tribal elders meet, and at what level or scale they are established. More analysis and description of this important process could help identify elements for the design of an adaptable process for arriving at agreements and alliances.

Pacts and alliances have always been set and broken in the history of Karamoja. The Toposa, having lived with the Dodoth under considerable strain during the original migration into Karamoja, left for southern Sudan after a major conflict. They later signed a peace pact on the Sudanese border that lasts to this day. They have often helped the Dodoth in their fights with the Didinga (Pazzaglia, 1982). The Jie and Turkana were friendly with each other until 1966, when a serious famine forced the Turkana to enter Jie territory, at first in peace; later, fights and robbery broke out, and government police were called in to quell them. Quite a number of Turkana were killed and captured, and the remainder fled back into Kenya. The enmity of the Jie and Turkana has existed since that time.

These traditional mechanisms are all flexible in how they can be used. Some are informal (e.g. social sanctions, reciprocity) and others are formal. Among those that are formal, some are transient and are expected to change with time (e.g. adaptations to stress, alliances) and others could be constitutional (e.g. long-term alliances). Traditional preventive mechanisms are still very viable in Karamoja, and should be seen as the framework for the establishment and strengthening of conflict-management mechanisms.

These traditional mechanisms are all flexible in how they can be used. A reduction in insecurity will in the long run be able to open up new pastures and disperse grazing pressure for a more optimal use of land. However, the

175

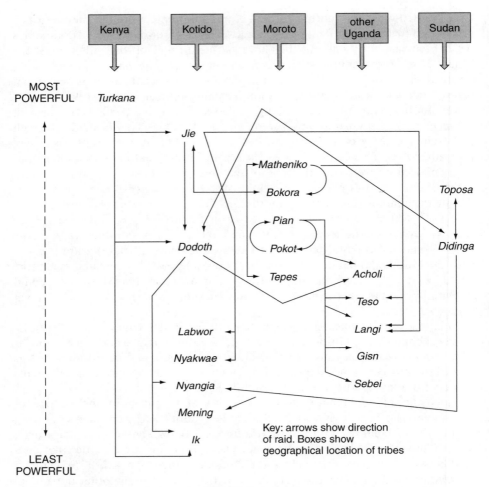

Figure 7.1 *Power relationships and raiding in Karamoja*
(Source: *adapted from Belshaw* et al., *1996: 13*)

Karimajong, as well as the Turkana, will continue to need to use their neighbour's lands in the event of drought or bad years. Therefore, the policy to stabilize will work in good or normal years, but will create hardships in bad years. Formal mechanisms that allow negotiated access in bad years are necessary, in order to avoid conflicts. Land-use planning and regional development, incorporating principles of inclusivity, negotiated access, flexible and opportunistic resource use, and equitable distribution of resources, would be able to contribute considerably to the prevention of land conflicts.

As long as there are alliances, conflicts between the 'friends' are minimal, and the breakdown or absence of traditional regulations for resource allocation and political power is not noticeable or destructive. But the absence of traditional regulations makes conflicts between 'enemies' appear sharper than ever before.

Traditional mechanisms for conflict resolution

Traditionally, elders exercise their power through neighbourhood, sectional or tribal assemblies, called *akiriket*, and in various specialized tribunals. The *akiriket* is a formal, prescribed event held periodically. It has jurisdiction over such issues as: raids, sacrifices for rain, returning cattle after raids, possible new grazing areas, etc. It is also the time to restore peace between individuals. The *akiriket* is a democratic assembly which even the uninitiated are allowed to attend, but elders always have the last word. The main speakers are never the same, but either are chosen or may volunteer. It is held periodically on an *ad hoc* basis whenever the need arises. Thus, whilst as an institution it is formal, permanent, and has set rules, its process is very flexible.

Tribunals are held when needed to resolve conflicts between individuals. There are separate tribunals for men and women. Women's courts are presided over by older women. Conflicts covered by the men's court can range from social ones to conflicts over land and resources (e.g. planting in another's fields, using another's water without permission). Clan tribunals, under the authority of clan elders, are also held in order to pass judgement over family problems and administrative matters (Pazzaglia, 1982).

The objectives of traditional tribunals in Africa are to reconcile disputants and to maintain peace, rather than to punish the wrongdoer (Rugege, 1995). Achieving consensus publicly (accompanied by social censure and ostracism) is more important than providing a 'lesson' through punishment. The 'winner-takes-all' judgements favoured by adversarial systems of law, are generally avoided in favour of a 'give-a-little, take-a-little' principle (Cousins, 1996). Rights of appeal exist, but are seldom used.

Recent mechanisms for conflict prevention

The current government policy is to prevent conflicts among Karimajong, Turkana and Acholi by continuing to push for the stabilization of each tribe inside its own provincial boundary. Estimates of the carrying capacity of the dry-season pastures in Karamoja – albeit 'guestimates' useful only for planning purposes – indicate plenty of 'space' within Karamoja, in a normal rainfall year, to accommodate all the Karimajong herds, *if, and only if*, the land could be used without undue restrictions on internal mobility (i.e. opportunistically). This is not possible at the moment because of insecurity. Insecurity forces households to congregate together (thus overgrazing in limited areas) or, by contrast, to disperse into marginal areas such as mountain slopes that can hide the herds and settlements. Insecurity also forces herds to vacate large areas, such as the eastern highlands.

A reduction in insecurity will in the long run be able to open up new pastures and disperse grazing pressure for a more optimal use of land. However, the Karimajong, as well as the Turkana, will continue to need to use their neighbour's lands in the event of drought or bad years. Therefore, the policy

177

to stabilize will work in good or normal years, but will create hardships in bad years. Formal mechanisms that allow negotiated access in bad years are necessary, in order to avoid conflicts. Land-use planning and regional development, incorporating principles of inclusivity, negotiated access, flexible and opportunistic resource use, and equitable distribution of resources, would be able to contribute considerably to the prevention of land conflicts.

The Karamoja Task Force, a body established within and by the Ministry for Karamoja in 1995, has been instrumental in lobbying and creating the necessary environment for peace talks, brokered negotiations and agreements, and fundraising through NGOs and donors for Karamoja. It works closely with the Karimajong parliamentarians. It is a transient institution and is expected to disband once its usefulness runs out. At the moment, however, it is able to step outside government politics, and act as a neutral initiator of equitable development. The KTF advocates that land outside trading centres and towns should be held communally, and should be assigned to specific groups of pastoralists. They would like to see this happen soon, so as to prevent expropriation by individuals (KTA, 1996). Formal recognition of communal ownership of land would also pave the way for establishing formal and informal mechanisms for mobility through tracking, and peaceful negotiation of access to neighbouring grazing resources.

Recently, Karimajong intellectuals have embarked on a major campaign to promote education among their people. They see education as necessary not only for rural development, but also for increasing security. The Karimajong, they believe, should be assisted to find alternative ways of securing livestock and livelihoods, and not to rely on raids and banditry. The 'Unearthing the Pen' ceremony of 1995 was instrumental in paving the way, and at the moment alternative methods of 'distance' education are being sought, such as mobile schools, radio, taped lessons, etc. Alternative livelihood systems need to be found not only for the disenchanted youth and bandits, but also for the herder and his family, in order to secure his livelihood. Development workers are working with a varied portfolio of experiments, such as: ox-drawn implements, small-scale irrigation, community forestry and private wood-lots, fruit production, rice, nitrogen-fixing plants, to name but a few in the production sector; and services such as woodworking, 'barefoot' water engineers for water pumps and irrigation systems, and paravets. More ideas need to emerge from the people themselves, and to be tested rigorously before being disseminated.

Conflict resolution: recent mechanisms
In the customary system, the guilty person is handed over to the elders, judged and punished in proportion to the crime committed. The colonial administration imposed its own, British-based judicial structure on this customary system. When they did catch raiders, for example, they applied their cumbersome procedures with their intricate subtleties, instead of the straightforward and swift Karimajong practices (Novelli, 1988). The Karimajong soon lost patience and respect for the modern judicial practices. The British system continues today in Uganda, and the Karimajong continue to distrust it.

Post-colonial governments were faced with serious conflicts along the international borders and tribal boundaries. By the early 1980s these conflicts were so severe that at least five meetings were held between administrators and people in all districts neighbouring Kenya, including Acholi and Turkana. None achieved a lasting solution (Ocan, 1992), primarily because the traditional leadership was not explicitly and formally involved.

A new government policy was instituted in 1994/5 in order to encourage independent warriors to become vigilantes under the command of the more respected *kraal* leaders. The vigilantes are allowed to keep their guns, and are provided with a monthly stipend and a shirt symbolizing their status. The programme has apparently had considerable success in attracting the warriors, and in reducing the incidence of banditry and, to a lesser extent, communal raids. There is widespread local knowledge of who committed banditry and raids, and it is much easier for the vigilantes, Karimajong themselves, to catch the perpetrators and recover their loot.

When perpetrators of raids are caught by the vigilantes, they are turned over to the local authorities, and not to the traditional elders system, because of the direct involvement of the government in this process. However, moves are underway to strengthen and rebuild the power base of the elders to take over this function. Elders' Councils in each tribe are now being convened on an *ad hoc* basis, and usually by parliamentarians and district commissioners, to resolve inter-tribal problems during 'peace meetings'. These Councils should be modelled after the *akiriket* and tribunals. If successful, this institution building process can strengthen the power the elders have over all internal problems (rangeland use, grazing rights, cropland, mineral rights, etc.).

Since Elders' Councils still do not exist, the peace talks normally invite all elders from the tribes concerned. Formal peace accords between tribes are still far in the future because of the disintegration of authority among the Karimajong, but the meetings have been instrumental in resolving conflicts on a case-by-case basis. With the help of the vigilantes, livestock have been returned, and bandits and raiders have been apprehended.

For all its perceived merits, the vigilante system may have the unintended effect of legitimizing the wealth gains and political power of the stronger groups who have more vigilantes and guns. This may lead to a weakening of the bargaining position of the smaller groups in negotiated peace agreements in tribal talks.

Design principles

An interesting dimension to development aid in Karamoja can be seen from the people's own eyes. The early generation of development workers, following on the heels of colonial administrators and missionaries, came with a top-down, paternalistic approach, hardly understanding the Karimajong mindset. Development workers were seen as strangers, and therefore neither *friends* nor *enemies*, but *all others*, who should be looked upon with indifference and exploited. What development aid lacked in the past was the ability to transform itself from *all others* to *friends*; such a transition would have assured human dignity to the development activities, and the possibility of 'reciprocating' assistance (Novelli, 1988). Only through a clear understanding of the Karimajong culture and world-view can true participatory, self-sustaining development be carried out.

Governments and traditional development aid alike view pastoralism as the problem not the solution. Raiding is seen as a fundamental backbone of pastoralism. Therefore it is thought that raiding can only be solved if pastoralism is transformed into sedentary forms of production (farming, ranching). Banditry in agricultural areas (e.g. south-west Uganda) is called 'thuggery' while banditry in Karamoja is portrayed as a 'natural behaviour'[9]. The misunder-standing of historical causes and dynamics of traditional raiding, and the bias against pastoralism, are two of the greatest hindrances to appropriate development in Karamoja.

Another constraint is the gradual breakdown of the 'community' spirit, and the disintegration of political authority. While customary conflict prevention and resolution mechanisms in Karamoja have been able to handle most conflicts in the past century, they have not been able to cover new situations that have arisen, such as banditry. Tribunals are rarely held these days, and the *akiriket* while still active, no longer has the political and moral authority to impose its decisions on the warriors. Recent experimentation with vigilantes, peace talks, and elders' councils looks promising, but should be approached in a more systematic way.

Conflict resolution is the most visible form of conflict management. To this must be added the importance of focusing on conflict-prevention modalities. In the case of Karamoja, there are traditional and modern conflict-prevention mechanisms that have been and can be used, such as informal sanctions, spon-taneous adaptations, alliances, land-use planning, regional development and education.

The typology of Karamoja conflicts provided in Table 7.2 shows that most mechanisms used to date are suited to dealing with conflicts at their early phases (formation and escalation), only a few for resolution of conflicts, and only one for the transformation phase. New mechanisms and institutional structures need to be developed if the gains from successful conflict-resolution are to be main-tained and sustained into the future. Examples of possible institutions are: inter-tribal committees for land-use planning and resource negotiation/allocation; vocational schools to attract young warriors; herders' associations and the 'professionalization' of herding.

Mechanisms are suited to different levels or scales of conflict. For example, the *akiriket* and tribunals are specific to tribal conflicts, while tribal mediators, Elders' Councils, and alliances are appropriate to both clan and tribal levels. Similarly, informal sanctions are usually more effective at the individual or section levels, while alliances, or their annulment, are more effective at the

---

[9] *New Vision Newspaper*, 6 Dec, 1990.

clan and tribal levels. Some mechanisms are appropriate to all levels, such as land-use planning and modern judiciary.

The process of conflict prevention and resolution is just as important as the goal. Agreeing to go to a negotiating table itself is an important concession and commitment. The process should be kept as flexible as possible to allow the emergence of new forms of institutions and mechanisms. Conflict management theory provides alternative processes and procedures for settling and resolving conflicts. Pendzich (1994) and Anderson et al. (1996) provide useful definitions (see Box 7.2). Each, or a combination of several, of these procedures should be tailored to specific situations.

Although conflicts are situation-specific, the process of conflict management can be applied systematically according to several guidelines. Some 'diagnosis and design' questions adapted from Cousins (1996) may be useful in Karamoja:

o what is the nature of the situation (type of conflict, causes, phases, nature of stakeholders and their political power, etc.)?
o how can the pre-negotiation process be managed (collaborative or participatory planning, empowerment strategies, equalization of power of parties, information gathering and communication, etc.)?
o what processes and procedures should be selected (culturally appropriate,

---

o **Fact finding** is the investigation of key issues in a conflict by a neutral third party, as an input into the negotiation process.

o **Facilitation** is the assistance of a neutral third party in running a meeting and making it productive (agenda, calling on speakers, bringing parties to the table, etc.)

o **Collaborative planning** is where the parties agree to work together to prevent a conflict.

o **Negotiation** is where parties meet vountarily to agree to an acceptable solution.

o **Mediation** is the neutral assistance offered by a third party to a negotiation process.

o **Conciliation** is an attempt by the third party to communicate separately with the disputants, in order to reduce tensions and agree on a way forward.

o **Arbitration** involves the submission of a dispute to a third party, who makes a binding or advisory decision after reviewing the evidence.

o **Adjudication** is a judgement rendered according to objective standards, rules or laws, by a person with the authority to rule on the issue (judges, administrative officers, or elders).

Box 7.2 Some procedures and processes for conflict-resolution
Source: Pendzich (1994) and Anderson et al. (1996)

---

gender and equality sensitive, integration of customary processes, role of mediators, sequence of processes, etc.)?

o what institutions and mechanisms need to be created or strengthened to sustain the process and the resolutions (transient or constitutional institutions, strengthening of customary structures, capacity building, etc.)?

One conclusion of the recent revival of peace talks in Karamoja is that inequality limits the usefulness of negotiation. District officials and parliamentarians aware of this fact make a point of inviting all stakeholders – even assisting minorities and those in remote areas who do not have as much political power as the rest. This observation has also been made elsewhere (Bradbury, et al., 1995; Ross, 1995). Attempts to equalize power can be made through various empowerment and capacity-building measures (Cousins, 1996):

o modifying the procedures used to manage or resolve the conflict, to ensure equal access by all (e.g. meeting in settings where the weaker party feels comfortable)

o legal advocacy and political action to change the legal framework of rights to resources (e.g. land-tenure laws, political representation in councils and parliaments, dissemination of legal information, etc.)

o mobilizing and organizing strategies to press claims and defend rights (e.g. organization of new associations, participatory decision-making, workshops and public debates. etc.).

Capacity building in an attempt to equalize power is an important but extremely sensitive process, that may backfire if the stronger parties raise objections or try to retain their advantage.

Conclusions

The most effective approach to conflicts in pastoral areas is to maintain flexibility, and to recognize the complexity of the problems. Changes in mobility patterns, constraints imposed by a reduction in territory, decreasing herd sizes, and increasing population, have all resulted in a lowering of the standard of living of the average Karimajong. Herders are no longer able to exploit the variable natural resources properly, leading to decreasing livestock productivity. Increasing poverty, the proliferation of guns, and the erosion of the power of elders, have all led to an increase in insecurity, banditry and other conflicts.

Resolution of the conflicts in Karamoja must benefit from both preventive and curative mechanisms. It is important to find ways to legitimize mobility of livestock in accordance with the non-equilibrium nature of the environment, to establish formal and informal mechanisms for negotiated access to grazing resources, to ensure equitable distribution of development benefits, and to allow effective participation of the Karimajong and their leaders in the development process.

Karamoja provides an interesting study because it is one of the few cases where the internal ingredients are present for spontaneous experimentation with conflict-resolution alternatives:

o the conflicts have approached a 'crisis' phase, thus prompting all stakeholders to take responsibility for their commitments
o the national political climate is favourable to democratic and participatory processes
o a core group of determined, committed, and more-or-less neutral mediators is present to push the process along
o a group of elders is able to analyse the extent of the loss of their authority, and is therefore able to make decisions, and
o the majority of the populace perceives the need for change.

Perhaps events are seemingly moving quickly in Karamoja because of the 'crisis' nature of the situation. However, the same factors operating elsewhere could also stimulate innovation and spontaneous change, albeit at a slower pace. Karamoja's case is still evolving and should be watched closely. It could provide important lessons for conflict management elsewhere in Africa.

The nature of conflicts is such that until they are resolved it is practically impossible to turn one's attention to a future strategy or plan. In Karamoja, participatory planning for development at any level (section, clan, tribe, province) is hampered by the lack of a viable local political authority and representation. In fact, local actors hope that the very process of conflict-resolution will stimulate the emergence of the proper conditions for participatory planning and development.

This chapter has focused on the prevention and resolution of conflicts within Karamoja. However, international conflicts continue to affect Karamoja's crisis. Cross-border disputes, civil war in Sudan, rebellion in northern Uganda, and arms proliferation will continue to have an impact on Karamoja. The success of innovative conflict-resolution mechanisms in Karamoja will also depend largely on the extent to which the region can be buffered from these external conflicts.

8

Stock-movement and range-management in a Himba community in north-western Namibia

Mouvements du bétail et gestion des parcours dans une communauté Himba du nord-ouest de la Namibie

ROY H. BEHNKE, JR

ABSTRACT

This Chapter describes seasonal herd movement and resource use in a pastoral Himba community in north-western Namibia. In this grazing system, livestock tend to occupy those areas best suited to their needs in a particular season. The self-interest of herd owners, the natural inclinations of the animals, and the dictates of the ecology are therefore routinely sufficient to enforce orderly patterns of stock movement and resource use. The perennial pastures used for dry-season grazing are the major exception to this rule. These pastures provide good grazing throughout the year, and group discipline is required to ensure their preservation. The local institutions that provide this discipline are described, and their capacity to cooperate with government authorities to develop and manage new water-points is assessed.

RÉSUMÉ

Ce chapitre décrit les déplacements saisonniers des troupeaux et l'utilisation des ressources dans une communauté de pasteurs Himba du nord-ouest de la Namibie. Dans le cadre de ce système de pâture, le bétail tend à occuper les zones les mieux adaptées à ses besoins pendant une saison précise. L'intérêt des propriétaires des troupeaux, les préférences naturelles des animaux et les impératifs du milieu suffisent donc en général à imposer un certain ordre dans les déplacements des animaux et l'utilisation des ressources. La principale exception à cette règle est liée aux pâturages permanents utilisés pendant la saison sèche; en effet, ces pâturages pourraient permettre de nourrir le bétail pendant toute l'année, et une discipline collective est nécessaire pour assurer leur sauvegarde. L'auteur présente les institutions locales qui assurent cette discipline en évaluant leur capacité à coopérer avec les pouvoirs publics pour créer et gérer de nouveaux points d'eau.

'By not moving you can kill an area'
Etanga livestock owner

IT IS OFTEN assumed that 'improved' forage-management is a universal prescription for the problems of extensive livestock production, one that should appeal to all environmentally responsible livestock keepers. This is unlikely to be the case.

Grazing systems in the communal areas of northern Namibia can be roughly sorted into two categories – those that are feed-limited and those that are water-limited. In feed-limited systems, animal performance, output and numbers are constrained by the unavailability of natural forage. In these systems increased feed-output has the potential to directly increase animal output and make herd-owners wealthier. Moreover, the need for more intensive natural forage-management and/or fodder cultivation is likely to be evident to communities whose daily experience suggests that natural feed-supplies are finite and must be husbanded carefully. Conventional range-management or other techniques to improve forage husbandry may have a future in these communities.

Water-limited systems are those in which livestock performance, output and numbers are constrained by insufficient feed adjacent to water, rather than insufficient feed overall. In northern Namibia, local interest in more intensively managing natural forage-resources tends to be minimal under these conditions. Producers instead concentrate their efforts on obtaining more widely distributed water supplies that will automatically give them access to more forage. In this setting, improved range-management is largely reduced to an exercise in appropriate water delivery.

The case study in this Chapter examines a water-limited grazing system in which livestock owners practise an indigenous form of range-management based on the seasonal use, resting and rotation of grazing areas. These communities also control access by outsiders to local resources and, as far as possible, adjust stocking pressure to annual fluctuations in rainfall and forage production. The indigenous husbandry practices used to achieve these objectives are sophisticated, cheap, and fundamentally different from those found on commercial ranches or in range-management textbooks. In this case, innovations in pasture management may be neither economic nor as important as a restrained water-development programme that is compatible with the current system of rangeland use.

Water development offers several advantages, as well as posing potential problems. Local herders unequivocally say they want more water points. Treating these communities as clients, honouring their requests and successfully executing such a programme would begin to build local appreciation for professional range-managers – a discipline that, to a remarkable extent, lacks a rural constituency among small holders both in Namibia and elsewhere in sub-Saharan Africa. The long-term benefits to be derived from a close working relationship between professional resource-managers and resource-users could be significant, though difficult to predict in detail.

But there are also risks. In a grazing system constrained by the availability of water, more water implies more animals and more prosperous herd-owners. It also implies more degradation in the 'sacrifice areas' around water points and settlements. Misgivings about the long-term implications of water development are expressed in an observation common among professional Namibian agriculturists: 'In the communal areas, where there is grass there is no water, and where there is water there is no grass'. Incessant, aggressive and frequently unreasonable local demands on government for water development do little to inspire confidence in rural communities as responsible resource-managers.

In these circumstances, mutual misunderstanding and poor communication between rural communities and outside agencies may pose a more serious threat to sustainable rangeland-use than any shortcomings in current husbandry practices. This Chapter examines the possibilities for community-based water development. It argues for improved range-management conceived of as a combination of institutional and infrastructural development, rather than technical innovation in the use of vegetation resources. The analysis suggests that changes both in government water-development policy and rural attitudes and institutions are both necessary and possible.

The setting

The settlement of Etanga consists of a couple of dozen seasonally occupied homesteads, a general store, school, clinic, and cattle sales pen. A Himba senior headman lives on the fringes of the Etanga settlement. His jurisdiction – also known as Etanga – is large but dry and sparsely settled. The parts of it that are routinely used cover about 3000 square kilometres, receive 150 to less than 50mm of average annual rainfall, and support about 2400 people dependent on cattle and goat husbandry. This Chapter looks at how the people of Etanga use – and, in using, manage – this vast area.

The Etanga area is part of Kunene Region in north-west Namibia. Kunene Region is bisected by a veterinary cordon fence that runs from east to west across Namibia and separates the predominately black communal areas of northern Namibia from commercial, largely white-owned ranches to the south. There are quarantine restrictions on the movement of livestock across this line, which also has deep historical and political connotations. The current veterinary cordon fence corresponds to the old 'police line' of colonial times – the northern limit not of white imperial power but of colonial settlement and land expropriation. Etanga lies well north of this line, but nonetheless felt the impact of colonial rule. For decades northern Kunene was cut off by administrative edict from external livestock markets, both north and south of the cordon fence (Bollig, 1996).

The vegetation of most of northern Kunene is *mopane* savanna. The people who live there are drawn from over half a dozen ethnic and tribal groups from predominately pastoral and agro-pastoral backgrounds (Malan, 1974). Most people in Etanga would identify themselves as Himba, an identity that developed as *OtjiHerero*-speaking people migrated south out of what is now Angola, and subsequently confronted the unsettled conditions that accompanied colonial penetration in the nineteenth century (Bollig, 1996; Bollig & Mbunguha, 1997). Most Himba now live in the north and west corner of Kunene Region or across the Kunene River in Angola.

Four senior Himba headmen control sections of the Kunene drainage on the Namibian side of the river from the Kunene regional border to the Atlantic (Bollig, n.d.). The main settlements for these 'riverine' headmanships are located at or near the headwaters of the feeder rivers that flow into the Kunene River, such that each headman's area combines both river frontage and upland grazing areas back of this frontage. The most westerly of these riverine headmanships shares a common border with the territory controlled by senior headman Ukorwavi Tjambiru at Etanga.

The Etanga jurisdiction is organized around similar principles but is adapted to very different topography to that of the riverine territories. Etanga is part of another set of five customary Himba jurisdictions, strung like beads along the drainage of the Etanga/Hoarusib ephemeral river, which flows into the Atlantic Ocean and provides the 'interior' Himba jurisdictions with their most reliable and abundant sources of stock water. The heartland of the Etanga jurisdiction lies at the headwaters of this drainage system in rugged highlands at about 1200 metres above sea level, with mountain peaks rising to 1600 metres. The lower, dryer and thinly settled western portions of the jurisdiction extend towards the Atlantic and the borders of Namibia's Skeleton Coast National Park (see Map 8.1).

This Chapter looks at the indigenous system of range-management in the Etanga area and the institutions that sustain it. It also examines the differences between the local system of resource-management and the range-management systems advocated by professional agriculturists and range-managers. We conclude that the local system, though distinctive, is effective given local conditions, and deserves external support. This support is being provided by a livestock-development project working through the local farmers' association. The future of this association and its potential role in resource-management and community development are also assessed.

The original purpose of this study was to provide guidance for livestock development programmes undertaken by the Namibian government with international donor support in the Etanga area. Project activities began in 1996 and thus far have included work on water development, veterinary services, livestock marketing and road maintenance. The most problematic aspect of this effort was, however, the project's work on range-management, which was the focus of the study summarized here[1].

---

[1] This study is based on six weeks of field research in April, May, August and October 1997, with an additional five weeks of field study undertaken by a representative of the local farmers' association. Material presented here was collected with the assistance of Kauta Koruhama and Jephta Kaurimuje and is presented in greater detail in our earlier report (Behnke et al., 1998). Anthony Hovey, Carol Kerven and Janson Mbunguha contributed both materially and intellectually to this study, and their support is gratefully acknowledged.

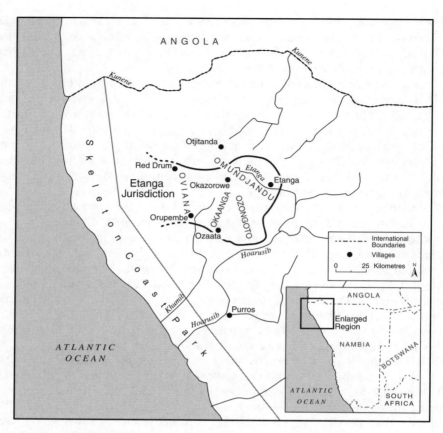

Map 8.1 *The Etanga jurisdiction in north-western Namibia*

The institutional basis for resource management in Etanga

By local reckoning there are 23 settlements (*otjirongo*) in the Etanga jurisdiction. The smallest of these settlements contain a couple of households and the largest – Etanga itself – contains 24. The average settlement is made up of about eight households. These are not nucleated villages, but dispersed clusters of compounds around water points, and it is not uncommon for a kilometre or so to separate one household from another. Apart from the elderly, most people do not live year-round in the main household compounds in the settlements. As will become clear later in the discussion, people and livestock come and go on a seasonal basis, most spending some of the year in impermanent cattle camps or *ohambo* – the sites of which tend to be reoccupied year after year by the same families, rainfall and pastures permitting.

Residential communities typically have authority over and are primarily dependent upon identifiable dry-season grazing areas, which are vacated and

188

rested in the rainy season. Outside herders may ask permission to live temporarily in a community and to use its water and grazing resources. If permission is granted, they may stay from a few weeks to a few years in their adopted location. Should they wish to relocate permanently, however, they must again seek permission from the long-time residents of the community. This stability results in communities that are close-knit social and kinship units (see Box 8.1).

Residential communities in Etanga are based on kinship. For example, one large settlement in the Etanga area consists of 18 households; 15 of these households are long-term residents, and three are newcomers who have lived in the community for two years. Of these three outside households, two plan to return to their native areas when rainfall and grazing conditions improve, and one household has sought and been granted permission to relocate permanently in its adopted village. In this settlement, the heads of the 15 'permanent' households share three (patrilineal) surnames. In another village of 11 households, seven household heads were the descendants of two brothers and a sister, one (or at most two) generations removed from the present heads of households. The remaining 4 households in this community were related either patrilineally (razu) or matrilineally (eanda) to the senior living male and female members of the settlement. In a third settlement made up of 9 households, 8 were headed by 4 pairs of brothers, and the last by their father's brother's son. Despite the mobility demanded by a pastoral way of life, these are stable communities, and their members are closely related to one another.

Box 8.1 The social organization of residential communities

Final authority over land-use decisions resides unequivocally with the senior headman who is resident near the Etanga settlement, and his councillors. But the headman does not seek out problems; he waits for disagreements to be brought to him. This means that many decisions may be reached amicably and quietly at the local level, with the headman actively involved only in contentious issues. How these more routine decisions are reached – and who reaches them – is less clear.

Old people – especially older men – are accorded respect. If an old man has significant cattle wealth, vigorous sons, is living near the graves of his ancestors (Bollig, 1997) and is related to the other senior household heads in his area, then he can probably claim a position of authority in local affairs (Malan, 1973; Crandall, 1991). Depending on how vigorous an elder is, this authority may be largely ceremonial, or exercised indirectly through kinship links to middle-aged heads of households and their spouses, who – in turn – control unmarried men and women.

In the Etanga area everyone knows and is in some way related to everyone

189

else, and it is unclear if specialist 'offices' with particular functions play a significant role in resource-management. The phrase *ovatjevere womario* or 'caretaker/guardian of the grazing area/pasture' is used to describe individual/s responsible for coordinating grazing movements in Etanga. But some herders claim not to know who these individuals are, or disagree about who they are. Bollig discusses the importance of the *omuni wehi,* the 'owner' or 'guardian' of land in the immediate vicinity of Himba settlements. Scattered throughout the Etanga jurisdiction there are at least five prominent families – and senior individuals in these family lines – who can claim this position because of their long residence in their area. Bollig describes this position as follows:

> The 'owner of the land' (*omuni wehi*) has little power to keep others off the land. However, the 'owner of the land', usually the oldest male member of a family that has lived in that place for generations, will have an important word in communal meetings. Occasionally, if resources are scarce, he may be influential enough to keep away outsiders who would bring further stress to resources (Bollig, n.d. 32).

Bollig also describes the existence in a riverine jurisdiction north of Etanga of 'grazing committees' with authority over pasture land (as distinct from settlement areas) and the capacity to impose fines on transgressors. In the Ozongoto zone of Etanga (described below) there is a group of four persons called a 'grazing committee', responsible for coordinating herd movements. But local residents admit that few people formally consult with the committee, and many parts of the Etanga area do not have a committee at all, or different individuals are suggested as committee members. It is therefore unclear to what extent committee members or holders of particular named positions might have additional authority independent of that which they already possess as kinsmen, and by virtue of their age, cattle-wealth and family status.

Herders using resources outside their home area often affiliate themselves with an individual 'host' household in the local area, sometimes camping together and sharing herding responsibilities. These 'host' households mediate between their guests and the wider local community. These links based on kinship, friendship and practical cooperation may be as important to the reciprocal exchange of resource rights as any appeal to special bodies or individuals with special authority in land matters (Bollig, n.d.; Crandall, 1991). No person spontaneously mentioned any formal government structures in relation to controlling the movement of herds or allocating land rights in the Etanga area.

Overview of Himba indigenous range-management

Figure 8.1 and Table 8.1 summarize stock-movements and resource-use in Etanga. Herders in Etanga tend to use the same category of resources every year in the same season. *Okuroro* and *Okuni* – the rains and the drought, respectively – are the main seasons. Depending on conditions in a particular

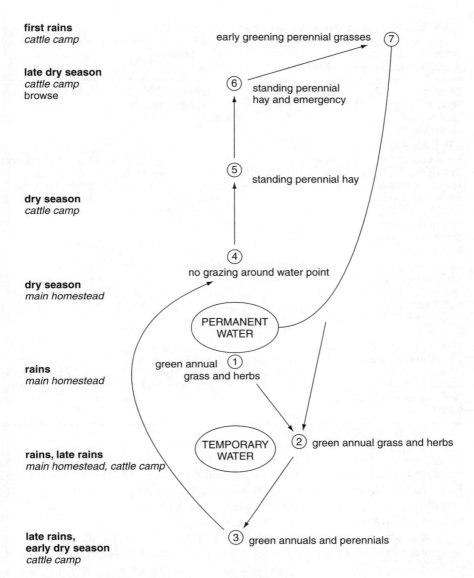

first rains
cattle camp

early greening perennial grasses ⑦

late dry season
cattle camp
browse

⑥ standing perennial
hay and emergency

⑤ standing perennial hay

dry season
cattle camp

④ no grazing around water point

dry season
main homestead

PERMANENT
WATER

rains
main homestead

green annual ①
grass and herbs

TEMPORARY
WATER

② green annual grass and herbs

rains, late rains
main homestead, cattle camp

late rains,
early dry season
cattle camp

③ green annuals and perennials

Figure 8.1 *Seasonal cattle-herd movements (Namibia)*

year, herders hope to keep their stock in the vicinity of their main village during *Okuroro* and early *Okupepera* (see Table 8.1), and anticipate using their cattle camps and associated grazing areas in the dry seasons of *Okuni* and *Orutene*.

The cycle begins in the wet season with most people and their stock in their 'home' settlements, where both water and pasture are readily available at this time. These settlements are located either near permanent (stage 1 in Figure 8.1) or near large and reliable temporary water points (stage 2). This concentration of

Table 8.1. Resource use in Etanga by season (Namibia)

Season in Otjiherero and local description	Approximate duration	Stock watering locations	Grazing locations
Okuroro: rainy season; good pasture and abundant milk – stage 1or 2 in Fig. 8.1	Jan–March	Ephemeral sources – pools, pans, standing water	Around permanent settlements
Okupepera: transitional cold season between rains and the dry season; trees change colour and the grass dries out – stages 2 and 3 in Fig. 8.1	April–June	Temporary sources, generally man-made or improved – dams, shallow hand-dug wells in smaller rivers	Around temporary water sources, starting near the water and working outwards
Okuni: hot dry season in which the trees have no leaves and there is little milk – stages 4 and 5 in Fig. 8.1	July–Sept	Permanent sources – boreholes, large hand-dug wells in major rivers, large springs	Start near major water points and work outwards as season progresses
Orutene: transitional from the dry season to the rains; the mopane trees get new leaves – stages 6 and 7 in Fig. 8.1	Oct–Dec	Permanent dry season sources until significant rains; ephemeral standing water after the rains	As above until the rains break, then shift to grazing areas where early rains are strongest

people and animals at the height of the rains is brief. When the rains have recharged temporary water points, most herds make use of these water sources. Reliance on temporary water sources gives the herds access to peripheral grazing areas (stage 3) which will become inaccessible later in the dry season, and it relieves grazing pressure around permanent water.

As the dry season progresses, however, herds are forced to fall back on permanent water points when temporary sources run dry (stage 4). In this season there are no good pastures left around heavily used permanent water points, and as the dry season progresses animals must walk increasingly long distances between permanent water points and peripheral pasture areas (stages 5 and 6). This pattern persists until the first strong rains of the new wet season, when the herds shift to far-flung pasture areas wherever the rain has been strong enough to promote early grass re-growth and provide ephemeral sources of water. Later in the wet season when rainfall has become more general and the grazing around permanent settlements has recovered, the herds will return to their 'home' areas, and the cycle begins again.

This simple schema is subject to adjustment according to the needs of different herds and households, the distribution of water and grazing in a particular locality, and the nature of the rains in different years – factors that will be discussed later in this Chapter. But it is clear that – conditions permitting – herders in Etanga tend to use the same category of resources every year in the same season, and it is the availability of water that dictates the recurrent pattern of movement and pasture use.

Water is unevenly distributed and so, therefore, is the intensity and timing of pasture use. There is resting of grazing resources in this system, but there is no systematic rotation of the resting and use periods through the seasonal calendar. A particular place (or kind of place) is rested or used for the same reasons at the same time each year. This pattern of exploitation exaggerates the natural heterogeneity in the natural vegetation as intrinsic differences between areas are reinforced by different histories of use. Some pastures are heavily grazed year after year while other pastures are lightly used or used only for short periods of time. The result is a grazed landscape that is not used like a commercial ranch and, even to the untrained eye, looks very different to a commercial ranch.

Pastoral-ecological zones in Etanga

The Etanga herding system exploits the productive potential of natural landscape features, and the geographical rearrangement of these features in different places forces herders to modify their patterns of movement to suit local conditions.

In Etanga local herders recognize two broad pastoral-ecological zones: the Omundjandu and the Oukoto/Koukoto. The Oukoto is further sub-divided into the Ozongoto, Okaanga and Oviana. Box 8.2 provides an overview of the characteristics of the four ecological zones in the Etanga territory. Grazing and water-management in two of these zones – the Omundjandu and Okaanga – are described for illustrative purposes.

Despite its relatively small size (by Kunene standards) the Omundjandu contains about half of the 'permanent' residences of the Etanga population (103 households) and in normal years sustains at least half of the territory's livestock. The Omundjandu is the heartland of Etanga. The main administrative settlement of Etanga – which contains the only government services and permanent store in the territory – is in the Omundjandu.

The Omundjandu also contains the senior headman's residence, the majority of large settlements, and the jurisdiction's most important source of permanent water – from hand-dug wells in the bed of the Etanga River. The Omundjandu is the only part of the Etanga jurisdiction in which arable farming (small-scale hoe cultivation of maize and cucurbits; see Crandall, 1991/2) is practised in most places in most years. Even for herders who reside outside the Omundjandu, the Etanga River is an important source of dry-season water, and by far the single most important source of water in droughts.

The Oukoto contains a resident population slightly less than the Omundjandu (88 households). Compared with the Omundjandu, the Oukoto supports more small stock, particularly goats, and fewer cattle. It also tends to receive rain later and to receive less than the Omundjandu, but sustains vegetation that is more nutritious than that of the Omundjandu and, according to local herd-owners,

Omundjandu *The Omundjandu is a wooded shrubland (dominated by Acacia ataxacantha, A. erioloba and A. fleckii, Commiphora glaucescens and Colophospermum mopane) along the upper reaches of the Etanga drainage. The Omundjandu is much smaller than the Oukoto but is equal to the Oukoto in importance as a pastoral resource.*

Oukoto *In an arc to the south and west of the Omundjandu lies the Oukoto/ Koukoto – literally the 'steep places' – a reference to the rugged topography of the area. The Oukoto comprises about three-quarters of the land area of Etanga, and is, in turn, subdivided by local herders into three zones: the Okaanga, Ozongoto and Oviana.*

Okaanga *Commiphorda mopane is shrubland on the south and western flank of Okamanga Mountain, principally along the Ozosaraombo River (at the headwaters of the Khumib ephemeral river), but also including the headwaters of the Ozongoto drainage (which eventually enters the Hoarusib).*

Ozongoto *An area to the south of the Omundjandu characterized by deeply incised canyons and rough terrain along the lower Ozongoto drainage.*

Oviana *The entire western portion of the Etanga, consisting of broad plains receiving little rain, supporting – as one moves from east to west – mopane shrublands, and then grasslands, gradually giving way to the flora of Namibia's coastal deserts. Most of this area lies outside the Etanga–Hoarusib drainage and empties either into the Kunene River, directly into the Atlantic Ocean via the Khumib ephemeral river, or into smaller seasonal rivers that never reach the coast. Farming is practised in the Oviana at a few sites, if not more broadly in years of sufficient rainfall. At the time of this study, the Oviana was lightly inhabited because poor rains had forced people and animals to leave the area.*

Box 8.2 Locally recognized pastoral-ecological zones in the Etanga area (see also Map 8.1)

supports higher levels of livestock reproduction and production if it receives sufficient rain.

The Omundjandu and the Oukoto therefore contain complementary resources for livestock production. The Omundjandu provides relatively abundant permanent water and high volumes of forage, which is routinely important in the dry season and critically important in drought years. The Oukoto has smaller permanent water sources and less dense but higher quality pastures and browsing, and is particularly attractive whenever rainfall is adequate. Most herders in Etanga, and especially those with large herds and flocks, use both these pastoral-ecological zones. In this sense the whole of the Etanga jurisdiction functions as a large and loosely integrated herding and resource-management system under the ultimate direction of the senior headman resident at Etanga.

The Omundjandu

Defined narrowly following common usage in Etanga, the Omundjandu consists of four interdependent geographical features: the Etanga River, Okamanga Mountain, the Omiramba plains, and the feeder rivers emptying into the Etanga River. Each of these zones fulfils a different agricultural/pastoral function (see Map 8.2).

o The Etanga River supplies the highest volume of reliable stock water in the territory, obtained from wells re-dug each dry season in certain portions of the sandy river bottom. For herders based in the immediate vicinity of the Etanga settlement, the river is their normal source of dry-season water. For herders whose main residence is farther afield, the river is an important fall-back watering point when other sources fail. However, because it is an important source of permanent stock water, grazing is not abundant around Etanga and, in a good year, many herd-owners prefer to obtain dry-season water for some of their animals from more peripheral sources where grazing pressure is lower and forage is more abundant.

o Scattered around the base of Okamanga Mountain to the north and east on flatter ground are permanent villages and arable fields sited along the major streams feeding into the Etanga River. These riverine settlement sites provide seasonal water from hand-dug wells in stream beds or from dams, and/or permanent water from boreholes that draw upon deeper sources of water underneath the streams. The stream valleys also provide relatively flat areas of alluvial soil suitable for small-scale arable farming, which is practised by most households in the Omundjandu.

o Okamanga Mountain is the most important emergency dry-season grazing resource used by livestock that are watering in the vicinity of Etanga.

o To the east and north east of Okamanga Mountain, roughly between the mountain and the Etanga River, lies a triangular, sandy, tree- and grass-covered plain cut across by numerous small dry river-beds. This area, known as the Omiramba, contains seasonally occupied cattle camps but no permanent water or village settlements. The Omiramba is the principal dry-season grazing area for herds watering in the Etanga River. Aside from the Omiramba, the most important concentration of reserved dry-season grazing in the northern Omundjandu lies near the northern limits of the Etanga jurisdiction, generally on high, rocky ground at the headwaters of streams that feed into the Etanga River from the north. These areas contain either no stock water or temporary wells, but are deserted in the rains when the herds relocate to permanent settlements.

Much simplified, seasonal grazing regimes in the Omundjandu conform to the following routine in years of adequate rainfall:

o In January (the beginning of *Okuroro*, the rainy season) most herds and

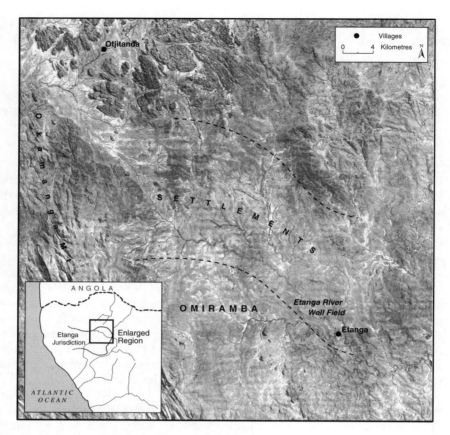

Map 8.2 *The central Omundjandu (Namibia)*

flocks return at night to kraals in the permanent village settlements (*otjirongo*), the bulk of which form a crescent along the track east from the Etanga settlement to the north end of Okamanga Mountain (see Map 8.2). Ideally, the herds will stay here through the main part of the rainy season (January to March) to allow dry season pastures to recover. When rain has been sufficient, herds in this season drink from pools or ponds in depressions and river beds.

o During the period April–June (*Okupepera*), ephemeral surface-water sources disappear as the rains draw to a close, and herds are watered at temporary but more substantial sources of water – typically, dams or small wells located in the beds of medium-sized rivers. Depending on the distance between permanent settlements and these temporary water points, herds may continue to return nightly to the permanent villages, or find it more advantageous to shift to satellite cattle camps (*ohambo*). In either case, herders attempt to use pastures accessible from temporary water sources until those sources become dry.

196

- As these temporary water sources run dry (in *Okuni*, July to September, or in *Orutene*, October to December, depending on the strength of the rains) herds fall back on the Etanga River, a source of water that will last through the entire dry season, except possibly in multi-year droughts. At this point, herders shift from watering their animals every day to watering them every second, third or (in emergencies) fourth day, depending on the distance between grazing areas and water. At this point most stock are managed from temporary cattle camps rather than the main homestead. Initially, herders locate their cattle camps close to the river, dig wells in the sandy river bottom, and use pastures adjacent to the water point.
- As the dry season progresses (late *Okuni* and *Orutene* until the rains come), herders move their cattle camps further and further out from the Etanga River in the direction of their dry-season pasture areas. This should be a co-ordinated movement so that individual camps do not push ahead of their fellows and take the best remaining pasture for themselves. The cattle camps are routinely located between the water source and the pastures. Stock walk one day 'in' for water and then on subsequent days 'out' for grazing, returning each night to the camp; in this way the animals do not trample fresh pasture on their way to water. Following this pattern, as the dry season progresses herders push their cattle camps and their grazing areas further and further out from the Etanga River and into the Omiramba plains, typically finishing the dry season at the foot of the mountain. In poor rainfall years the cattle push up Okamanga Mountain itself, in hopes that this last emergency grazing reserve is not depleted and that animals will not die in large numbers before the rains come. Alternatively, in the dry season, cattle (especially those without calves) may not be herded or return to kraals at night. Herders simply position themselves at the wells in the river and supply water for those animals that present themselves. Cattle are then free to seek the best pasture they can find, and herders concentrate their attention on providing water and locating strays.
- The rains ending the dry season are likely to break any time between October and December (*Orutene*). When this happens, herds and cattle camps relocate opportunistically, depending on the distribution of early rains. Until this happens, however, pasture continues to deteriorate and water becomes increasingly scarce as the dry season wears on. This is the time of human hardship and livestock mortality, the harshest season, and the one around which the rest of the year is planned.
- When the rains do come, Okamanga Mountain, the Omiramba, and other smaller, less notable dry-season grazing areas are vacated and left to recover, the herds return to the main settlements along the river courses, and the cycle begins anew.

Because of their size and regional importance, Okamanga Mountain and the Omiramba are managed directly by the Etanga senior headman and his advisors, although smaller dry-season grazing areas are controlled by the senior men in the nearby settlements associated with these areas.

197

The Okaanga

'Okaanga' refers to the area on the western and southern flank of Okamanga Mountain (Map 8.1). Major settlement sites in Okaanga, as in the Omundjandu, occur along the larger dry river-beds where it is possible to dam, dig or drill for water. Grazing patterns in Okaanga revolve around an oscillation between permanent/dry season and temporary/wet season water sources, and their associated grazing areas.

Small and/or temporary water points The abiding objective of the Okaanga community's system of resource-management is to preserve for use in the late dry-season pastures around the region's most important and centrally located permanent water sources. To this end, many households do not have a permanent homestead or locate their primary wet-season homestead near impermanent or low-capacity water sources. In any case, since most households do not practise crop agriculture, the distinction between permanent settlements (*otjirongo*) and cattle camps (*ohambo*) – which is quite clear in the Omundjandu – becomes blurred in Okaanga.

Okaanga contains one large temporary water point, a dam, as well as scattered impermanent or small sources of water that, if taken together, make an important contribution to the overall supply of water in Okaanga. The dam and temporary wells are critical at the end of the rainy season and the beginning of the dry season. During this time, large numbers of livestock congregate to use pastures around the dam before it runs dry. This helps to preserve the grass in the grazing areas linked to permanent water that are used later in the dry season[2].

Temporary sources of water permit herders to delay using the major water points until later in the dry season. Small but relatively permanent water points also allow herd owners to water their weak animals away from the large sources of water where forage is scarce and animals must be strong enough to walk considerable distances to find adequate grazing. These water sources are often too small to sustain large cattle herds, or are situated in rocky areas where access is difficult for cattle, but are commonly used well into the dry season by flocks of goats and sheep.

Permanent water points Okaanga contains three relatively large, permanent sources of water: a diesel-driven borehole (at the settlement of Okazorowe, Map 8.1) and a nearby wind-driven bore, and a cluster of strong, permanent springs (at Ozaata, Map 8.1) about 35km away at the southern edge of the Okaanga. These water points are the backbone of the dry-season grazing system. In Okaanga, as in the Omundjandu, dry-season grazing begins near the water point and moves gradually outward as pastures are depleted. In a year

---

[2] This pattern of movement was not reflected in the household herding schedules collected in this study, because there was insufficient rain in 1996–7 and the dam did not fill. In this situation, movement patterns became complex as herders attempted to delay their return to permanent water sources by lingering at the other, widely scattered but often very small water points.

of good rainfall, many larger herds finish the dry season watering at the large permanent springs at the southern edge of Okaanga. Typically, it is the more mobile male cattle and the castrated goats that are sent to the springs under the care of boys and young men.

The regional Etanga herding system

The preceding sections of this Chapter have documented local herding arrangements in two of the four pastoral-ecozones of Etanga. This section examines the Etanga-wide regional system of herd movement and resource-use, which both links together individual ecozones within Etanga and connects the Etanga territory with surrounding Himba jurisdictions. Three kinds of data are discussed:

o a 'snapshot' of resource use at one point in time in one ecozone – in this case the Ozongoto at the end of the Okuni dry season in 1997
o a description of wells and water use and ownership in the Etanga River bed
o herd movements over the past year for one Okaanga herd owner.

Resource-use in the Ozongoto ecozone in September 1997

The following discussion documents the need for reciprocal grazing rights in a variable climate, illustrated by conditions in the Ozongoto ecozone, one of the three subdivisions of the Oukoto.

Because of poor rains, movement out of the Ozongoto was extensive in the late dry season of 1997. By local calculations, 23 households occupy the Ozongoto on a more-or-less permanent basis and thereby have unquestioned rights to use resources in the area. Of these 23 households, more than half (12 households) were using water points outside the ecozone at the time of the survey. Three households were using water in the Okaanga ecozone, and movement in this direction might have been more extensive had rains in this zone also not been poor in 1996 to 1997. The majority of absentees – nine households – were using watering points entirely outside the Etanga jurisdiction. In all cases these households were using the closest neighbouring Himba jurisdiction to the Ozongoto along the main Hoarusib River channel. These Ozongoto residents had requested permission for and been granted temporary watering and grazing privileges by local residents of this jurisdiction. These privileges were negotiated on a household-by-household basis from the Hoarusib communities that controlled and normally used these resources.

Despite the failure of the rains in the Ozongoto and the exodus of over half of the zone's households, eight 'guest' households were simultaneously present in the eco-zone, and most had been there for two or three years. The bulk of these guest households came from the Omundjandu, either within Etanga or from an adjacent headman's territory immediately to the east of Etanga. These households were in the Ozongoto to take advantage of its more nutritious pasture and browse (but especially browse for goats), relative to that which could be found in their home areas. One household, on the other hand, was fleeing drought in its

199

home area, the Oviana, which had received even poorer rains than the Ozongoto in the preceding year.

Patterns of residence and resource use in the Ozongoto illustrate how Etanga herders adjust animal numbers and grazing pressure to annual variations in rainfall and the availability of feed resources in a particular zone: they move. Especially mobile in late 1997 were large cattle herds with the capacity to relocate and the need to obtain adequate pastures. Moving into the Ozongoto at the same time were households with an interest in goats, which require browse that is less sensitive than pastures to annual fluctuations in rainfall. As the cattle moved out, the goats moved in. Sometimes, households that were moving in with goats, also moved their cattle out.

Mobility is the primary way in which livestock keepers in Etanga adjust stocking densities to annual fluctuations in rainfall and forage production. The process of adjusting forage supply and demand is not coordinated by any central authority. Each household makes its own decisions based on its assessment of the relative costs and benefits of different locations, and its ability to negotiate grazing rights with a desirable 'host' household or community. Like prices that are set by market forces, grazing pressure in a particular locality is the summation of numerous decisions made by individual actors in their perceived self interest. Contrary to the standard position of professional range-managers, it is the absence of a central authority capable of controlling herd numbers and movement that permits stocking rate adjustments in this system.

Wells and watering in the Etanga River in late October 1997

The permanent well-fields in the Etanga River bed are the single most critical resource controlling inter-zonal herd movement within the territory. As a dry season progresses, the number of hand-dug wells in operation and the number and origin of herds using these wells is in continuous flux. Initially, herders begin watering at standing pools (the larger of which have individual names) in the river bed. Some of these pools are located along stretches of the river that provide permanent water, but most are located on tributaries of the Etanga or along stretches of the main river that do not provide water throughout the dry season. Typically, these pools are located under rock faces at the side of the river channel. When surface water in these pools is depleted, herders gradually deepen them, chasing the water as it retreats until they hit solid ground through which they cannot dig, and the well must be abandoned. When this occurs, the herds must shift to deeper well sites in the middle of the river channel, either digging a fresh well or teaming up with users of an existing well. It is at this point that the well census summarized here was taken – shallow wells in pools were either being deepened or abandoned, and new mid-channel wells were being dug daily.

These deeper, mid-channel wells also have a variable life expectancy, depending on the strength of flow in the river in a particular year and the depth of water-bearing sand that lies on top of impermeable strata. Just like the shallower wells dug at the bottoms of pools, mid-channel wells also come into and go out of production as the dry season progresses. At the end of a two-year drought there are said to be only three well sites in the Etanga River that will reliably produce water.

About 71 herds were using the Etanga well fields at the end of October 1997, but additional herds were arriving daily as more peripheral water points ran dry and animals converged on the river. A few weeks later, in November 1997, at least four new cattle camps from the Oviana had started watering at the Etanga wells, there having been poor rains in the Oviana in the preceding wet season. If we assume that cattle herds (management not ownership units) have an average size of about 100 head and goat and sheep flocks about 200 head, then the Etanga well-fields were supporting about 7000 cattle and 14 000 small stock along a 7 to 8 kilometre stretch of river. And pressure was still building.

At least at the time of the well census – when water was relatively abundant – there was little tension over the distribution of water. Each mid-channel well takes at least two people to operate it – one to lift the water and another to control the animals at the trough. As the season progresses and the wells deepen, three or four people may be needed to lift water, in addition to a stock manager at the head of the well. A considerable labour force is therefore needed to meet the demands of these routine tasks and the difficulties of well excavation and maintenance. Help is also occasionally required to water animals when a herder is sick, chasing after strays or otherwise unavailable. For these reasons herders seemed to be much more interested in establishing working partnerships for effective well-management than in excluding other users.

Wells owned in three different ways according to the type of well. Natural pools and the shallow impermanent wells that evolve out of natural pools are un-owned and are used on a 'first-come-first-served' basis. The only alternative to these arrangements would be to fence natural water sources. Since thirsty cattle are difficult to keep out of open water, fencing around numerous impermanent water points would be expensive and probably not be worth the effort. Everyone has some cattle that are stray or are simply left to take care of themselves for a period of time, and all owners probably benefit more or less equally from the current *laissez-faire* system. Besides, by the time water gets really scarce and must be closely managed, open water and shallow wells are long since depleted.

Sites for the more valuable mid-channel wells are owned by individual households, inherited and managed on that basis. Because of the need for extra labour, these wells are not used exclusively by one household, but one household or family group has the authority to make management decisions regarding the facility. Finally, the three permanent well sites that are of critical interest to the whole community are said to belong collectively to 'the people of Etanga' or 'the people of the Omundjandu' and are managed under the authority of the Etanga senior headman.

In sum, the more valuable a water source the more tightly it is controlled, and the more important a source is for the whole community, the greater the tendency for senior political authorities to be involved in its management. At least with respect to water, this is not simply a 'communal' system of resource ownership. When individual ownership rights are technically feasible and rewarding to maintain, they tend to exist – as in the case of mid-channel well sites. At the extremes, less valuable or less easily controlled natural water points (such as pools) tend towards open access, while the most important watering points are regulated by the political authorities in the interest of the community at large.

The well census conducted for this study was taken relatively early in the dry season of a normal to good rainfall year, for those parts of Etanga east of Okamanga Mountain. If the new rains are delayed, we could expect to see many more 'outside' herds falling back on the Etanga River, as peripheral water points go dry. At the end of a serious drought we might expect to see many more herds using just three well sites in the entire river. At the time the census was taken, however, by far the bulk of the herds watering in Etanga were settled in adjacent communities, and used the river as their routine dry-season source of water. Because the river is a linear topographical feature, herders were able to disperse grazing pressure over a distance of more than 7 kilometres, rather than concentrating around a single point like a borehole. Only one of the three main well sites was in use at all during the census.

Individual herd movements

The generalizations about herd movement contained in this Chapter are substantiated by about 50 household case studies of retrospective herd-movement over an entire year. Presented below is the herding schedule of one wealthy herd-owner resident in Okaanga (Table 8.2). Thus far we have examined the regional grazing system in terms of the use of a particular zone (the Ozongoto) or a particular kind of resource (water in the Etanga River). The material now presented here allows us to see the diverse resources of the Etanga area from the perspective of an individual household.

Because they can afford to split different kinds of livestock into specialized herds and flocks, the movement patterns of wealthier households are usually more complex than those of households with smaller numbers of animals. While unusually rich in detail and explanation, the herding schedule summarized in Table 8.2 is typical of the complexities of herd movement for wealthier families.

Outside technical experts and agricultural extension agents routinely assume that Himba livestock husbandry is 'traditional'. That it may be, but it is also exceedingly professional and technically competent. Table 8.2 displays a detailed knowledge of both regional environmental conditions and the requirements of different herd species and age and sex groups, which are managed differently and migrate on different schedules. Reforming or improving this system is not going to be easy, nor are the changes likely to be dramatic, because the level of technical knowledge already possessed by herd owners is high and it is used in practice.

This conclusion applies particularly to herd movement, which some professional observers would like to see regulated and restricted. In the course of a normal year, the herds and flocks (or herd and flock components) of a single household may be using widely distributed resources. In Table 8.2 the main homestead was usually located in the central Okaanga area, but the majority of the family's livestock gathered there only in certain seasons (*Okuroro* and

202

Table 8.2. Movements of the herds and flocks of one Okaanga household (Namibia)

Season	Movement
Okuni 1996	The family herd split during the early rains in 1996. Immature cattle and old milk cows were kept on soft sand in the central Okaanga, and watered at the nearby diesel- (at Okazorowe, Map 8.1) and windmill-driven boreholes. A second cattle camp was located 35km south near permanent springs on stony ground at the edge of the Okaanga (Ozaata, Map 8.1), and all oxen and all goats were kept there.
Orutene 1996	The bulk of the cows and oxen and nanny goats joined the immature cattle and old milk cows in the central Okaanga, but switched to a new cattle post and now began using the impermanent water from a local dam, plus the diesel borehole. This is their normal early wet season location, and with better rains they would have stayed here longer.
	Some early showers occurred in another grazing area associated with the permanent springs at the southern edge of the Okaanga. All the castrated goats and enough milk cows to supply the shepherds moved their kraal to this location, but continued watering at the springs.
Okuroro 1997	At first all livestock returned to the main homestead in the central Okaanga, but the rains were poor, the dam did not fill and when it ran dry they shifted to watering again at the diesel borehole. At this time all stock were managed from one camp using several kraals for different categories of animals.
	The herds and flocks then split. Castrated sheep and goats remained where they were in the central Okaanga. The bulk of the cows, oxen and nanny goats and ewes shifted to a cattle post normally reserved for *Okuni*, which is why they stayed there only a short time and then moved away (see below). While at this cattle post they obtained stock water from ponds formed by rainfall. When the ponds dried out, they first dug small wells in the bottom of the ponds, and then fell back on the windmill for water.
Okupepera 1997	All animals gathered together at the windmill. The area had received insufficient rain and the (nearby) diesel-driven borehole had many animals already using it, so the stronger animals were sent away to leave room for the cows. This is a common pattern. The big herd owners who live around the diesel bore at Okazorowe might have a small number of oxen at the borehole, but the greatest number have been taken away. Residents with smaller herds would have complained, there having been poor rainfall last year and a shortage of grass in the area.
	The livestock were split three ways at this season. The bulk of the oxen went to the Omiramba in the Omundjandu. Castrated small stock went back to the southern springs with enough cows to supply the shepherds with milk. The breeding goats were managed from the main household at Okaanga because nannies begin to kid between *Okuni* and *Okupepera* and the owner wanted to observe their management and make sure that the shepherds allowed the young kids to suckle properly.

Table 8.2. Continued

Season	Movement
Okuni 1997	The main household is currently at the windmill in Okaanga. Disposition of herds and flocks is as follows:
	○ All castrated goats and some milk cows for the shepherds are at the southern springs (Ozaata).
	○ Oxen and immature cattle are in Etanga, in the Omiramba under the care of the owner's nephew, without any women accompanying. They are still drinking from ponds in Etanga River.
	○ Nanny goats and the majority of the cows are at the windmill.
	In *Okuni* they routinely divide the goat herd into male and female components. This is done to spread the grazing pressure from the goats more evenly; otherwise, the combined herd would strip an area clean and leave nothing for the cattle. The castrated goats are separated from the nannies because the former are stronger, and if they are combined in one flock the males finish the grass, leaving none for the females.
Future	They expect the animals at the southern springs to remain there all through *Okuni*, but the first rains generally begin in the Etanga vicinity in *Orutene*, and for that reason they might move to that area if the drought breaks on that side. Those animals now at the windmill will – stage by stage – work their way out towards one of the family's usual dry-season cattle camps. The Etanga animals will shift from pools to wells in the Etanga River, as the dry season progresses.

Okupepera). Otherwise, parts of the family herd (certain categories of cattle) were routinely 20 kilometres to the north of the main homestead in another ecological zone – the Omundjandu around Etanga – or 35 kilometres to the south and west (castrated goats). All these moves were reasoned responses to the shifting distribution of critical grazing and water resources and the requirements of different kinds of animals.

These are not 'closed' or tightly bounded resource management systems; people and stock routinely move outside their home areas in response to localized variations in rainfall, but these movements must be negotiated with other local resource-managers. In the Etanga area there are boundaries between different traditional jurisdictions and between communities, but all these boundaries are permeable.

Grazing and forage management: the rationale behind production and resource use in Etanga

Thus far this Chapter has focused on the dominant variable that controls grazing patterns in Etanga – water. We have examined the local distribution of stock water within two of the pastoral ecozones that make up the Etanga jurisdiction

– the Omundjandu and Okaanga – and the impact of low rainfall on the use of a third ecozone, the Ozongoto. We have also examined the organization and use of the single most important permanent water source in the territory – the Etanga River well-fields – and the impact of this water source on inter-zonal herd movements. Unavoidably, this discussion has briefly touched upon the other factors – the habits and needs of livestock, forage quality and quantity, and human management-decisions – that interact with water availability to create the current grazing system. In the following sections we examine more systematically how these 'other' factors are manipulated by herders to exploit efficiently their scarce water resources.

Livestock grazing habits

Cattle are disinclined to walk great distances, to walk on stones or to walk uphill, all else being equal. Left to themselves, cattle first graze low-lying sandy areas close to water, and only work their way uphill when more accessible sources of forage are depleted.

The Etanga grazing system takes these inclinations into account. It could not do otherwise since cattle without calves are routinely left unattended and do not return to their kraals at night in the late dry-season. In this season, the cattle that do not need to come home for milking, are not herded. This is for several reasons:

o there are no standing crops for them to damage
o night grazing is advantageous when forage is scarce, and cattle are disinclined to eat during the hot days
o free-ranging cattle can be monitored periodically when they return to drink at the few permanent water sources
o the labour required to provide water interferes with more routine herding duties.

In this critical season, cattle are making many of the decisions about where they go to feed, and any unfenced management system that was contrary to their bovine inclinations would collapse.

Dry-season herding patterns make sense in terms of these constraints. Given the dictates of the topography, most of the major water points are located in dry river-beds at low elevation on sandy ground. Dry-season grazing begins around these points and moves outward, generally using sandy areas first. It therefore makes little difference if the cattle camps gradually move outwards and the cattle are kraaled at night (a practice known as *okuvonga*) or if the cattle are left to care for themselves (termed *osamununa*)[3]. Just as the cattle will not bother to walk further out from water until the pastures closer to water are finished, herders should only move their camps forward when the grass in front of them is depleted[4]. In either case, as herders point out, this system minimizes the

---

[3] Almost all herds in the Omundjandu were free ranging (*osamununa*) in the dry season of 1997 because poor rains had resulted in low calving rates.

[4] The following interchange (which refers to grazing areas in the Okaanga on the west flank of Okamanga Mountain working up a valley to the headwaters of the Ozongoto River) illustrates this system, and suggests some of the social controls that keep it functioning:

trampling of useable pasture since cattle walk back to the water point over areas they have already grazed. In this system, rocky high areas tend to be used last.

The most important example of this pattern is Okamanga Mountain, the emergency grazing reserve for the Etanga well-fields. In normal years, herders hope that their cattle will be required to spend little time on the mountain, being instead able to finish the dry season on the Omiramba (literally 'the sands'), the normal dry-season pastures at the base of the mountain and situated between the mountain and the river (and therefore closer to the water).

It is informative to contrast the Etanga system of dry-season grazing-management from that promoted by the range-management profession. From the perspective of mainstream range-management, Etanga dry-season grazing practices are hardly a system at all. Professional range-managers would, generally, advocate that dry-season grazing reserves be situated adjacent to watering points to minimize walking and energy loss by the cattle when they are weakest, rather than at the furthest point from water. The consumption of nearby pastures during the height of the wet season would, from this perspective, be viewed as an unnecessary waste of forage needed later in the dry season, and the 'degraded' areas around water points would be viewed as prime candidates for 'rehabilitation'.

Though sensible for commercial beef breeds that are forage- and water-demanding, this assessment does not take into account the physiological poten-tialities of the indigenous breeds of cattle kept in Etanga. As the dry season progresses, the Etanga cattle gradually shift from drinking every day to every other day or every third day. If they did not do this they would soon starve to death because their pastures are distant from water and they would spend all their time walking between water and pastures rather than eating.

However, these necessary changes in watering regime also induce several advant-ageous metabolic changes in the animals. Research has shown that indigenous African cattle breeds like those in Etanga have the capacity when they are water-and/or feed-deprived to reduce their energy expenditure quickly to a fasting metabolic rate that is two-thirds of the rate of animals on a full maintenance ration. They can also reduce the energy cost of walking as body mass declines in the dry season, and conserve energy by allowing larger fluctuations in body temperature at lower body weights (Payne, 1965; Finch & King, 1979; King, 1983; Western & Finch, 1986). In other words, as these animals are deprived of water they not only have time to eat more but they also use more efficiently what they do eat. The Etanga dry-season herding system might, in sum, spell disaster for

---

Researcher: What are your future plans?

Herder: I will push my animals further up [the valley and away from the borehole]. I expect the grass to be finished here in August, and I will move my cattle camp in November.

Researcher: How are these moves arranged?

Herder: We are all here around the borehole and using it for water. We decide together to move, and then we all move together.

Researcher: If you expect the grass to be finished in August, why are you waiting until November to move camp?

Herder: We are afraid that if we move earlier we will also finish the grass there [further up the valley] before the new rains.

206

a commercial rancher, but is well suited to the physiological characteristics of the indigenous cattle breeds that have co-evolved with the system.

Goats are, of course, different from cattle. They require smaller volumes of water, are primarily interested in browse rather than pasture in the dry season, and are unimpeded by rough ground. Goats are also shepherded year round – usually by younger boys – because they are preyed upon by hyena and jackal. When they do compete directly with cattle for pasture, goats are able to graze more closely, stripping an area clean and leaving nothing for the cattle. For all these reasons, goats are herded separately from cattle. Even when they are managed from the same camp or settlement, goats are frequently watered from separate sources than the cattle, and graze in different directions. Often, a few milking goats or females with young will share the larger water points with cattle, but there is considerable social pressure on the owners of large goat flocks to move their animals out to peripheral and smaller water points, which are often springs tucked away in inaccessible or rocky areas. Because goats are shepherded in all seasons, human rather than caprine logic determines their movements. But the goats are taken to precisely those areas which meet their specialized feed preferences, exploit their small stature and agility, and keep them out of the way of the cattle.

Herders in Etanga have partitioned their environment according to the needs of their dominant herd species – cattle and goats. Human decisions reinforce and systematize the pairing of herd species to favourable environments within the total landscape, and there is little attempt to get the animals to behave in ways that contradict their natural inclinations.

Forage production

Etanga herd-owners have a sophisticated and detailed knowledge of the vegetation of their area and the usefulness of particular plants as forage. Where there exists sufficient information from which to draw comparisons, local descriptions of the distribution of forage species and vegetation types conform closely to the observations of field botanists (Burke, 1998).

Goats eat grass, especially annuals and *ongangahozu* (*Eragrostis nindensis*), a perennial found in stony, upland areas[5]. But goats also browse in all seasons, relying on a wide range of tree and shrub species and moving from fresh leaves, flowers and shoots to dry leaves and fallen fruits and seeds as the year progresses[6]. Herders from the Omundjandu consistently cite the better quality browse of the Oukoto as a reason for taking their goats to this region.

Cattle, on the other hand, are little interested in browsing when green grass is

---

[5] Plant identifications based on Roux, 1971 and Malan and Owen-Smith, 1974.

[6] Important tree species for goats include (list far from exhaustive): *omukaru* (*Ziziphus mucronata*, especially fallen dry leaves), *omuhama* (*Terminalia prunioides*) and *omukaravize* (*Catophractes alexandri*, eaten year round, green and dry), *omuvapu* (*Grewia bicolor*) and *omumbuti* (*Rhigozum virgatum*, green leaves in *Orutene*), the seed pods of *omue* (*Acacia albida/Faidherbia albida*, the pods of which should not be shaken down but left as a feed reserve to fall naturally at the end of the dry season), and *omutati* (*Colophospermum mopane*, not dry leaves but the early regrowth before the rains of *Orutene*).

available. When the grass is dry, they do feed on both green and dry leaves and the seeds of certain trees and shrubs. Consistently cited as important emergency cattle feed were the green leaves of *omuvapu* (*Grewia bicolor*) and *omumbuti* (*Rhigozum virgatum*), both eaten in *Orutene* when forage supplies are lowest.

By far the bulk of cattle diet comes from grass, and it is the availability of grass that influences the seasonal movements of cattle herds. Herders recognize three phases in the yearly feeding cycle of cattle – the wet season, the dry season, and a short but critical period at the end of the dry season just after the rains break. Herds use distinctive pasture types in each of these periods, usually moving their location to obtain these feed types.

In the wet season (*Okuroro, Okupepera*) all kinds of forage are readily available, and all are generally acceptable. However, the cattle seem to prefer *omupito*, a generic term referring to all grass species when they have just sprouted and are difficult to identify. The term applies more especially to annual grasses, described as plants which do not have roots that stay in the ground and are propagated from seeds 'just like when you plant maize'[7]. *Omupito* is commonly found around settlements and/or on sandy soils (when asked to list wet-season forage plants, livestock owners in Etanga spontaneously list species said to be common in sandy areas).

The usefulness of *omupito* as forage ends with the closing of the rains and the drying of the annual grasses that are weak and 'fall down and disappear' in the wind. Thus, in the dry season *omupito* is transformed into *ongundju*, loose hay of mixed-species composition tossed about in the wind, which is of little use to cattle. Two important forage species of sandy areas – *omurondji* (scientific name unknown) and *orwejo* (*Eragrostis porosa*) – are both annuals and prone to breakage and loss in this way.

Standing hay from perennial grasses is the foundation of cattle diets in the dry season (i.e. *Okuni* and *Orutene* prior to the rains). *Ongumba* (*Stipagrostis uniplumis*) is the most important single species, universally lauded as a grass that when dry 'is never blown away by the wind.' Descriptions of the distribution of *ongumba* (*Stipagrostis uniplumis*) suggest that it is most commonly found neither in the mountains nor the sandy river bottoms, but on what is variously described as soft to somewhat rocky soil in reasonably flat areas. Only in the Oukoto (Oviana, Okaanga and Ozongoto) is *ongumba* found on deeper sandy soils because (in the estimation of herders) the soils in these areas are particularly strong. Rarely is *ongumba* (*Stipagrostis uniplumis*) found around wet-season settlement sites in the Omundjandu. Also important are at least two other perennials characteristic of upland areas – *ongangahozu* (*Eragrostis nindensis*) and *otjimbere* (scientific name unknown). These grasses are not classed as

---

[7] Species of *omupito* include: *ombanga* (*Eragrostis superba*), *eriangwari* (scientific name unknown), *orwejo* (*Eragrostis porosa*, a valuable forage plant) and *ohoke* (*Aristida adsensionis*, described as a weak grass easily broken in the wind, eaten by cattle when green but dangerous when dry because it has sharp tips that blind the animals), *omurandji* (scientific name unknown, found in sandy areas but scarce around settlements) and *okatjirakonduno* (*Enneapogon cenchroides*, commonly restricted to the Oukoto). Also eaten in this season is the spreading herb *ohongo* (*Tribulus terrestris*, which later produces sharp burrs when it is mature).

omupito since they are all perennials and readily identified in all growth stages. Areas containing these perennials are especially targeted for wet-season resting followed by use later in the dry season.

Late *orutene* – just after the rains have broken following the long dry season – is a particularly critical season. At this point cattle tend to be in the poorest condition of any time in the yearly cycle. This is also the only period in which cattle herders 'chase the rains' – opportunistically relocating their herds in areas where early rainfall has been strong. Even at this time, herders are seeking particular kinds of pastures consisting of perennial grasses that rejuvenate quickly with 'only a few drops of rain' because they are sprouting from established root systems. These early greening perennials (*ongumba* – *Stipagrostis uniplumis*) and *ongangahozu* (*Eragrostis nindensis*) are repeatedly cited, though other perennial species are also mentioned) provide the first flush of fresh grass well before the annual grasses (*omupito*) have had an opportunity to sprout from seeds.

In sum, Etanga herders crop each category of pasture – fresh annuals, perennial standing hay, and perennial regrowth – when it is most valuable relative to other forage sources. Minimal management-controls are required in a system of this sort, which directs the cattle to feed in areas where they are already inclined to do so, given the relative value of alternative pastures.

Local concepts of degradation and 'range-management'

Etanga herd-owners recognize that an area can become degraded. In their estimation, long-term degradation occurs because of permanent settlement and continuous grazing around such settlements. Commonly cited as the species lost in this process were *omurondji* (scientific name unknown), *orwejo* (*Eragrostis superba*) and *ongumba* (*Stipagrostis uniplumis*) – all preferred forage plants that occur either in low, sandy areas around settlements (*omurondji* and *orwejo*) or on flatter areas readily accessible to livestock (*ongumba*). No perennial forage grasses characteristic of mountainous or hilly areas were cited in this regard. Herders noted that even when adjacent to settlements, stony or hilly areas tend to preserve threatened species because it is difficult for the livestock to graze these areas too closely or pull the roots of perennials out of the ground. *Omundumba* (*Pechuel-Loeschea leubnitiae*, which is apparently browsed by goats) and *oruvahu* (scientific name unknown) were cited as invasive species indicating that an area had been badly abused.

Like professional range-managers, Etanga herders recognize pasture degradation and identify critical indicator species with this process. Unlike professional range-managers, Etanga herders do not view settlement sites and large water points as inevitably degraded. The plant communities of these areas have a high annual grass component, but within limits this poses few problems since these areas are used for grazing during the rains at the only time of the year when annual grasses are the preferred forage type. Later in the dry season when the annuals around these sites are depleted and would in any case have disappeared with the wind, livestock are using these areas only for water and walking out to more distant pastures dominated by perennials. As one group of informants put it:

> Later in the dry season [the interview took place in early May] when you observe cattle here you will see that there is no grass to be eaten. The animals in the sandy area around this river bed will be just waiting for water and will leave as soon as they are watered.

Annual-dominated pastures around settlement sites and water points may be theoretically 'degraded' – if degradation is defined solely in terms of changes in botanical composition, and if it can be demonstrated that human use is responsible for these changes[8]. But does the label 'degraded' mean very much in this instance? Etanga herd-owners obtain numerous valuable benefits from these sacrifice areas – water, garden sites, places to periodically congregate, residences for elderly people, forage in the wet season, etc. That they do not also obtain dry-season forage seems to strike them as a predictable and acceptable trade-off for these other benefits[9].

Etanga herders do not spontaneously cite changes in species composition and long-term rangeland degradation as problems in the perennial pastures reserved for dry-season grazing. Problems of overgrazing take a different form in these areas[10]. In *Okuroro*, the height of the rains, forage is plentiful and all forage types are universally palatable, including the perennial grasses that will later become the only available feed source in the dry season. Herders are supposed to vacate areas dominated by these perennials and leave them for later use. Some livestock owners, especially ambitious young men intent on herd growth, may be tempted to linger in these closed areas, capturing for their own stock forage that

---

[8] A recent study of vegetation resources around Etanga noted few obvious signs of human and livestock use in the areas surveyed. The study did pick up 'distinctive signs of utilisation . . . tree and shrub stumps' which might explain the 'comparatively high grass-component and reduced shrubby undergrowth' in some portions of the settlement zone north of the Etanga River (Burke, in press). The study did not comment on whether or not these changes might constitute 'degradation'. Other observers have been less reticent. Generalizing about northern Kunene as a whole in the early 1970s, Malan & Owen-Smith concluded that: 'Today most rangeland on the highlands is dominated by annual species . . . which appear for a few months only during and just after the rainy season. The much healthier state of those pastures where a lack of permanent water prevents intensive exploitation, leaves little doubt that the elimination of perennial species can be directly attributed to overgrazing by cattle . . .' (1974: 167). 'Although no quantitative surveys have been conducted, there is much visual evidence to suggest that selective browsing by domestic stock is gradually eliminating the more palatable shrubs from areas of high animal density' (1997: 168).

[9] Etanga stock owners are not alone in making these kinds of judgements. Consider the Namibian capital city of Windhoek, which has irreparably degraded the once-pristine pasture areas upon which highways, offices and homes now stand. However lamentable Windhoek may be as an example of failed range-management, these high levels of botanical degradation are unlikely to change because so many local residents profit from current arrangements. Efforts to botanically 'rehabilitate' settlement sites and the areas around large water points would elicit similar levels of disinterest and incredulity in Etanga.

[10] Local herd-owners are probably correct in their assertion that current levels of grazing pressure have little permanent impact on dry-season pastures. In the late wet-season of 1998, a field assessment was made of vegetation resources in the Etanga area. The field team visited the Omiramba, the most heavily used dry-season pasture area in the Etanga jurisdiction. The field team found the Omiramba but discovered so little evidence of use by livestock that they concluded that they were at the wrong site (Burke, pers. com.).

should be preserved for later general use. In these cases, it is the responsibility of older men to control the self interest of their juniors, or to refer the matter to the senior headman and his councillors.

Sanctions can also come into force later in the grazing cycle. It is the moral obligation of Etanga livestock owners to admit 'guest' herders to their home areas whenever this is feasible, and especially when people are fleeing from drought. But herders are under no obligation to accommodate people who have mismanaged their home areas through short-sightedness or greed. As long as this system remains in force, cheaters cannot prosper, for any advantages they have gained in the wet season when conditions are good are more than offset by losses in the dry season when conditions are difficult, and co-operation is essential for herd survival.

Management

Management of the Etanga grazing system is decentralized and unobtrusive. Many routine decisions are left to individual households or small clusters of households with minimal interference from any outside authority. Stocking-rate adjustments to fluctuations in rainfall are a clear example. Individual households decide whether it is in their interest to occupy their home area or beg resources elsewhere. At the receiving end of these population transfers, again it is individual households or small clusters of households that must decide whether local resources are sufficient to host outsiders. The stocking rate in an area arises out of the summation of these individual decisions, rather than through any centralized decision-making process.

Resource-management in Etanga can be unobtrusive because the management system matches the natural proclivities of the livestock to the productive characteristics of the rangeland. Cattle, for example, begin grazing on flat, sandy areas adjacent to water sources, and only move from these areas when their grazing is depleted. This is how the cattle are moved when they are herded, but it is also how cattle behave when left to themselves. Similarly, cattle are moved to the best pastures in the seasons of maximum productivity for different pasture types – annuals in the rains and perennials in the dry season or just after the rains begin.

For much of the year, patterns of herd-movement and resource-use are dictated by ecological considerations that make an elaborate management code unnecessary. The needs of stock and the self interests of informed herd-owners are enough to send the animals in predictable directions and to keep the system functioning. Rules come into play only at those points in the yearly cycle when movement is not constrained by environmental conditions and individual self interests might undermine collective welfare. It is biologically feasible, for instance, for cattle to use dry-season pastures in the wet season, for goat flocks to graze the same areas as cattle, or for large herds to hang around the most attractive and permanent water sources. At these critical junctures social control comes into play.

211

The system rests on a small number of simple and generally accepted rules of behaviour. These rules are frequently negative – stating what should not be done but leaving to individual ingenuity the choice as to what to do. The most notable of these rules are as follows:

o *Okuni* (dry season) grazing areas should not be occupied in *Okuroro* (the rains)
o whenever possible, large goat flocks should use inaccessible water points and leave the larger and more accessible water sources for cattle
o large cattle-herds should vacate central water points that are reserved for smaller herds, particularly milk animals
o in relocating their dry-season cattle camps, individual herders should consult with and not push ahead of their fellow herders
o impermanent water sources and associated grazing areas should be used before falling back on permanent water and on the pastures reserved for the dry season.

The system works because the rules are simple, everyone knows them, and visiting resource-users agree to submit to the authority and husbandry practices of their hosts. Within these broad limits, herding decisions are left to individual herding units, an efficient arrangement in a complex, variable and continuously changing environment. Hard and fast territorial boundaries, set stocking-rates and the centralized regulation of stock movements are unnecessary. The functioning of the Etanga grazing system relies instead on the social cohesion of small residential communities that make up the jurisdiction, backed up by the authority of the senior headman and his councillors.

Herd movement in conventional and indigenous range-management

The benefits of herd movement are exploited very differently in Etanga and on commercial ranches. Formal pasture-rotation systems subject part of a ranch to special stress or resting, and then reverse the process. No section is grazed or rested year after year in the same season. In this way, the intrinsic differences between paddocks within the grazing rotation are minimized by subjecting all of them to roughly equivalent levels of grazing pressure and compensatory relief, while rotating the periods of relief and stress annually or through the seasonal calendar.

In Etanga things work differently. Pastures are used and rested on a set seasonal rota; long-term resting occurs but on an unplanned basis when water points fail or rainfall is insufficient to support stock in a particular area. In this system the landscape is used unevenly and heavily grazed 'sacrifice zones' around settlements and permanent water are a potential hazard. The capacity of conventional rotational grazing systems to redress this problem is, however, unclear.

While they are theoretically plausible and practicable for the commercial rancher, the advantages of deferred, rotational and multi-paddock grazing systems have not been consistently demonstrated despite considerable efforts

212

to do so (O'Connor, 1985a; Stoddart et al., 1975)[11]. Rotational grazing schemes implemented in the communal areas of South Africa have, on occasion, yielded perverse results: resting an area can encourage both bush encroachment and the overgrazing of preferred grass species when undesirable species escape grazing, mature and become unpalatable (Forbes & Trollope, 1991). Often the advantages from rotational systems are so modest as to be difficult to discern statistically, even in carefully controlled experiments (APRU, 1978–1990). Irrespective of the final scientific verdict on the efficacy of these systems, communities like Etanga are unlikely to realize from them the obvious benefits needed to promote their voluntary acceptance.

These difficulties are compounded by the high costs that communities would incur in shifting to a radically different system of forage use and management. Ignorant of the rationale behind village-level management practices, outsiders frequently assume that improved schemes are simply filling a 'management void'. All that is needed is to educate the locals, free them from the shackles of custom and get local communities properly organized. In reality, improved management schemes are competing against established husbandry and resource-use systems and must out-perform these systems by providing more benefits or lower costs.

Etanga already has a system of resource-management that addresses the same problems as conventional range-management – the sustainable use of a hetero-geneous resource in a variable climate. But the indigenous system pursues medium-term management objectives and employs husbandry practices that are radically different from those of commercial ranchers. This means that the dislocation and cost of adopting conventional systems of range-management would be high and the rewards would be minimal, if the indigenous system already performed appropriately despite its different methods of operation.

Finally, there is the question of access to industrial inputs, particularly man-made water installations. When they employ rotational grazing systems, ranchers have the means to fence and to develop the water supplies that create paddocks and render them uniformly habitable, overriding the erratic natural distribution of water sources and the propensity of livestock to use the land unevenly. In the past, Etanga stock-owners did not have this option. Their response was to adapt to the exigencies of the natural environment, rather than attempt to alter it. This situation is now changing and Etanga – in common with many other African pastoral communities – now has access to modern water-development tech-nology. The future of livestock and rangeland husbandry in Etanga depends on how this option is exercised.

---

[11] 'From the data at hand it cannot be categorically stated that a rotational-grazing system will invariably improve the range or give greater livestock production than moderate, continuous seasonal grazing. Existing evidence is contradictory' (Stoddart et al., 1975: 297, referring to North American research).

'It is also notable that grazing system trials have shown no significant effect on composition over a wide range of savanna types . . . even though many of these studies compared continuous grazing and some form of rotational grazing. Furthermore, there is no evidence that controlled selective grazing has any influence on botanical trends' (O'Connor, 1985a: 42, in a review of the southern African literature).

Water and development: 'screaming at the bosses'

In 1996 the Government of Namibia initiated a range and livestock development project for its northern communal areas, the Northern Regions Livestock Development Project or NOLIDEP. Etanga was one of the pilot communities included in this programme, and the relationship that has evolved between NOLIDEP and the community is an additional factor influencing the future of natural resource management in the Etanga area.

Under the auspices of the Etanga Senior Headman, a committee was formed in 1996 to mediate between the project and the community, and negotiations began to carry out several construction projects requested by the community. At that time Etanga was over 100 kilometres away from the nearest cattle-sales point. Work started on a cattle sales pen near the Etanga settlement and on a hand-built masonry dam in the Okaanga area, and negotiations began for the mechanical excavation of several additional dams. The project undertook to supply construction materials and technical support for building the masonry dam and sales pen, if the community supplied the labour. For the mechanically excavated dams, the project provided engineering services, arranged for commercial contractors to carry out the work, and paid the bulk of construction costs. But the community agreed to provide incidental labour and pay a portion of the contractor's fees.

The committee formed in 1996 successfully identified development priorities but was less effective in implementing them. After about a year some Etanga residents became impatient with the lack of interest by committee members in executing their projects. The old committee was dissolved and a local farmers' association was formed. Members of the association's governing board were young, ranging in age from their late 20s to late 40s, and predominately literate – precisely the kinds of people that Etanga had to integrate into the community's traditional authority structures if it was to successfully deal with government and other modern institutions. The association drew up a formal constitution, opened a savings account, and now obtains funding independently of NOLIDEP through a rebate from the Namibian parastatal abattoir company to maintain the sales facility.

Though initially set up to deal with NOLIDEP, the association now has an independent existence. Its governing board contains members from outside the NOLIDEP pilot area and hopes to expand further. It also initiates activities on its own, most recently (unsuccessfully) banning the sale of alcohol by itinerant traders who attend the small 'fairs' that spring up around the sales pen whenever cattle buyers are present.

These are potentially important developments for the future of range-management in Etanga. The 'customary' institutions that regulate resource-use in Etanga are still strong, as this Chapter has shown, but they have survived because the area has been cut off from outside development. This situation is changing, and the prognosis for the continued survival of these institutions in their present form is not good, based on precedents elsewhere in semi-arid African and in more commercialized pastoral regions of Namibia.

214

These institutions require support and modernization if they are to continue to perform their current functions. This will be a gradual process in which engagement with government will familiarize local communities with administrative procedures, and local institutions will change in response to new opportunities and responsibilities. In this respect, the local farmers' association provides an outlet for the talents of younger and better educated rural residents and thereby complements the strengths of the customary authorities, who have thus far supported and closely monitored its affairs.

Much remains undone. The following interchange summarizes some of the challenges facing the farmers' association as it undertakes community-based water development in Etanga.

Herder I want a dam to be built here so that I can settle and do not need to migrate in the dry season to the Kunene River.

Researcher You must take your request to the farmers' association, and they will decide whether the dam you want or some other request is more important. They will contact the project. I cannot personally help you.

Herder This way of doing things will not work. The Chairman of the Farmers' Association will provide his father's place and the place where he stays with benefits. All the boreholes in this area result from the struggle of my father, the councillor of the old Etanga headman, and I tell you that there is no one on this earth who will provide another community with water without first taking care of his own place. I am sitting here now at my father's place. If I find bosses I must scream at them, and if they give me help, then they give me help. I was cursed by my father's dead brother when I tried to resettle in another area because I could find no water here. My father's brother wanted me to come back and take over from my father. I must stay here and scream for the bosses to put in a dam.

As this interview indicates, resource-management in Etanga is supported by spiritual sanctions, as well as local institutions and material interests. But the kinship loyalties that underpin Himba society cannot be automatically transferred to a public institution like a farmers' association. The herder quoted above assumed that association leaders were unlikely to act in the public interest unless they also set aside deep attachments to their ancestors and home communities. In Etanga a sceptical local audience will need to be convinced that an institution like a farmers' association will really work for the general good. This will take time, and problems – real or perceived – are almost certain to arise at some point.

Equally informative is the herder's perception of the prevailing system for siting new water developments, summarized in the phrase 'screaming at the bosses.' For local residents, playing the 'water-development game' is rather like buying a lottery ticket. In both instances, the initial investment is small – very little money is required for the lottery ticket, and all that is required to participate in the water-development game is a bit of time and energy spent importuning outside technicians or government representatives. Nothing comes of most of these encounters. Most lottery tickets are torn up, and whatever they may be forced into promising, most government representatives do not deliver. But if the possibilities of a favourable payout are infinitesimally small, the payout – should it ever arrive – is huge. Persistent, unrealistic demands for water development make sense in these terms – small initial investment, small probability of success, but large unpredictable rewards on rare occasions. Why not have a go?

NOLIDEP'S water development policies in Etanga have attempted to counter these expectations, by asking communities to prioritize their water needs, contribute to the costs of water development, and accept formal contractual obligations in return for the reliable provision of promised inputs. Sensible and environmentally sustainable water development requires this kind of community involvement, rather than a screaming contest.

But old habits and expectations may be slow to change, and if local communities and technicians are to make this system work, government agencies must also reform the way they do business. Politicians must relinquish some of their capacity to garner support by dispensing largesse in their constituencies. Both government water-development agencies and commercial contractors must learn to deal with rural communities that have paid for construction and view themselves as clients rather than supplicants. Finally, there is a temptation for donors and government departments to build large-capacity water points without community contributions, in easily accessible areas where the returns on investment are likely to be better, the demands on staff time are reduced, where money can be spent more quickly and a short-term programme has more to show for its efforts. Environmental problems would emerge long afterwards and would, no doubt, be attributed to the inability of rural communities to manage resources effectively, rather than irresponsible central planning.

Water is the single most important variable influencing rangeland use in Etanga, a fact that local herders clearly understand. Etanga residents already possess an impressive understanding of their local environment. Putting this knowledge at the service of a restrained water-development policy will require both new local institutions and a national policy environment that is conducive to their operation. As things now stand, this is possible. But it is not assured.

9

Ecological dynamics and grazing-resource tenure: a case study from Zimbabwe

Dynamique de l'écosystème et régime d'utilisation des pâturages: étude de cas au Zimbabwe

IAN SCOONES

ABSTRACT

Ecological dynamics, and particularly spatial heterogeneity in grazing resources, interact with tenure regimes and other institutions in complex ways. An historical review of the shifting patterns of rights over resources during the past century in rural Zimbabwe shows how changes in the relative value of different resource types can be linked to changes in tenure and institutions. The types of institutional arrangement for managing different resource types are examined, with contrasts made between *ad hoc* institutions for ephemeral, low-value resources and more formal institutions for relatively stable, high-value resources. The experiences of intervention in grazing management in southern Zimbabwe illustrate how simplistic assumptions about resource dynamics have been made, leading to inappropriate institutional forms – with the resultant failure to address the complexities inherent in community management of grazing resources.

There is thus a need for a disaggregated view of environment, community and institutions for effective project and policy design. This requires a different conceptual approach, rooted in an understanding of both ecological and social dynamics.

RÉSUMÉ

La dynamique de l'écosystème, et en particulier la variabilité des ressources fourragères dans l'espace, ont des liens et rapports complexes avec les régimes d'occupation des terres et d'autres institutions. Un rappel historique de l'évolution des droits sur les ressources au Zimbabwe depuis un siècle montre comment les variations de valeur relative des différentes ressources peuvent être liées aux changements des institutions coutumières et du régime d'occupation des terres. Les types d'institutions susceptibles de gérer des différentes catégories de ressources sont examinés: les institutions *ad hoc* sont adaptées aux ressources éphémères de peu de valeur et les institutions plus structurées aux ressources relativement durables de valeur élevée. Les interventions réalisées dans la gestion des pâturages dans le sud du Zimbabwe montrent comment des hypothèses simplistes concernant la dynamique des ressources ont abouti à créer un cadre institutionnel mal adapté, de sorte que la complexité que présente fatalement la gestion communautaire de ces ressources n'a pu être prise en compte. Il est nécessaire d'avoir une vue détaillée de l'environnement, des communautés et des institutions pour que les projets et les politiques soient conçus de façon rationnelle et novatrice à partir d'une bonne compréhension de la dynamique de l'environnement et de la société.

Introduction

ANY GRAZING LANDSCAPE is made up of different patches of different value for livestock feed. Such patches vary from wide expanses of relatively low-value grazing, dominated by annual grasses, trees and shrubs, to relatively high-value patches which are highly productive and provide high-quality feed. The patterning of resources is critical for our understanding of use-patterns, perceptions of resource-value, and therefore incentives for management through tenure regimes and other institutional forms. In debates about resource tenure the heterogeneity of landscapes is often forgotten, with simplistic categorizations of tenure types (open access, common property, private or state-owned) being used to describe situations over wide areas. Yet spatial patterning and temporal shifts in relative value of different landscape components are critical in structuring the likely nature of tenure regimes.

This Chapter explores the dynamics of grazing management in southern Zimbabwe, investigating the way ecological dynamics, and particularly spatial heterogeneity in grazing resources, interact with tenure regimes and other institutions. Following a brief review of some of the pertinent theoretical debates concerning property rights and rural resource-management institutions, a historical review is offered of the shifting patterns of rights over resources during the past century in rural Zimbabwe. This focuses on the shifts in the relative value of different resource types and the resultant changes in tenure and institutions. The next section examines the contemporary situation by looking at the range of tenure regimes making up the grazing resource. The types of institutional arrangements for managing these resource types are examined, with contrasts made between *ad hoc* institutions for ephemeral, low-value resources and more formal institutions for relatively stable, high-value resources. This provides the background for an examination of the experiences of intervention in grazing management in southern Zimbabwe. The Zimbabwe experience illustrates how simplistic assumptions about resource dynamics have been made, leading to inappropriate institutional forms, with resultant failure to address the complexities inherent in community management of grazing resources. The final section concludes by noting the need for a differentiated view of environment, community and institutions for effective project and policy design, requiring a different conceptual approach, rooted in an understanding of ecological and social dynamics.

Property rights and social processes: tenure and institutions for resource-management

The economic theory of property rights provides one route into a differentiated understanding of tenure regimes (Demsetz, 1967). Property-rights theory posits that the incentives to invest in forms of exclusive tenure, most obviously private regimes, but also types of exclusive common-property arrangement, increase with the increased value or potential value of the resource. Thus, when land is scarce or when opportunities for reaping significant economic benefit arise through changes in market conditions, for instance, then there is an increasing likelihood of exclusive forms of tenure emerging. However, these will only arise when the stream of benefits derived is perceived as secure, and exceeds the transaction costs of investing in a more complex, exclusive-tenure regime. Transaction costs arise during coordination of contracts as a result of negotiation, monitoring and enforcement; through acquiring and integrating information in the process of establishing contracts; and through a range of strategic behaviours (Ostrom et al., 1993). Exclusive tenure is therefore more likely under conditions of increasing resource pressure, when the political and policy environment is stable, when exclusive-tenure regimes are socially accepted and when there are existing institutions which can take on the role of exclusion at relatively low cost.

The problem with much of the property-rights analyses, and associated theories of institutional economics applied to questions of resource-management, however, is that, as with other simple classifications of tenure regime, little attention is paid to the ecological dynamics of the resource base. In a highly variable environment, typical of most dryland pastoral and agro-pastoral areas, where a wide range of different landscape components exist, each with different resource-values which vary in different ways over time, the patterns of tenure will not be uniform (Behnke, 1994). In the language of game theory, appropriation problems arise whereby assignment of patches of different resource-value must be carried out in the negotiation of contracts (Ostrom et al., 1994).

Another problem with property-rights analysis is the set of assumptions made about individual behaviour. An ahistorical and asocial analysis based on in-dividual actors with (bounded) rationality typifies many analyses, particularly those based on gaming simulations of decision options (cf. Ostrom et al., 1994). Interpretations more effectively grounded in field complexity recognize the importance of the 'embedded' nature of tenure regimes and institutional forms (cf. Peters, 1987; Mearns, 1996). As Sara Berry notes (1993: 104) 'People's ability to exercise claims over land remains closely linked to membership of social networks and participation in both formal and informal processes'. This social and political negotiation over land rights means that people are not operating simply as individuals independent of context, they are social actors engaged in processes of negotiation with a wide range of social and political implications. Thus, tenure regimes and rural institutions may not reflect the result of repeated games between individual rational actors seeking to maximize utility. Instead, tenure regimes and other resource-management institutions are the result of complex social processes which are poorly explained by simple rational-actor models. Due to peoples' continuous investment in the 'means of negotiation as well as the means of production' (Berry, 1993: 15) 'rural institutions often operate as arenas of negotiation and struggle, rather than as closed corporate units of accumulation and resource-management' (Berry, 1993: 20–21), as is conventionally assumed. The result is often ambiguous rules, flexible membership of organizations, and overlapping and contested boundaries.

In highly variable dryland ecosystems, characterized by non-equilibrium dynamics, this is particularly likely to be the case. As Roy Behnke (1994: 7) observes:

> In pastoral African tenure systems the natural landscape is seldom carved up into neat territorial packages owned by distinct groups or individuals. Instead, any area is likely to be used by a myriad of different ownership groups of variable size and composition, with overlapping claims to territory derived from particular claims to different categories of resources within it.

Therefore, with a disaggregated view of the grazing landscape and the social processes governing its management, we can expect to see quite a complex pattern of tenurial arrangements, with some patches being exclusively managed, while others are managed intermittently as exclusive resources and at other times more loosely. Tenure regimes will thus likely be overlapping in both time and space, with a variety of different institutions, operating at different scales and with different degrees of intensity, being involved in the management of different portions of the landscape.

Table 9.1 provides a summary of the relationships between resource type, tenure regime and institutional arrangement and investment levels that can be expected in two contrasting landscape components[1]. These are of course the extreme cases, and in any grazing landscape a gradation between these types can be expected, with associated difference in tenurial and institutional arrangements.

Table 9.1 offers a spatially differentiated view, but does not address the temporal dimension. There are two important aspects to this. Over the longer term, the relative value of different parts of the landscape will change. Thus, in situations of high land abundance, extensive use of low-value areas at low levels of intensity may be sufficient to sustain the needs of livestock.

However, if land becomes scarce relative to the demands of livestock, then particular high-productivity patches may become more and more valuable. Thus, it is the relative value in relation to demand for different patches, not the inherent value in terms of resource productivity, that is important. The same principle applies over a shorter time frame. The relative value of different landscape patches may change seasonally and between years, such that in droughts or during dry seasons the value of available resources increases, thus increasing the incentive to institute more exclusive tenure regimes.

---

[1] This parallels the contrast between the dynamics of equilibrium and non-equilibrium systems which may exist at broader landscape or regional scales (Behnke et al., 1993). Within these systems, resource types or patches exhibiting these characteristics may be found.

Table 9.1. Resource characteristics and tenure regimes: two contrasting situations (Zimbabwe)

Resource type	High-quality, high-value resource spread over relatively small area; stable and high production	Low-quality, low-value resource spread over relatively large area; unstable and low production levels
Tenure regime	Exclusive: privately or commonly held among relatively small group	Open: loose common-property regime or open access; wide access to resource
Institutional arrangements	Strong institutions (rules, regulations); high individual transaction costs	Weak institutions; limited transaction costs
Investment levels	High	Low

This complexity in tenure regimes and institutional forms may appear to some chaotic and inherently inefficient. Indeed many attempts at intervention have been focused on somehow 'rationalizing' this apparently disorderly mess. During the colonial period many attempts were made to codify so-called 'customary' law and land rights (Channock, 1991). Yet, because of the flexible and ambiguous nature of actual tenure systems, this proved complicated. As Elizabeth Colson (1971) argues, 'customary law', which formed the foundation of much of the legal framework for British indirect rule in many parts of Africa, can be seen as a creation of colonial policy. Equally, attempts at creating structures for local administration based on tightly defined geographical territories and strict patterns of authority, typified many interventions from the colonial period to the present. Chiefs, headmen, councillors and committee members are thus often creations of administrative intervention by the State wishing to intervene, organize and develop. The search for tradition on which to base such interventions has been a powerful guiding principle. Somewhere, it is argued, behind the apparent mess of conflicting, overlapping and contested institutions, there must be an original, legitimate and 'traditional' form. The argument for 'resurrecting traditional institutions', which have somehow 'broken down' or 'disintegrated' has, of course, become a rallying cry for development agencies today in search of sustainable development solutions. But the quest for tradition may lead to its reinvention (cf. Hobsbawm & Ranger, 1983) by élites and others who may gain by being granted new authority and resources under recreated customary forms or other new institutional forms.

But are complex, overlapping tenure regimes, regulated by vague or ambiguous rights and governed by flexible institutions with competing claims, so inappropriate in the context of dryland grazing management? A good argument can be made for a better recognition of such *ad hoc*, flexible approaches to resource-management. In situations where a variety of resources coexist within the landscape – grazing is juxtaposed with arable; water points and salt licks are scattered across the landscape; and high-value key resources are found alongside

221

low-value extensive resources – it is not surprising that a complex system of multiple-use rights evolves (cf. Peters, 1987; Behnke, 1994). Superimpose upon this, the temporal dimension of changes in relative value of different resources and the need for flexible responses to local resource-scarcity, and the likelihood of competing claims increases.

In dryland grazing settings, rapid response and flexible approaches are key if opportunistic strategies are to operate efficiently (Sandford, 1982; 1994). Where negotiations must happen in response to a sudden change in the local situation, and new and flexible contracts must be formed due to unexpected circumstances, institutions derived from a complex interplay of individual and group interaction based on the negotiation of rights within and between social networks are likely to be the most effective at managing resource access in such dynamic ecological settings. Formalized structures may be too inflexible to adapt to changing circumstances, being too cumbersome and unwieldy due to the constraints of procedure, bureaucracy and legalistic approaches. The result may be excessively high transaction-costs and failure to respond quickly and opportunistically to risk and uncertainty.

However, there may be a downside to such flexible, adaptive arrangements. Some commentators argue that where systems of customary tenure are not clearly defined and effectively enforced, land users are reluctant to invest in boosting the productivity of their land (cf. Feder & Noronha, 1987), and so more concrete delineation of tenure (with the ideal explicitly or implicitly being assumed to be freehold title) is necessary. However, a vast amount of evidence suggests that where tenure is perceived to be secure, investment occurs under a wide range of tenure regimes (Place & Hazell, 1993; Platteau, 1995). Arguments for titling thus appear to be driven more by ideological leanings than empirical considerations.

However, flexible tenure systems and complex institutions may be transformed through external intervention. For instance, development projects often like to support well-defined 'community' institutions, and governments like to plan and govern with well-defined administrative units. The re-creation of local organizational forms to fit the needs of project plannners, and so have access to the fruits of development aid, is a common route by which institutional arrangements become formalized, and the benefits are captured by élites able to exercise their power in such settings.

Another reason for the creation of more formalized institutions is the protection of rights to resources for more vulnerable groups. While complex, flexible arrangements may work well among connected social networks with a history of interaction, negotiation and adjudication, they may not be so effective when confronted by, for instance, a sudden influx of outsiders wishing to usurp rights over resources, or where local élites seize the opportunity of resource capture. In such settings access to more formal processes through widely recognized legal and administrative structures may be necessary. Thus, in highly variable grazing environments, local, flexible and *ad hoc* institutions with low transaction costs and rooted in existing social formations, may effectively manage resources day to day, but may need to be complemented by investment in more formal, and thus costly, administrative, legal and co-management structures in order to cope with the range of contingencies likely to be encountered (Swift, 1995; Sylla, 1995). This may take the form of co-management arrangements, where the interests of weaker groupings are promoted in the pursuit of greater equity in resource tenure. However, such co-management arrangements too may have a downside, as formalized structures supported by the State are inherently political in character, and again may be open to co-option by élites (Cousins, pers. comm.).

The following sections explore the range of propositions suggested by the theoretical debates described above. Through a historical examination of grazing management practices in the communal areas of southern Zimbabwe (Map 9.1), the ways in which tenure regimes and institutions have responded to the changing patterns of the grazing landscape are explored.

Grazing management in southern Zimbabwe[2]

During the latter part of the nineteenth century relatively few livestock were held by the Shona living in southern Zimbabwe. These were largely grazed close to the homes, many of which were sited on hilltops, secure from raids by the Ndebele. Lack of security meant that people did not have an opportunity to make use of the extensive grazing lands in the plains area and instead, limited grazing was secured from wetland (*dambo*) patches. This combined with *miombo* woodland browse to make a reasonably good diet for the limited number of stock held. By contrast the Ndebele herds grazed the lowlands, with large-scale migrations being undertaken to secure grazing.

All this changed following the colonial conquest in the 1890s. With the defeat of the Ndebele, raiding ceased and Shona farmers were able to colonize the plains areas. This opportunity increased with the arrival of the plough when extensive clearance and opportunistic ploughing of large areas gave reasonable returns. No longer were livestock restricted to hilltop settlements and their immediate surroundings – herds and flocks could now expand and make use of the areas once too unsafe to enter. With the clearance of land and the settlement of the plains area, wild animals, too, were eliminated and livestock thrived, growing many-fold from the low point following the rinderpest epidemic of 1896. During the early part of the century no major grazing shortages are reported in oral or archival histories. Indeed, this is regarded as a time of plenty in the communal areas, when frontiers were expanding, agriculture was booming and livestock populations were increasing.

However, a number of factors changed this situation. The increasing political pressure to implement the colonial Land Apportionment Act, together with a wide range of environmental legislation, meant that State interventions from the 1930s, but particularly following the Second World War, resulted in a growing resource squeeze in the communal areas. Population increases combined with

---

[2] This section is based on historical research carried out in southern Zimbabwe and reported in Wilson (1990), Scoones (1990) and Scoones et al. (1996).

Map 9.1 *Study area in southern Zimbabwe*

restrictive land-use legislation and changed the nature of agricultural land-scapes and so the availability of grazing resources. The pattern of extensive grazing, amongst scattered fields and settlements, was firmly disrupted with the introduction of the centralization policy from 1929 onward. This was later further reinforced by legislation in 1951, when a large-scale programme of State-led land reorganization was initiated following the enactment of the Native Land Husbandry Act. Landscapes were thus redesigned by State intervention, with a typical pattern involving linear settlements dividing a grazing area and arable-land area. During this process, large blocks of land were thus allocated for particular land uses with associated tenure regimes. The arable lands were allocated to individual (male) heads of households, while the grazing lands were meant to be held in common according to reinvented 'traditional' rules centred on ward (*matunhu* pl.) -based management and control.

From this time onwards, the value of different parts of the landscape was dramatically changed. The large-scale conversion of upland woodland to the new arable lands meant that low-lying areas became vital for grazing, in combination

224

with an increasing reliance on crop residues for early dry-season fodder supply. The banning of *dambo* and river-bank cultivation prevented conversion to arable use and such areas became vital 'key resources' for grazing. Over the following decades as livestock densities increased[3], this pattern became even more evident, with relatively small patches of high-value grazing in lowland resource types providing a large proportion of the total grazing for cattle. For instance, a study carried out in 1987 showed how cattle had significant grazing preferences for riverine strips, drainage lines and *dambo* patches, especially during the rainy season and late dry season when fodder resources from the arable areas were not available (Scoones, 1995).

This pattern of spatial use is particularly apparent in periods when livestock populations are high. In such situations, the stable, high-productive grazing patches may be the only resource left for hungry animals and foraging patterns tend to be highly concentrated (Scoones, 1995). However, if forage resources run out in an area, such as when drought strikes, animals are forced to move further afield. For instance, during the 1982–84 and 1991–92 droughts, cattle were moved up to 100km in search of feed (Scoones, 1992; Scoones et al., 1996). If drought results in large-scale mortality, foraging patterns switch to a more dispersed grazing pattern with more even use of different patches. Thus boom and bust population dynamics driven by drought combine with spatial heterogeneity of the forage resource to produce variations in foraging behaviour over time.

The history of grazing management over the past century has therefore seen a number of phases, switching from intensive grazing of relatively limited areas with expansion restricted by the dangers of Ndebele raiding, to a period of relatively extensive resource-use during the early colonial period when farming and settlement expanded into the plains. This trend was reversed through the combination of population growth and colonial government policy, with again a grazing strategy focused on key patchy resources being employed. This general pattern has been maintained to this day, with oscillations between more and less extensive grazing-use depending on the interaction of livestock populations and drought conditions. These shifts between extensive and intensive use, between low- and high-resource pressure and between restricted and relatively free use-patterns, naturally has major implications for the type of tenure regimes found. A discussion of the contemporary situation is the subject of the next section.

Resource types and tenure regimes

In a heterogeneous landscape different resource-types can be identified (Figure 9.1), each with different ecological characteristics. Table 9.2 highlights the major resource-types in the contemporary grazing landscape of Zimbabwe's communal

---

[3] Over the period between 1955 and 1980 the cattle population density in Chivi, for instance, increased from *c.* 0.15 to 0.45 animals per hectare (Scoones et al., 1996).

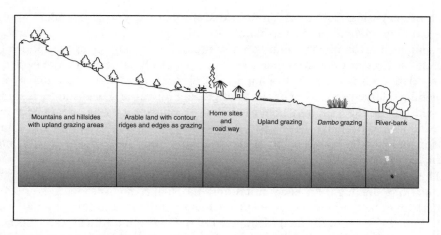

Figure 9.1 *Transect diagram illustrating the major resource types*

Table 9.2. Resource types and tenure (Zimbabwe)

Resource type	Fodder characteristics	Seasonality and variability	Tenure regime
Upland grazing	Low grass-productivity; extensive browse	Early rainy season; highly rainfall dependent	Common property, with increasing exclusive control in dry season/drought years
Bottom land grazing (*dambos*, river banks)	High grass-productivity	Year-round; stable production	Common property (exclusive), sometimes private
Arable land grazing	Crop residues, contour bund and field-edge grazing	Dry season, following harvest	Common property, sometimes private if fenced or if residues/grass harvested
Home sites	Limited fodder; some crop residue feed from stores	Year round	Private
Road sides	Strip grazing	Year round	Open access (except main roads)
Mountains and hillsides (sometimes including sacred sites)	Refuge grazing and browse	Dry season	Open access, except for sacred sites

areas and how these are associated with different tenure regimes and different institutional arrangements for governing their use.

Institutional arrangements are as varied as the resources they govern. Institutions exist as 'regularized patterns of behaviour between individuals and groups in society' (Mearns, 1993b: 103); they are part of social practice, created and recreated through people's actions (Giddens, 1984). In the Zimbabwe context, access to grazing resources is dependent on the interaction of a range of different institutional and organizational forms, including formal organizations such as local government structures, ranging from the district council to the ward to the village committee, as well as more informal arrangements based on lineage and other networks and associations. In order to gain access to resources, people must invest in the social relations which allow institutions to persist (Berry, 1989). As discussed above, different institutions and organizations have been important in mediating grazing-resource access and control over the past century. This section will examine the variety of institutions which are significant for the present setting of southern Zimbabwe.

Four broad types of grazing resources (and many variations) associated with different types of institution governing access to grazing can be distinguished. First, in some low-value areas there is effectively no control initiated directly for grazing-management and rules governing use are instituted for other reasons. Road sides are examples of this type of arrangement. These areas provide small strips of grazing, which often remain green longer than other areas due to the run-off effects of road surfaces. Such areas are important last-resort grazing, but the danger of animals suffering road accidents if grazing is along a busy road and the legal restrictions banning road-side grazing may be disincentives for some.

Second, there are areas where resource access and use are governed by sets of broad cultural norms. Such areas are regarded as sacred, sites where lineage ancestors reside and therefore where appropriate respect should be paid. Sacred areas are also linked to broader territorial land cults in southern Zimbabwe, and land spirits exert control over use (Mukamuri, 1988; Schoffleers, 1979). Grazing use in sacred areas is not necessarily specifically controlled, although tight regulations may apply to water sources, although less likely in sacred woodlands or on sacred mountains. While it is often permissible to graze animals in some sacred areas, herd boys may not wish to enter such areas unless animals are in severe need of grazing during a drought. In such cases, these sites sometimes provide important refuge grazing.

Thirdly, there are areas which have flexible, intermittent, *ad hoc;* arrangements for managing control and access to grazing resources that vary between seasons and years depending on the relative value of the resource. Flexible, *ad hoc* arrangements are typical of the upland grazing areas. These are extensive grazing areas making up perhaps 75 per cent of available non-arable grazing land. For most of the time, these grazing resources are only used at low intensity. While early rainy season grazing is important, productivity is low and much of the fodder value is exhausted during the first few months of the rains. Although people recognize rough grazing territories associated with particular clusters of homes or village sites, these are generally not tightly defined. Indeed, many such grazing territories overlap and boundaries outside key resource areas, although spoken about, are generally not contested.

Thus a loose form of common-property management exists, whereby membership is defined through residence and territories are roughly known, with the potential for exclusion of outsiders. This is not always exerted as, in periods when local livestock populations are low, rainfall is good and grazing resources are plentiful, there is no need to establish firm rules of use. In these periods an effectively open-access situation reigns, where rules and regulations are not applied. However, this is not always the case, as in times of local resource-scarcity, exclusion rules are more firmly applied. The way this control is exerted is variable. In some cases, local by-laws are invoked (these may be real, or very often imagined) and councillors, elected representatives of local government, are urged to mediate. This may involve bilateral negotiations with neighbouring wards and their representatives or may involve the petitioning of a higher authority at district level, attempting to get the District Administrator to intervene.

In many areas, particularly following the decreasing power of the village and ward authorities and the reassertion of power by so-called traditional leaders in the late 1980s and 1990s, more informal networks were activated. Mediation by lineage leaders is the most common form of informal negotiation. Such negotiations tend to be very localized and informal, with complex reciprocal arrangements made, involving personal deals and counter-deals. The type of arrangement arising will depend on the relationship between the leaders. Sometimes they are close blood relatives, or at least hold the same family totem; on other occasions they are friends; in other situations the two parties may be represented by ruling and immigrant lineages with long-running feuds, in which case disputes may be less easy to resolve. In these informal situations, existing institutions and networks are employed, ones that exist for reasons well beyond grazing-management.

Finally, there are 'key resource' areas which are more tightly controlled by more elaborate institutional arrangements. This type of institutional arrangement is more permanent. In areas with high-value resources which are relatively stable, the incentives to capture control are high. In dryland regions such areas include *dambos*, river banks and other run-on sites. Although there are government restrictions through environmental legislation that prevent conversion to arable land and so effective privatization, there are a variety of ways in which people are able to exert considerable control over such high-value resources as both groups and individuals. At one extreme is the annexation of common resources through the extension of private holdings. The 'creeping fence' phenomenon is often found on *dambo* edges. Fences are often found to progressively extend into the *dambo* area, capturing valuable grazing or creating vegetable gardens for private use. Not everyone can get away with such a strategy and those farmers who are successful in appropriating land in this way are usually representatives of the local élite, rich and politically powerful members of key lineage families. In areas where the ethic of common ownership remains strong, other people would not be allowed to get away with it[4].

In other instances, control over such key resources is not exerted by individuals but by groups. If it can be argued that an important key resource exists within the broad territory of a village or lineage cluster, then claims may be exerted. A lineage leader may establish a series of rules about use which his immediate family, and any other immigrants living in the village area, are obliged to follow. Such rules may set patterns of use of water sources, the timing of grazing of the sites, the type of herding arrangements to be used at different times of year, and so on. These, however, are very often contested, particularly as many key resource sites are found along notional boundaries (e.g. rivers and streams, drainage lines, etc.). In many instances, two (or more) groups may claim control and disputes arise. Thus many of the grazing disputes over broad territorial areas discussed above, often centre on key resource sites, particularly at times of resource scarcity.

The general pattern seen shows how with the increased value of the resource for grazing, more exclusive patterns of tenure are seen and more elaborate institutional forms are employed to regulate use. However, this pattern is not static, with overlaps and shifts between different tenure regimes and institutional arrangements characterizing grazing-resource management in the communal areas. Tenure regimes are therefore the result of complex struggles at the local level, and arise as a product of the interaction of a variety of social networks and relationships. Thus:

> Existing land rights involve a diffuse pattern of overlapping rights, including those of land spirits, chiefs, ward heads, village heads, local patrilineage heads, and individual homesteads. Rights at any one level never fully exclude rights at another (Scoones & Wilson, 1989: 108).

This complexity makes intervention in grazing management highly problematic, as is illustrated below.

Experiences of intervention: grazing schemes in southern Zimbabwe

Since the 1960s grazing-management interventions have focused on the implementation of grazing schemes in the communal areas. With the failure of earlier livestock policies which concentrated on highly unpopular destocking of what were perceived to be excess livestock numbers, a more positive developmental approach was adopted and grazing schemes began to be promoted. By the mid-1970s hundreds of such schemes existed across the country, with 315 alone in Masvingo Province (Danckwerts, 1974; Froude, 1974). During the period of the liberation war such interventions ceased, but the grazing-scheme idea was

---

[4] The overall impact of such annexation on grazing resources varies from place to place. Legal restrictions on *dambo* cultivation reduce the likelihood of wholesale privatization, but in some areas, especially where such key resource grazing is limited, the impact may be significant.

revived as a centrepiece of rural development activities soon after Independence, with a proliferation of schemes being initiated or revived. By 1986, over a hundred schemes were active in the country (Cousins, 1987; 1992), and many more have since been planned, funded and implemented.

The design of grazing schemes derives from a series of assumptions about technical and management objectives. The technical design is premised on the assumption that increased productivity of livestock can be derived from rotational grazing using paddocks. This model is based on the approach used in large-scale beef ranching, where low stocking-rates, limited herding-labour and reticulated water systems mean paddocking offers a useful solution (even though it may not actually increase productivity over continuous grazing[5]). In communal areas, paddocks are demarcated within the upland grazing area, often with little consideration being paid to existing patterns of use and the heterogeneity of the resource base. Not surprisingly, most grazing schemes are not used in the way the planners imagined. In times of drought, for instance, when upland grazing areas have very limited available biomass, animals are removed and grazed outside the scheme boundary. Equally, when resources are differentially available between areas, reciprocal arrangements are made between individuals and groups, and livestock are able to move to new pastures, beyond the boundaries of the scheme. These opportunistic patterns of livestock management, responses to the variability of the resource, are necessary parts of livestock-keeping in the communal areas, yet are not always recognized in the design of schemes (Cousins, 1992).

Assumptions behind management design can also be seen to be flawed. The idea of a management unit based on a paddocked grazing area derives from over-simplistic notions of community cohesion and resource territoriality. This dates back to analyses of administrative structures in the communal areas in the 1960s which gave primacy to the idea of the ward (*dunhu*) as the major administrative and management unit in Shona society (Holleman, 1969). This 'traditional', 'customary' form of administrative control was thus a product of colonial invention, rather than any widely recognised entity (Scoones and Wilson, 1989). The ward was a convenient construct for development activities and thus provided the ideal geographical unit for grazing management, with the ward community (and its apparently stable membership and defined leadership structures) being the appropriate group for collective management. However, as already discussed, rural communities in the communal areas are much more diffuse, with less-defined and often overlapping boundaries, more complex leadership and authority structures and more fluid group membership, than assumed by such idealized notions of a ward community.

Despite the many experiences where 'community'-based approaches to development have faltered, the ideal of the rural community pervades much development thinking and practice. Interventions in the post-Independence era replicated many of those from the colonial period. During the mid-1980s

---

[5] A review of over 300 studies from the southern African region have shown how there is no consistent benefit of rotational over continuous grazing (Gammon, 1978; O'Connor, 1985b).

attempts were made to create a decentralized system of local government based on ward and village committees. Villages and wards were designated according to population estimates, with around 100 households making up a village and around six villages making up a ward. Boundaries were then demarcated on official maps, often with scant regard for what already existed on the ground. Indeed, the new wards often did not coincide with those created during the colonial era. Finally, posts for councillors and committee members were created in an attempt to impose a systematic and ordered administrative hierarchy. Administrative expediency and politically induced haste resulted in widespread confusion and an administrative and political nightmare in many areas that has taken the last decade to unravel[6] (Mutizwa-Mangiza, 1991).

Grazing schemes in the post-Independence era have largely made use of this administrative structure, with scheme committees being drawn from the village and ward levels. Grazing schemes have involved the definition of a scheme boundary, often marked with fencing; the adoption of a series of management recommendations, including rotational grazing, bush clearance and controlled breeding; the election of a scheme committee; and the adoption and implementation of by-laws (Cousins, 1992; 1993). Grazing schemes can thus be seen as an attempt to fulfil, at least notionally, the requisite conditions for effective common-property management (cf. Oakerson, 1986; Lawry, 1990). Joint use of a valued resource is initiated; boundaries are defined; group membership is established; resource-management organizations are formed; and use and access regulations are defined and enforced. In other words, grazing schemes have been an attempt to transform a 'minimum' common-property regime, based on flexible and informal arrangements, into a 'fully fledged common-property regime' (Cousins, 1993).

How successful has this been? The verdict is mixed. While there have been some undoubted successes, where groups have come together to defend their (usually relatively well endowed) grazing area from encroachment by others, as well as investing in the improvement of the resource itself, there have also been notable failures, or at least outcomes that were not predicted, resulting in conflict of various sorts. Three types of conflict can be highlighted: conflicts over technical design, conflicts arising between scheme members and conflicts between groups, especially over boundaries.

First, the imposition of a uniform design of rotational grazing on a highly heterogeneous landscape has highlighted the inadequacies of the technical design. A more flexible approach, more attuned to local ecological complexity and in line with existing herding management systems, has been suggested as an alternative (Table 9.3).

Second, conflicts between scheme members have arisen when the project has been used as an opportunity for the capture of resources by particular indivi-

---

[6] In Mototi ward in Mazvihwa communal land, six villages were marked on the official map kept by the Topographic Section of Agritex in Gweru. However, in practice eight villages were in existence by 1986, each with their own committee and their own set of perceptions about where boundaries should be.

Table 9.3. A comparison of grazing-scheme design (Zimbabwe)

Aspects of design	Multi-paddocked, fenced grazing schemes	Key-resource grazing schemes
Stocking rates	Low	High
Objectives	Increased beef production; reduced draught availability	Maintain multi-function herd
Costs	Fences	Herding
Ecological issues	Rotational resting improves grass in extensive grazing areas	Concentration on most degradation-susceptible and valuable areas
Planning issues	Focused on scheme boundaries and fence lines	Focused on identification of key-resource areas

(From Scoones, 1989.)

duals or sub-groups. For instance, Ben Cousins reports on the experience of a grazing scheme in Mhondoro communal area, where the effective manipulation of the project by local élites allowed the capture of resources for private gain in the form of a paddocked-pen fattening scheme. Cousins observes:

> The 'grazing scheme' in Chamatamba denoted much more than a project to manage grass and livestock; it was at the centre of a carefully nurtured image, or representation, of a self-reliant and dynamic 'resource managing community'. This image was being used as a vehicle for the establishment of purely private economic ventures undertaken by the Chamatamba élite. (Cousins, 1993: 30).

In such cases, the ideals of community cohesion leading to collective action are shattered, and the reality of resource-management as a process involving intense contest between different interest groups, with different power and authority, is revealed.

Third, attempts to define strict boundaries and limit group membership have resulted in disputes, particularly when scheme boundaries exclude others from key grazing resources (see Box 9.1).

In these cases, disputes arose over highly valued resources and the scheme plan broke down, with overlapping, multiple and shifting rights being reasserted. The result was that the fences for this grazing scheme were erected in areas that people did not really care about, across areas which were of limited value for grazing and not subject to overlapping claims and dispute. The result was that the scheme fulfilled a limited function, that of reducing herding by providing some holding paddocks for animals to be placed when children were at school. The scheme therefore failed to become a full common-property regime in that exclusion from the most valued resources could not be effected.

The experience of grazing-scheme intervention in Zimbabwe shows how in most cases only a 'minimum' form of common-property management emerges; instead of a strong, exclusive management system based on a single institutional set-up with a single set of rules, a more complex situation arises. The new grazing

Box 9.1 Boundary disputes and resource patterns: the case of Indava Ward grazing scheme
Sources: Field notes: July 1987; April 1988; March 1990; February 1991

institutions – the committee, by-laws and regulations – became part of the complex of existing institutions, sometimes negotiating a prominent space in resource-management issues, at other times taking a much lesser role and getting subsumed in existing institutional and organizational forms. Despite the rhetoric and paraphernalia of community-development projects, most schemes in practice are only marginal adaptations of existing patterns of rights over land. Although the scheme intervention may have afforded the opportunity for shifts in the degree of control by different actors, the same principles apply. What is found is a system of land rights which remain diffuse, ambiguous, overlapping and very often disputed (Cousins, 1992, 1993; Murombedzi, 1992; Scoones & Wilson, 1989).

The more general question arises: is this type of complex tenure the inevitable result of spatially heterogeneous and temporally variable resources being managed by groups with diverse membership through social networks influenced by differential power, authority and status? The answer appears to be, 'yes'. This suggests a number of important practical implications for approaches to external intervention in grazing management. These are discussed in the concluding section.

Conclusions

What are the appropriate resource-management interventions for highly variable grazing resources, such as are found in many of the dryer communal areas of Zimbabwe and elsewhere across pastoral Africa? How can opportunistic use and management, particularly of key resources in highly heterogeneous landscapes, be supported? What role, if any, is there for external intervention?

The experience from Zimbabwe suggests some preliminary answers to these questions. The problems arising from the implementation of grazing schemes have pointed towards some fundamental design and management issues. As Ben Cousins notes (1992: 145):

> Grazing schemes and opportunism are not necessarily antithetical. The central issue is rather the design of management systems and institutional regimes which have a better fit with the objectives of livestock herds and herd-owners on communal rangelands in Africa.

What elements of design and management must be given priority? A vital first step must be the description of resource types within the grazing landscape according to the criteria and at a scale of relevance to local herders. This would result in the identification of key grazing resources – those patches of critical value to livestock at different times of year – for different types of livestock. Mobility mapping would also focus on patterns of use during extreme events, such as droughts, and take a longer time perspective and wider geographical scale for looking at resource use. Exploring institutional forms involved in regulating the use of different landscape resource-components would be a parallel exercise[7]. A focus on perceptions of boundaries and histories of disputes would again highlight key resource sites. A deeper understanding of the roles of different actors and the composition of social networks involved in resource-management could then be developed through the mapping of connections between actors involved in different resource disputes. Such an analysis would reveal the nature of local institutional forms, and if carried out with different groups, the degree to which they are widely accepted or contested. Such a participatory analysis would ideally emerge as part of a co-learning process, whereby different actors – herders, researchers, extensionists and others – would

---

[7] A range of participatory rural appraisal techniques are potentially useful (Chambers, 1992; Waters-Bayer & Bayer, 1995). For instance, resource mapping and aerial photograph analysis with herders may help in the identification of key resources and perceptions about resource boundaries; network diagramming with different interest groups may help in identifying important networks of actors, key organisations and influential individuals; and role playing of dispute situations may reveal the range of institutions and organizations involved in regulating resource access, and their effective-ness (Sithole, 1995).

explore options together. But a co-learning process is not without its pitfalls. Each actor involved in such analysis and planning exists as part of their own social networks; knowledge thus emerges in the context of often unequal power relations between different actors (Long & Long, 1992; Scoones & Thompson, 1994).

With limited resources for external intervention and the need to keep administrative intervention to a minimum, a concentration on key resources would be justified. Such areas are at the centre of the production system and technical interventions there would have effects far beyond. Thus, productivity increases in areas which provide feed at critical times of year may have major impacts, whereas returns on investment in resources with low or erratic productivity may be very low. As key-resource areas are usually the centre of dispute and are the areas where most management effort is spent, they also provide the obvious focus for management interventions. Due to the complex and negotiated nature of local arrangements, external intervention is unlikely to succeed unless focused on issues where external support might benefit local mediation of resource rights. However, the essential political nature of external intervention, whether through the state extension agency or an NGO project, must also be recognized (Long & van der Ploeg, 1989). Negotiation over and capture of development gains by certain individuals or groups is a real possibility, as the experience of grazing-scheme implementation in Zimbabwe shows.

One important area might be in the area of legal provision, not so much in the form of prescriptive land law or lists of regulations enforced through by-laws, but in the form of procedural law. Procedural law would specify the framework within which different groups could put forward claims to different resources, provide facilities for mediation and arbitration of disputes and would enforce decisions made (Vedeld, 1993a; Behnke, 1994)[8].

Rather than trying to intervene everywhere all the time, the role of external intervention would thus be limited to support for focused technical interventions at key resource sites and the provision of a broader framework for dispute mediation and arbitration which would be important only at certain times. While recognizing the political context and the potential for capture by certain interest groups, it appears that some form of external intervention may be required in such settings. For instance, a co-management arrangement (cf. Lawry, 1990) between the State and a whole complex of local institutions could be envisaged – where most activities are managed and regulated at the local level, according to locally specific, often ambiguous and certainly overlapping rights to different resources, but with certain functions retained by the State in the domain of limited technical support and the provision of a conflict-negotiation framework.

---

[8] Procedural law dealing with the type of disputes described in Box 9.1 would likely focus on the mediation of conflicts over key resource sites. Such disputes inevitably escalate during times of resource scarcity and drought and so would require the establishment of a forum for discussion during such times. In the Zimbabwe setting this could be facilitated by rural local-government structures at village, ward and district levels. However, recent experience of local government interventions in resource-management issues has not been good, and such fora would have to gain legitimacy from local people.

10

Gestion de la mobilité et résistance des organisations pastorales des éleveurs du Haut Atlas marocain face aux transformations du contexte pastoral maghrébin

Mangement of mobility and the persistence of pastoral organizations of the Moroccan High Atlas in the face of transformations of the pastoral context in the Maghreb

ALAIN BOURBOUZE

RÉSUMÉ

Comparé au reste des pays méditerranéens, le pastoralisme maghrébin reste fondamentalement marqué par la mobilité des troupeaux et des hommes d'une part, et par la persistance de vastes territoires à usage collectif d'autre part. En intégrant différents facteurs tels que mode de vie pastoral (semi-nomadisme, transhumance, sédentarisation) et types de déplacements (horizontaux ou verticaux), il est possible d'identifier une dizaine de systèmes différents de l'utilisation de l'espace pastoral. Mais de nombreux évènements concourent depuis les années soixante à transformer ces modes de vie. Les changements de statut foncier remettent en cause les bases du système collectif pour promouvoir partout où c'est possible l'exploitation individuelle. Les déplacements se simplifient et diminuent d'intensité, mais surtout, la motorisation (ailleurs que dans les montagnes) fait naître un modèle différent d'utilisation de l'espace qui profite surtout à une classe d'individus puissants qui ne voient dans les règles coutumières qu'un frein à leur expansion et exercent une pression très forte sur les ressources. Tous ces phénomènes concourent à faire disparaitre les institutions traditionnelles. Elles résistent pourtant dans certaines régions isolées comme le Haut-Atlas marocain.

Deux modèles d'organisation pastorale s'y rencontrent. Le premier, plus officiel, issu d'un long processus engagé du temps de la colonisation, est né des arbitrages menés par les autorités et les marabouts locaux. Bien que très imparfait, il représente le produit métissé d'une gestion coutumière et d'une volonté politique pour en fixer les règles sur le papier. La règlementation, consignée dans une charte, comprend un certain nombre de règles inspirées directement des pratiques locales. Le second modèle d'organisation est celui qui s'appuie sur les institutions coutumières, plus discrètes, et donc moins connues des autorités. La qualité essentielle de ces institutions est la simplicité qui fait reposer l'organisation sur quelques principes; des territoires délimités et des ayants-droits identifiés, des restrictions et non des interdictions sur les droits de construire des abris, de mettre en culture, de prendre des animaux en association, des droits d'abreuvement. L'institution de l'*Agdal*, ou mise en défens saisonnière à des dates fixées par la coutume mais négociables chaque début d'année, est la formule la plus simple qu'un technicien puisse proposer. . . mais la seule acceptable en l'absence de contrôle des effectifs. La richesse des détails d'une gestion quotidienne, les règles de bienséance entre bergers, la tolérance sur le franchissement des limites – doublée d'une grande intransigeance sur les droits formels – introduisent une grande souplesse de fonctionnement. Les institutions de contrôle sont d'une grande légèreté et diverses: grande ou petite assemblée des chefs de famille (*Jemaa*), avec ou sans délégué, le plus souvent sans recours aux autorités officielles (le *Caïd*) en dehors des conflits les plus graves.

Mais ces organisations coutumières sont fragiles. Aucune limitation d'effectif n'est appliquée, les prises d'animaux en association se révèlent de moins en moins contrôlables, la multiplication des abris qui préparent la privatisation d'une partie des parcours n'est plus maitrisée par la *Jemaa* que les puissants manipulent. C'est donc un système très peu égalitaire, sans esprit coopératif au sens moderne du terme, car le principe de gestion n'est plus la mise en valeur en commun mais le contrôle de la concurrence.

On peut cependant retenir de ces modes de gestion traditionnelle quelques principes pour un développement futur tels que:

o l'identification des territoires et de ses usagers, donc des études socio-foncières fines qui enregistrent les pratiques réelles

o des restrictions sur les privatisations occultes avec une intervention plus déterminée de la part des autorités

o un meilleur contrôle de l'accès aux ressources en étendant l'institution de l'*Agdal*

o des institutions au niveau local mieux reconnues et plus officielles.

ABSTRACT

Compared with other Mediterranean countries, pastoralism in the Maghreb is still characterized by mobility of herds and people, and vast areas under common-property use. It is possible to distinguish more than a dozen different systems of pastoral land-use, depending on the degree of mobility (semi-nomadism, transhumance, sedentarization) and the type of movement (horizontal or vertical). However, since the 1960s, livelihood systems have changed due to numerous factors. Changes in land-tenure laws have encouraged individualized exploitation of collective resources. Mobility has diminished and simplified. A minority of powerful individuals move their large herds by trucks over long distances in the steppes and plains, imposing strong pressure on fragile resources. Customary regulations are a constraint to these individuals' expansion. All these factors contribute to a weakening of traditional institutions. However, in certain isolated regions, such as the High Atlas of Morocco, customary institutions continue to function.

Two models of pastoral organization can be found in the High Atlas. The first originated during the colonial period after negotiations between the authorities and local *Marabouts*. Although far from perfect, this 'Transhumance Charter' represents a compromise between the needs of customary management, and that of the authorities to document and formalize regulations. The second model derives from customary institutions, is less formal, more discrete, and therefore less known by the authorities. The essential quality of these customary institutions is their simplicity, based on a few principles: delimited territories, clearly identified membership, and restrictions rather than prohibitions on land- and resource-use, such as on construction of animal shelters, on cultivation, on watering, and on joint-herding. The traditional institution of *agdal* is still practised, where certain pastures are seasonally closed and opening dates are negotiated at the beginning of the season. This practice, based on timing of grazing, is also the most simple technique that government technicians can propose in the absence of ways to control stocking-rates.

The details of day-to-day management of pastures, including informal rules of decency and fairness among shepherds, are extremely rich. The tolerance of a certain degree of infringement of informal rules is coupled with intransigence on formal rights. These customary principles provide a high degree of flexibility to the management of the pastoral system and collective property. Customary institutions are very 'light' and diverse. These include a council of chiefs of the lineages (*Jmaa*) which, depending on the need, can be a large or small assembly, with or without the presence of the elected representative to the provincial government. In the case of serious conflicts, this council can have access to the official judicial and administrative authority (*Kaid*).

But these customary institutions are delicate and easily destroyed. In recent years, they have not been able to control the practice of joint-herding (where outsiders entrust animals to members of the collective, thus increasing grazing pressure), nor the gradual appropriation of collective land by those who settle. The *Jmaa* has proven to be easily manipulated by powerful individuals. The current system is not very egalitarian and lacks a cooperative spirit (in the modern sense of the term), because the basis is no longer common-property management but the control of individual competition.

Nevertheless, certain principles can be distilled from these traditional forms of management for the design of new models:

o clarification of current territorial and use-rights, through in-depth socio-geographic land tenure studies
o greater restrictions on surreptitious privatization of collective land, particularly through more determined government intervention
o improved control over access to resources, by expanding the institution of *agdals*; and
o better official recognition and understanding of local institutions.

Introduction

PAR LE PASSÉ, les sociétés rurales du nord de l'Afrique ont toujours su mettre en oeuvre des formes de gestion communautaires de l'espace pastoral, de l'eau d'irrigation ou des terres cultivables que des générations de paysans se sont transmises oralement et qui, pour certaines, furent transcrites dans des coutumiers (*orf*, *izreg*). Les témoignages de ces organisations traditionnelles sont multiples pour qui prend la peine de les découvrir.

Mais le veut-on vraiment? Cet héritage du passé est trop souvent tenu pour négligeable puisqu'il est tiré d'une histoire précoloniale ou coloniale dont on n'attend *a priori* rien de bon. Il est vrai que ces modes traditionnels de gestion ont été fortement désorganisés sous les effets d'un puissant essor démographique et surtout d'interventions massives des États qui ont imposé leur vision verticale du développement agricole.

Ces nouveaux modes de gestion décidés d'en haut ont sans conteste transformé les paysages agricoles, notamment dans les régions les plus riches et sur les terres irrigables. Ils ont eu un impact considérable sur les modèles de production et sur les pratiques agricoles, intégrant sans trop de difficulté les formules modernes d'organisation (coopératives, groupements de producteurs, syndicats d'usagers). Par contre, dans les régions marginales où prédominent les systèmes pastoraux et agro-pastoraux fondés sur l'utilisation de ressources en partie collectives, la plupart des projets de développement initiés par l'État se sont heurtés à l'impossibilité de faire émerger des institutions spécialisées ou des structures socio-politiques locales nouvelles, capables de gérer ces ressources sur un mode suffisamment flexible qui puisse se plier aux contraintes particulières de ce milieu. Il y a donc un besoin impérieux de mettre en place des institutions adaptées au contexte écologique, économique et social.

Il se trouve que dans certaines régions reculées où les terres de statut collectif appartenant à des collectivités ethniques occupent de vastes superficies, certaines de ces institutions pastorales ont survécu et montrent des formes résiduelles d'une organisation cohérente. C'est notamment le cas au sein des systèmes pastoraux du Haut Atlas marocain qui présentent des exemples reliques de ces organisations pastorales, portant témoignage d'un passé récent d'une gestion traditionnelle qui semblait montrer une certaine efficacité.

Notre démarche s'articulera par conséquent sur une présentation générale des principaux modes d'utilisation des espaces pastoraux maghrébins et l'analyse des transformations profondes qui les perturbent, pour ensuite replacer dans ce cadre l'exemple très particulier des organisations pastorales coutumières du Haut Atlas. Nous tenterons de démontrer comment et pourquoi de tels systèmes traditionnels de gestion ont pu survivre et comment ils se sont adaptés à ces changements. La connaissance de ces mécanismes traditionnels, même résiduels, devrait ainsi offrir la possibilité d'imaginer de nouvelles formules et de mettre en relief les quelques principes clefs qui participent à la réussite d'une bonne gestion des ressources naturelles par une communauté. Sous certaines conditions, de telles règles seraient-elles susceptibles d'être adaptées à différents contextes, en Afrique du Nord ou ailleurs?

Le contexte maghrébin et les transformations des systèmes pastoraux

Milieu physique et ressources naturelles au nord de l'Afrique

En dépit de la diversité de leurs conditions naturelles, Maroc, Algérie et Tunisie présentent bien des similitudes. Le climat est partout de type méditerranéen, caractérisé par des manifestations météorologiques très contrastées:

- concentration des précipitations sur quelques jours, voire quelques heures, le plus souvent en périodes fraîches, d'octobre à avril, quelques orages d'été en montagne;
- pluviosité ne dépassant que rarement 400mm sur la plus grande partie des territoires (Carte 10.1);
- coefficient de variabilité des précipitations de l'ordre de 30–40 pour cent, c'est à dire que le maximum égale quatre à six fois le minimum;
- températures dépendant de la latitude, de l'altitude et de la continentalité avec des moyennes des minima de janvier de 9–10°C sur le littoral atlantique, 7–8°C sur le littoral méditerranéen et de 0°C dans l'intérieur au-dessus de 1200m d'altitude. La moyenne des maxima de juillet est de 27–30°C sur les côtes et de 30–35°C dans l'intérieur (Figure 10.1).

Hormis sur certains reliefs montagneux, les formations géologiques comprennent essentiellement des affleurements sédimentaires du secondaire au quaternaire. Les zonalités climatiques et géologiques induisent ainsi des types de végétation et de sols comparables d'un pays à l'autre (Le Houérou, 1995).

Pour s'en tenir aux seules zones marginales qui portent actuellement la majeure partie des terres de parcours, la végétation est de caractère essentiellement steppique sur les plaines des bioclimats arides et désertiques et un peu plus diversifiée en montagne. Cette végétation qualifiée de steppique se caractérise par l'importance des espèces vivaces, ligneuses ou graminées, couvrant 10–80 pour cent de la surface du sol, haute de 10–50cm, avec un développement très variable des espèces annuelles liées aux pluies.

Dans les zones à hiver rigoureux des régions montagnardes de l'Atlas maghrébin, la végétation steppique s'étend naturellement au climat semi-aride sous une forme arborée ou arbustive. Les montagnes portent ainsi deux grands types de végétation: l'un, de haute altitude essentiellement asylvatique, est

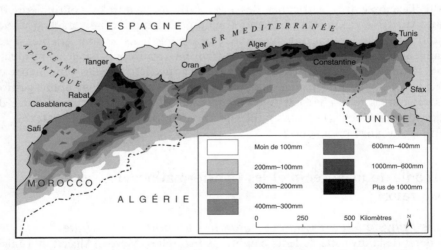

Carte 10.1 *Répartition des précipitations annuelles en Afrique du Nord* (Source: *Despoix & Raynal, 1975*)

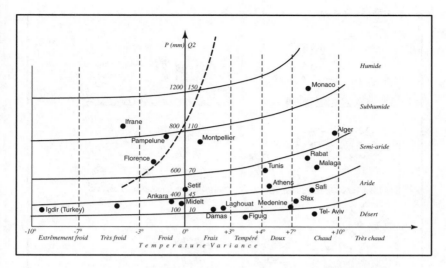

Figure 10.1 *Climagramme schématisé d'Emberger pour la zone méditerranéenne*
(Source: *Quezel & Barbero, 1982*)

composé de xérophytes épineuses sans grande valeur et de quelques plages de pelouses qui offrent un meilleur aspect; l'autre est composé d'un 'matorral' arboré ou de forêts (*Quercus ilex, Juniperus, Fraxinus dimorphis, Pinus alepensis*) installés sur les pentes basses des bassins versants.

Depuis quelques dizaines d'années, cette végétation pastorale est soumise à une pression de plus en plus forte de la part des populations usagères et bien des pastoralistes portent un jugement sévère sur la régression de l'état des ressources et la non-reproductibilité du modèle de production pastoral. Qu'en est-il dans les faits?

Les principaux modes d'utilisation de ces espaces

On sait que les élevages sur parcours se partagent en trois grands types, traduisant des modes de vie et des systèmes techniques bien différents, l'élevage semi-nomade, l'élevage transhumant et l'élevage sédentaire. D'autres critères peuvent enrichir cette typologie succincte, tels que déplacements horizontaux ou verticaux, type d'itinéraires, amplitude du mouvement, types d'animaux exploités, place de l'agriculture, modes de commercialisation, etc.

Pour nous en tenir à des généralités à l'échelle du Maghreb, il est possible d'identifier une dizaine de types (ou modes) d'utilisation des espaces. Le Figure 10.2 Carte Map 10.2 tentent d'en résumer les principaux traits. Le semi-nomadisme, qui implique le déplacement de toute de la famille (ou d'une grande partie) sur des distances en général importantes (plus de 100km), résiste dans sa forme traditionnelle dans les régions les plus austères de très faible productivité et à l'écart des courants de l'économie. Au sein de ces communautés ce sont

241

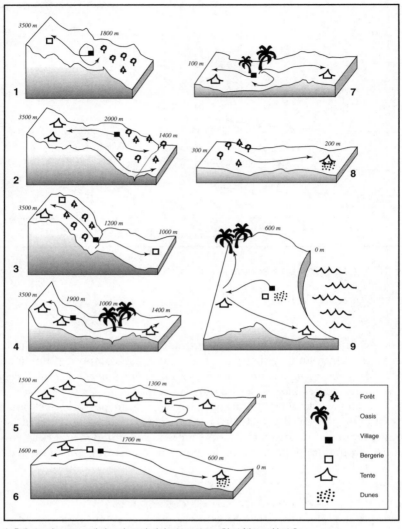

1. Petite transhumance estivale en bergerie de haute montagne (Haut Atlas occidental).
2. Grande transhumance estivale ou transhumance double (estivale et hivernale) ou semi-nomadisme de montagne, sous la tente (Haut Atlas Central et Oriental).
3. Transhumance estivale ou transhumance double en moyenne montagne, sous la tente et en bergerie (Moyen Atlas).
4. Semi-nomades des régions arides, l'hiver sur les plateaux sahariens, l'été en montagne, sous la tente (versant saharien de l'Atlas).
5. Semi-nomades des steppes, l'hiver sur les plateaux alfatiers sahariens, printemps et été sur les parcours à Armoise au Nord, sous la tente (steppes de l'Oriental marocain).
6. Semi-nomades des steppes algériennes, l'hiver en steppe, l'été sur les chaumes des plaines céréalières (Achaba), sous la tente.
7. Semi-nomades, éleveurs de dromadaires, des régions arides et désertiques, l'hiver au Sahara, l'été près des oasis (régions sahariennes).
8. Eleveurs en voie de fixation, transhumance estivale exceptionnelle vers les plaines céréalières du nord (Friga), sous tente (zones arides et semi-arides du Sud tunisien).
9. Bergers semi-nomades gestionnaires de troupeaux collectifs, l'hiver au désert et l'été près des oases et des zones de cultures.

Figure 10.2 *Diagramme sur la mobilité des troupeaux et les modes d'utilisation des espaces (Maroc)*

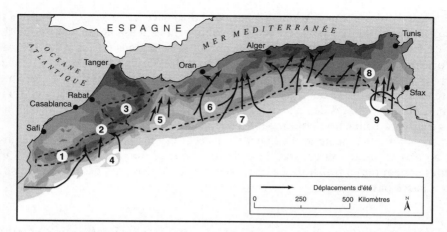

Carte 10.2 *Mobilité des troupeaux et grands modes d'utilisation des espaces en Afrique du Nord*
(Source: Bouirbouze, 1998)

plutôt des éleveurs moyens et les bergers des riches propriétaires qui le pratiquent encore.

En montagne, et notamment dans l'Atlas marocain, les transhumances verticales de type simple (estivale) ou double (estivale et hivernale) restent très actives. Dans les steppes, les amples mouvements horizontaux des transhumants qui suivent un *transect* nord–sud, hiver au Sahara et été sur les parcours et cultures des plaines du Nord, résistent au Maroc, régressent en Algérie et disparaissent en Tunisie.

Quant à l'élevage sédentaire sur parcours, il est présent partout, *sédentaire* signifiant ici que les troupeaux se déplacent, souvent sur de longues distances, mais qu'ils reviennent chaque soir au village. Ce mode est sans doute plus représenté dans les systèmes agro-pastoraux que pastoraux, mais il est banal de rencontrer côte à côte dans un même village des éleveurs sédentaires, transhumants et semi-nomades, ayants droit d'un même territoire, mais l'utilisant chacun selon ses possibilités de déplacement. L'élevage sédentaire est donc une formule technique toujours présente, notamment pour les petits troupeaux, quelle que soit la difficulté du milieu. On peut s'en étonner, mais en milieu méditerranéen, le pâturage est, en tout lieu, possible toute l'année au prix de quelques apports complémentaires tirés de la forêt et d'achats d'aliments de sauvegarde (orge et paille) pour passer les pires moments.

Mais ces classifications sont de moins en moins pertinentes, car ces sociétés pastorales sont maintenant agitées par de puissants changements qui les marquent profondément.

Transformation des systèmes pastoraux et mobilité des éleveurs dans les régions pastorales du Maghreb

Comparé au reste des pays méditerranéens, notamment de la rive nord, le pastoralisme maghrébin reste fondamentalement marqué par la mobilité des

243

troupeaux et des hommes d'une part et par la persistance de vastes territoires à usage collectif d'autre part. La tente, auxiliaire indispensable du semi-nomade ou du transhumant, survit dans de très nombreuses régions (Haut Atlas central et oriental, pays Zemmour et Zaer, et steppes de l'Oriental au Maroc, hautes steppes et régions désertiques en Algérie, régions arides tunisiennes). Et lorsque la tente a été remisée, ou dans les régions de vieille sédentarisation où elle n'a jamais existé, les longs déplacements n'en restent pas moins indispensables pour la survie des troupeaux, notamment des plus grands.

Cependant, en à peine un peu plus d'un siècle, et plus particulièrement depuis les années soixante, on constate que de nombreuses forces sont à l'oeuvre qui toutes concourent à transformer les modes de vie et les modes de production sur ces espaces pastoraux.

Les changements de statut foncier sur l'espace pâturé   L'intégration du Maghreb dans l'empire colonial français (colonisation de l'Algérie en 1845, protectorat sur la Tunisie en 1881, puis sur le Maroc entre 1906 et 1937), a entraîné dans chacun des trois pays un processus de basculement économique lié en particulier à la mise en place d'une politique foncière qui s'est appliquée à redéfinir les espaces agricoles y compris dans les zones les plus marginales afin d'y installer les colons (Abaab et al., 1995; Boukhobze, 1976). Ces législations qui avaient surtout pour objet de faciliter aux colons l'accès aux terres collectives vont marquer d'une empreinte très forte les paysages de ces trois pays. La colonisation des grandes plaines va réduire les complémentarités qui existaient entre régions céréalières et régions steppiques et freiner les déplacements saisonniers des troupeaux. L'immatriculation des terres, le partage de certains collectifs, la fixation des limites des grands territoires tribaux vont engager un processus irrésistible de sédentarisation des éleveurs et leur mutation en éleveurs/ agriculteurs (ou agro-pasteurs). A l'avènement des Indépendances, une nouvelle ère s'ouvrait pour assurer un développement plus harmonieux de la société et mieux intégrer les régions marginales dans l'économie nationale. De nouvelles politiques foncières furent alors menées à des rythmes différents selon la législation en vigueur dans chaque pays.

Au Maroc, où le statu quo persiste sur les terres collectives, on observe partout dans le paysage des mises en culture dans les zones les plus difficiles, opérées par les éleveurs eux-mêmes au détriment des parcours et des forêts. Les stratégies d'appropriation sont multiples: dans les montagnes du Haut Atlas central, les éleveurs commencent par installer des bergeries en dur, puis cultivent en céréales quelques parcelles alentour sur le collectif en se réclamant de la coutume, soutenus le cas échéant par quelques témoins qui garantissent l'ancienneté de l'installation.

Dans les steppes à Alfa de l'Oriental, l'utilisation du camion citerne et du 'pick-up', qui permet aux éleveurs ayant des moyens de séjourner sur des sites éloignés où les petits ne peuvent rester, entérine l'individualisation des droits d'usage; ces grands propriétaires (plus de 500 brebis) pratiquent des défrichements savamment dispersés qui bornent les limites d'un territoire pastoral qu'ils finissent par annexer. Les cultures – on sème même les années sèches sans grand espoir de récoltes – ne servent ici qu'à affirmer le droit exclusif au parcours.

C'est donc dans le sein même de la société pastorale que naissent les compétitions sur l'espace entre éleveurs et éleveurs convertis à l'agriculture. Il n'y a pas ici conflit entre deux communautés, l'une d'éleveurs et l'autre d'agriculteurs, mais plutôt émergence au coeur d'un même groupe de stratégies divergentes, qui s'expriment le plus souvent par des oppositions entre grands et petits.

Mais ce n'est pas toujours le cas. Des acheteurs extérieurs mobilisant des capitaux urbains peuvent aussi accéder à la propriété. Ainsi, sur les steppes algériennes, la loi portant sur 'l'accès à la propriété foncière agricole' (APFA) ouvre des possibilités d'investissement sur les *terres arch*[1], mises à profit par de nombreux détenteurs de capitaux totalement étrangers à la steppe. Mais en général, les stratégies d'appropriation sont à peu près les mêmes qu'au Maroc et s'opèrent sans réel contrôle.

Il en va très différemment en Tunisie où les nouvelles procédures administratives ont considérablement accéléré la privatisation des collectifs qui s'accompagne non seulement d'une mise en culture (creusement de puits, plantations d'oliviers), mais aussi d'un afflux de transactions foncières (Abaab et al., 1995). Dans la plaine de la Jeffara par exemple, marquée jusque dans les années soixante par un système agro-pastoral steppique sous une pluviométrie de moins de 200mm, la privatisation s'est appuyée sur des programmes de mise en valeur conduisant à une réduction des meilleurs parcours au profit de plantations d'oliviers et d'une diversification accélérée de l'activité économique soutenue par les revenus de l'émigration.

La mutation est encore plus totale quand se restructurent autour de petits pompages des unités de production plus intensives comme dans les steppes de Gafsa ou de Sidi Bouzid. Plus de 7000 puits y ont été creusés en quinze ans provoquant un inquiétant rabattement de la nappe de plus de 20m et poussant l'administration à mieux contrôler ce développement. A Gafsa, sitôt les opérations de lotissement terminées, l'éleveur, nanti de son titre de propriété (le 'certificat de possession'), a vendu en général les trois quarts de son troupeau pour financer le puits, la pompe, le matériel agricole de base et les plants d'arbres fruitiers. Il a alors réorganisé son exploitation autour d'un petit périmètre irrigué (un à deux hectares) avec une arboriculture semi-irriguée (pistachiers, oliviers) de la céréaliculture en sec . . . et un élevage ovin progressivement reconstitué à partir de la race algérienne à queue fine plus exigeante (c'est-à-dire plus agricole) que la Barbarine à grosse queue.

Du fait de cette privatisation accélérée, de nombreux éleveurs qui pratiquaient encore dans les années soixante-dix la vaine pâture sur chaumes dans les terroirs cultivés des villages, se voient maintenant contraints de passer des accords de gré à gré avec des propriétaires privés au travers de locations coûteuses.

---

[1] *Terres arch*: terres anciennement collectives de statut à présent domanial depuis la révolution agraire, mais qui restent fortement revendiquées par les ayants droit d'origine.

Un bilan chiffré de ce mouvement d'appropriation des terres collectives montre ainsi des différences considérables d'un pays à l'autre:

o Au Maroc, sur 10 millions d'hectares de terres collectives, seuls 3,5 millions sont immatriculés dont 1 million cultivé à la suite de partages officialisés sur ces trente dernières années. Mais ce relatif statu quo cache un puissant mouvement souterrain de privatisations occultes qu'on peut évaluer à plus de 1 million d'hectares auxquels s'ajoutent les nombreux défrichements et mises en culture en forêt domaniale. La pression semble s'exercer avec une même intensité tant en montagne (au-dessous de 2300m) que dans les steppes (au-dessus de l'isohyète 200mm).

o En Algérie, les opérations d'APFA ont permis l'attribution de près de 100000 ha dont 10000 seulement sont mis en valeur. Par contre, entre 1970 et 1994, les terres cultivées et les parcours dans la steppe sont passées respectivement de 1,1 à 2,4–2,9 millions d'hectares et de 14,3 à 12,8–13,3 millions d'hectares sous la pression de défrichements illégaux.

o En Tunisie, les terres collectives occupaient à l'Indépendance (1956) dans la partie sud du pays 3 millions d'hectares. Actuellement, la moitié est en passe d'être attribuée à titre individuel (1,2 sur 1,5 millions d'hectares attribuables), l'autre moitié devant être soumise au régime forestier malgré l'hostilité déclarée des populations usagères. Les partages s'opèrent même dans les régions subdésertiques entre les isohyètes 100 et 150mm.

Une même dynamique est donc à l'oeuvre qui remet en cause les bases du système pastoral collectif pour promouvoir chaque fois que possible l'exploitation individuelle.

Régression de la mobilité et redécoupage de l'espace pastoral Du fait de ces profondes modifications portant sur les statuts des parcours, les modes d'occupation de l'espace et les déplacements des éleveurs ont changé de nature. Chez les semi-nomades dont le nombre est incontestablement en régression, la motorisation a fait son apparition. Partout au Maghreb, là où les pistes sont carrossables, la camionnette (le pick-up) rend des services inestimables et modifie profondément les façons de faire: l'eau et les concentrés viennent maintenant vers les troupeaux et non l'inverse, les ventes s'organisent plus souplement, les déplacements se décident plus vite et l'on va éventuellement plus loin. Mais ce sont les gros troupeaux individuels, amenés par camions, qui conquièrent l'espace au détriment des élevages moyens. Les steppes, plus lourdement et plus complètement exploitées que par le passé, marquent des signes évidents de surpâturage. Dans l'ensemble, les déplacements se simplifient et l'on démonte moins souvent la grande tente.

Dans le Sud tunisien, les années de sécheresse déclenchent encore le départ de gros troupeaux vers les terres céréalières du Nord après la récolte, déplacement nommé *Friga*. Mais le plus souvent, les troupeaux sont dans ces circonstances immobilisés à proximité des habitations et nourris 'à coups' de concentrés fournis à bas prix par l'État et payés par la vente progressive d'une partie des animaux quand la sécheresse se prolonge.

En Algérie, *l'achaba*, qui n'est autre que la *Friga* tunisienne, reste très pratiquée par les éleveurs des steppes et elle intéresse encore plusieurs millions de brebis. Elle régresse cependant depuis le partage des domaines autogérés en exploitations agricoles privées qui pratiquent maintenant des tarifs de location de chaumes ou de jachères moins avantageux, poussant les éleveurs à recourir de plus en plus à des achats de compléments qu'ils font venir par pleins camions du Nord. Le transport des fourrages remplace le transport des moutons.

Dans les régions montagneuses, les transhumances doubles se sont simplifiées par disparition de la séquence hivernale au cours de laquelle les troupeaux descendaient dans les terres basses maintenant cultivées. Il en va ainsi des transhumances du Moyen Atlas marocain où, après que les pactes avaient été rompus en 1970 sur l'initiative des grands agriculteurs de la plaine, les éleveurs ont pris l'habitude de monter plus tôt sur les parcours collectifs d'altitude, de redescendre plus tard en hiver dans les forêts domaniales ou sur les rares parcours privés . . . et se sont mis à construire des bergeries et produire ou acheter du foin de vesce-avoine pour abriter et nourrir les animaux une partie de l'hiver. De l'organisation pastorale ancienne, il ne reste que quelques règles que l'administration de tutelle (le Ministère de l'Intérieur) s'efforce de faire respecter sur les parcours collectifs d'altitude: tracés des limites des parcs pastoraux affectés à chaque communauté, identification des ayants droit (mais les contestations sont de plus en plus vives à propos des 'étrangers', même installés depuis plus de vingt ans), interdiction de construction de bergeries et de défrichement pour mises en culture, contrôle des associations.

Il faut donc souligner ces deux idées: il y a bien régression de la mobilité des troupeaux, mais parallèlement, la nature même de ces déplacements évolue et s'adapte à ce nouveau découpage de l'espace. La motorisation, ailleurs que dans les montagnes où les pistes sont rares, est en train de faire naître un modèle différent de l'utilisation de l'espace: concentration de l'élevage au profit de grands éleveurs, recours à des bergers salariés, exploitation systématique de toutes les ressources, transport d'eau et d'aliments, émergence d'un marché de l'herbe qui concerne tout le territoire national.

Dans les régions marginales de ces écosystèmes maghrébins, l'utilisation de zones complémentaires qui implique de déplacer les troupeaux, reste une nécessité. On enterre trop souvent ces modes de production jugés anachroniques, sans assez mettre en valeur leur rationalité et leur capacité à s'adapter.

Conséquences sur la gestion de l'accès aux ressources La plupart des systèmes d'organisation collective de l'accès aux ressources pastorales qui tentaient de gérer les déplacements des troupeaux ont été fortement déstabilisés par les délimitations administratives qui n'ont pas toujours respecté les découpages traditionnels. La constitution du domaine forestier et la difficile mise en oeuvre de l'immatriculation et de la réglementation forestière, la généralisation des cultures partout où c'était possible et le partage, officiel ou occulte, des terres collectives, ont bouleversé les modes traditionnels de gestion. Mais surtout, ces mutations sur l'espace pastoral se sont accompagnées d'une forte montée de l'individualisme d'entreprise et d'un recentrage sur l'individu au détriment du groupe et de toute forme d'organisation collective.

De ce fait, les nouvelles relations que les éleveurs entretiennent avec leur espace, annoncent une certaine déresponsabilisation des acteurs vis-à-vis de leur patrimoine. Désormais, les déplacements se décident individuellement, indépendamment de ceux des voisins. Les décisions du groupe comptent peu et le chef d'exploitation se détermine beaucoup plus en fonction de contraintes propres au fonctionnement de son unité de production que sur les usages en vigueur: main-d'oeuvre disponible, prix du marché, stock fourrager, relations avec les autorités, location de pâturages de gré à gré, etc.

Bien sûr, la notion d'ayant droit a encore un sens et conditionne l'accès aux ressources collectives, mais le sentiment d'appartenance à une communauté élargie, construite autrefois sur un projet défensif – protéger les intérêts du groupe contre les voisins – s'estompe au profit de cet appel à l'individualisme lié à la pénétration des valeurs de la démocratie libérale, de l'ouverture de ces régions sur l'économie nationale et de la monétarisation des échanges. Le principe de gestion des terres collectives n'est donc plus la mise en valeur en commun de ressources mais le contrôle des voisins pour se ménager une niche individuelle (Chiche, 1992).

Plus récemment, de ces communautés s'est dégagée une classe d'individus puissants dont les intérêts se tournent vers l'extérieur et qui ne voient dans les règles coutumières qu'un frein à leur expansion. C'est à leur initiative et sous leur pression que bon nombre des institutions traditionnelles ont été vidées de leur contenu et sont tombées en désuétude.

Le processus est invariable chez ces nouveaux notables: multipliant les sources de revenu (émigration de parents, achats de commerce, charges officielles), résidant temporairement en ville où leurs enfants font des études, ils étendent leur emprise sur leur terroir d'origine en transgressant les règles coutumières, gonflent les effectifs de leurs troupeaux en recrutant des bergers (contrats au quart, *rebaa*) et en achetant des animaux à engraisser avant l'ouverture des parcours, construisent des bergeries et sèment des céréales en zone interdite avec la complicité bienveillante des autorités, s'entendent entre eux d'une communauté à l'autre pour transgresser les règles d'accueil, interdisent la vaine pâture sur leurs champs de céréales après la moisson, etc. Méprisant les règles coutumières, ils donnent l'exemple aux éleveurs plus modestes qui tentent leur chance à leur tour. Mais ces notables contrôlent étroitement le processus qu'ils ont déclenché.

Tel est le constat assez négatif que l'on peut porter sur les mutations que connaissent ces systèmes agro-pastoraux du Maghreb. Mais notre propos n'est pas de nous contenter de ce sombre bilan. Il est plutôt, comme annoncé plus haut, d'analyser en contrepoint, à partir de cas concrets, la résistance que certaines communautés ont su développer face à ce contexte pour continuer de gérer leurs ressources collectives par le biais d'institutions traditionnelles. De tels exemples subsistent qui tous témoignent de la capacité de groupes humains importants à organiser l'utilisation de vastes espaces sans l'aide d'une structure étatique, hiérarchisée et forte. C'est certainement dans le Haut Atlas marocain que de telles organisations montrent un degré de gestion exemplaire malgré les menaces qui pèsent sur leur fonctionnement. Dans la partie suivante, nous nous proposons donc d'analyser en détail cette gestion dite coutumière pour en saisir les mécanismes et les limites.

Organisations pastorales et accès aux ressources dans le Haut Atlas marocain

L'écosystème montagnard et les systèmes agraires

Le Maroc se distingue des autres pays d'Afrique du Nord par le caractère dominateur des montagnes qui le recouvrent sur plus du tiers de sa surface. Au sud du pays, le Haut Atlas est une chaîne particulièrement puissante culminant à plus de 4000m d'altitude et allongée sur 700km depuis l'Atlantique jusqu'aux confins algériens. D'accès difficile, cette montagne, au lieu d'être pénétrée par les paysans des piémonts et de la plaine qui auraient pu l'annexer à leur économie, reste assez fermée et traversée par de rares pistes carrossables. L'enclavement est donc un trait déterminant de la partie centrale de ce massif.

Les systèmes agraires répondent à un modèle caractéristique que l'on pourrait définir comme celui de 'vallées irriguées de montagnes sèches sous influence méditerranéenne'. Le système agraire type comprend trois sous-ensembles. Le premier est intensif et occupe les fonds de vallées au travers d'un lacis dense de terrasses irriguées sur de modestes superficies de statut privé. Le second est le vaste domaine de statut collectif, domanial ou 'présumé domanial', très extensif, qui s'étend sur les pentes des bassins versants et qui supporte pelouses, parcours et forêts que pâturent des troupeaux de petits ruminants. Le troisième, enfin, est intermédiaire et comporte des cultures en sec, le *bour,* d'importance très variable et dont le statut foncier est fort ambigu (terrains en voie d'appropriation plus ou moins contestée). Les exploitants agricoles combinent ainsi ces trois espaces au sein d'unités de production de petites tailles (de 0,2–2 ha en irrigué, à peine plus en sec) où ils cultivent des céréales, du maraîchage, des fourrages et de l'arboriculture fruitière, et conduisent des troupeaux de bovins, ovins et caprins sur les parcours.

Il faut cependant considérer dans ces montagnes des types différents de mise en valeur car, sur le plan de la population, plusieurs vagues d'émigrants venues au cours de l'histoire ont introduit des spécificités culturelles et façonné des systèmes agro-pastoraux différents. Alors qu'à l'Ouest, dans le pays dit *chleuh,* toute la vie montagnarde s'organise dans des vallées refuges de paysans sédentaires qui ont remarquablement aménagé leur terroir au moyen de réseaux d'irrigation complexes, à l'est, où les montagnes sont calcaires et plus sèches, l'élevage est la clef de voûte du système agraire. L'agriculture s'y trouve toujours associée, mais les terroirs irrigués sont plus modestes et la vie pastorale prédomine. Malgré la sédentarisation d'une grande partie de la population, beaucoup d'éleveurs pratiquent encore une transhumance d'été de forte amplitude dans le cadre d'organisations traditionnelles qui s'appliquent à gérer la mobilité des hommes et l'accès aux ressources. La grande tente, inconnue à l'Ouest, demeure toujours l'auxiliaire indispensable des déplacements.

Sur ces espaces pastoraux de l'est du Haut Atlas, les institutions pastorales traditionnelles ont été fortement marquées par l'organisation sociale particulière de ces communautés et l'histoire agitée qu'elles ont vécue. On retiendra deux modèles principaux, l'un plus officiel que les autorités s'efforcent de gérer au travers d'une charte, l'autre plus discret, plus souple, d'origine strictement coutumière et que les autorités ignorent.

Genèse d'une organisation 'officielle': l'exemple de Zaouia Ahansal
Partant d'un exemple qui traite d'une gestion à une échelle intertribale large, celle des parcours de Zaouia Ahansal dans l'est du Haut Atlas Central, nous nous efforcerons d'expliquer la genèse et la logique de cette organisation pastorale particulière sanctionnée par une 'charte de transhumance'.

L'élaboration de la 'charte de transhumance'   La commune rurale de Zaouia Ahansal occupe un territoire situé sur le versant nord du Haut Atlas. Le fond de la vallée ne dépasse pas 1500m d'altitude et est dominé par de vastes hauts plateaux s'étalant entre 2500 et 2800m, eux-mêmes couronnés par des sommets importants (Azurki, 3680m). Les grands traits de ce milieu physique décident du cadre général des activités humaines et dictent aux systèmes de production un mode d'organisation spécifique. Ainsi, les troupeaux montent précocement sur les versants bien exposés au fur et à mesure de la disparition de la neige, séjournent en altitude l'été sur les pelouses (*almou*) et redescendent l'hiver dans les vallées pour y pâturer les chênes verts et les buissons aux feuillages pérennes. On constate une faible extension des terres irriguées en fond de vallée, un développement des cultures céréalières sur certains versants favorables et de nombreux défrichements en forêt.

Deux groupes sociaux se sont depuis fort longtemps partagés l'utilisation de ce territoire pastoral:

o les nomades Aït Atta du Sud qui viennent chaque année du versant saharien, et notamment la fraction Aït Bou Iknifen qui va jouer aux avants postes un rôle particulier;
o les marabouts Ihansalen fixés selon la légende depuis près de cinq siècles dans leurs villages du versant nord (Figure 10.3).

Profitant de l'aide que ces marabouts leur accordaient et des arbitrages qu'ils rendaient en échange d'une protection efficace, la puissante confédération Ait Atta du versant saharien parvint en effet, il y a un ou deux siècles, à imposer sa présence l'été sur les pâturages d'altitude du versant nord, repoussant devant eux les tribus voisines. De cette association légendaire entre le pouvoir spirituel des descendants du Saint et le pouvoir temporel des guerriers Ait Atta naquit un *modus vivendi* pastoral entre les éleveurs du sud et du nord de l'Atlas.

Au début de ce siècle, ces éleveurs du Nord, rattachés à la fraction maraboutique Ihansalen étaient ainsi confinés dans leurs vallées boisées et

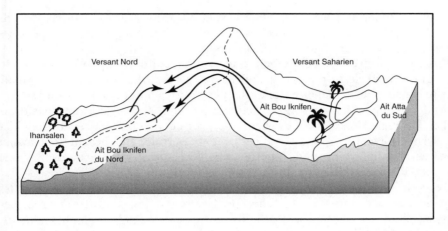

Figure 10.3 *Occupation du territoire par les groupes sociaux, Commune Rurale de Zaouia Ahansal (Maroc)*

venaient occuper à la fin du printemps les parcours d'altitude qu'ils partageaient ensuite en mai–juin avec les nomades Aït Atta venus du Sahara. Une fraction de ces derniers, les Aït Bou Iknifen de Talmest, s'était même fixée sur ce versant nord autour de ses greniers fortifiés (*tighremt*), dans un poste avancé qui témoignait ainsi de la présence Aït Atta.

L'implantation coloniale dans le Haut Atlas fut progressive et bouleversa les déplacements des troupeaux et les usages en vigueur. Certains Aït Atta du Piémont Sud se rallièrent dès 1919, d'autres refusèrent de se rendre jusqu'aux derniers combats de 1932, de telle sorte que les transhumances furent interrompues de longues années et que les éleveurs du Nord (Ihansalen et Aït Bou Iknifen du Nord) en tirèrent largement profit.

Lors du retour au calme, les autorités coloniales espérèrent dans un premier temps que les transhumances redémarreraient dans le respect des usages passés. Mais de quel usage s'agissait-il? Dans le passé, les usages n'exprimaient rien de plus qu'un rapport de forces. A présent que de nouvelles habitudes étaient prises, pourquoi ne pas les considérer comme des usages?

Les autorités coloniales s'efforcèrent alors, avec l'aide des marabouts, d'élaborer une charte de la transhumance. Les officiers des Affaires Indigènes voulant imposer une codification claire et précise en total contraste avec la souplesse et la fluidité des usages précédents, manipulant grossièrement les procédures coutumières, recourant aux notables qui se livraient à des luttes politiques confuses . . . ne firent en vérité que relancer les conflits pastoraux. Ils purent cependant, au fil d'une multitude d'accords, d'arbitrages et de confrontations, et malgré des erreurs, des injustices et des à peu près, établir 'la grande charte de la transhumance de Zaouia Ahansal' qui fait maintenant référence dans cette région et qui garantit les droits de pâturage de la totalité des éleveurs de la commune rurale de Zaouia Ahansal et des autres usagers Aït Atta venant du Sud.

Depuis l'Indépendance, les conflits se sont poursuivis sur leur lancée, mais ils ont changé de nature. L'essentiel des problèmes vient maintenant de ce que les intérêts de ceux du Nord, dont le nombre augmente, et qui s'installent, défrichent, construisent des bergeries, mettent en culture, s'opposent à ceux du Sud qui, venant de trop loin, ne veulent que pâturer. La solidarité intertribale qui prévalait pour la défense des droits pastoraux n'a donc plus aucun sens.

Malgré ces conflits, mais aussi façonnés par eux, une organisation pastorale cohérente règle donc les mouvements de près de 100 000 têtes et d'un millier d'éleveurs. La charte, quoique rédigée en 1941, sert de référence, mais chacun a conscience qu'il serait bien de 'mettre les papiers et les cartes en conformité avec les faits' (Capitaine Ransan, 1952, Archives locales).

Avec beaucoup de prudence, et fortifiées dans leur conviction par les expériences malheureuses de leurs prédécesseurs, les autorités marocaines laissent plutôt les éleveurs se débrouiller entre eux et régler les conflits mineurs. Quant aux conflits majeurs, rapports Nord–Sud et agressions de bergers, mises en culture, construction d'abris, ils s'enlisent dans les méandres d'une justice peu décidée à trancher.

La réglementation en vigueur Ceci posé, l'organisation, mi-coutumière mi-officielle, existe. Elle repose sur un certain nombre de règles dont la plupart sont consignées dans les cahiers de la charte, déposés dans le bureau du *Caïd*:

○ Première règle: Le pâturage est découpé en unités, attribuées aux ayants droit de l'un ou l'autre des groupes, selon un découpage qui respecte en gros 'l'état des forces en présence', dressé par les officiers des AI dans les années trente. Ces collectifs ne sont pourtant pas encore immatriculés.

○ Deuxième règle: Ces unités ou parcs pastoraux sont délimitées très soigneusement à partir de points de repère reconnus par tous et inscrits sur les différents procès-verbaux issus des arbitrages.

○ Troisième règle: La limite des cultures est repérée par des tas de pierres. Sur le parcours proprement dit, il est donc interdit de cultiver mais aussi de construire des bergeries, car en droit coutumier, construire en dur, c'est revendiquer la propriété.

○ Quatrième règle: Partout où l'eau est rare, les droits d'abreuvement sont arrêtés dans le détail: couloirs d'accès, temps de séjour au point d'eau ('sans prendre le temps de boire un thé'), ordre de passage des troupeaux.

○ Cinquième règle: Les meilleures parties des parcours, notamment les pelouses d'altitude, sont mises en défens chaque printemps. C'est la pratique de l'*Agdal*[2], élément essentiel de la gestion des parcours. L'*Agdal* est marqué par des dates précises de fermeture et d'ouverture; la période de mise en défens se situe au printemps et dure de 1–3 mois, jamais plus; l'*Agdal* n'accueille pas d'étrangers, sauf par le biais des associations; la surveillance de l'*Agdal* est

---

[2] *Agdal* ou *Agoudal*, nom berbère, pluriel *igoudlane*. Nous le garderons, même au pluriel, sous la forme au singulier par souci de clarté.

assurée par un gardien choisi pour son sérieux et rétribué par la collectivité; des amendes en argent ou en nature sanctionnent les délinquants.

Le Tableau 10.1 ci-joint, repris de la charte, résume quelques unes des caractéristiques de cette organisation dont la plus remarquable est qu'elle soit le produit métissé d'une gestion coutumière et d'une volonté politique pour en fixer les règles sur le papier. Mais cette organisation ainsi présentée paraît bien formelle et se révèle très insuffisante pour prétendre gérer dans le détail les contentieux et les conflits. En effet, les rédacteurs successifs de la charte ont tenté d'intégrer tout un ensemble de règles, de pratiques et d'interdictions dont certaines sont archaïques, d'autres injustes et ne répondent plus aux exigences d'un aménagement rationnel de l'espace. De fait, sur le terrain, les pratiques des éleveurs sont en constante évolution et tellement en décalage avec les textes que, pour une bonne part, le système actuel fonctionne en autogestion.

D'autres exemples souligneront la diversité des pratiques dans le cadre de modèles moins officiels.

Variété, richesse . . . et fragilité des organisations coutumières

La charte des parcours de Zaouia Ahansal est donc une tentative intéressante de gestion contrôlée, mais unique car, partout ailleurs dans le Haut Atlas, les systèmes de gestion sont si modestes et si peu officiels que les autorités n'interviennent pratiquement pas dans leur fonctionnement, en grande partie parce qu'elles les ignorent, ou tout simplement parce que les conflits, quand il y en a, se résolvent de façon interne. Ce sont ces règlements multiples qui émaillent la vie pastorale du Haut Atlas que nous voudrions maintenant évoquer afin de porter témoignage de la richesse et de l'inventivité dont les éleveurs de ces communautés montagnardes ont su faire preuve.

Il est frappant de constater que toutes ces organisations coutumières, plus ou moins réajustées par les autorités coloniales ou nationales, ont certes un cadre commun, mais sont marquées surtout par l'importance des facteurs de variation et la grande souplesse d'application. On nous pardonnera d'insister sur ce qui pourrait passer pour des détails de fonctionnement, mais la force de ces organisations semblent tenir à eux.

Il faut de plus impérativement faire la différence entre ce qui est affirmé (ou revendiqué) par les usagers et les faits tels qu'on les constate car il y a un perpétuel hiatus entre ce qui est présenté comme la norme et ce qu'il en est sur le terrain.

Le découpage des territoires pastoraux   Les éleveurs utilisent un espace pastoral particulier qu'on peut désigner comme leur 'territoire', composé de parcours dont le statut est collectif et domanial. La domanialisation dans ces régions n'a pas changé les habitudes des usagers qui continuent à considérer ces parcours forestiers comme les leurs. Ces parcours sont pâturés et attribués à des ayants droit précis. C'est l'appartenance à un groupe ethnique, que les découpages administratifs modernes ne respectent pas nécessairement, qui fonde ce droit au parcours sur ce territoire. Jusqu'au début de ce siècle, tous les groupes

Tableau 10.1. Extrait de la charte de la transhumance de Zaouia Ahansal sur l'utilisation des Agdal

Zones de pâturage	Tribus bénéficiaires du pâturage		Date de fermeture	Date d'ouverture	Observations
	du nord de l'Atlas	du sud de l'Atlas			
Afella n'Izran	Aït Bou Iknifen du Nord				Ne l'utilisent pas
		Ilamchane / Aït Bou Iknifen du Sud / Aït Aïssa ou Brahim / Ignaouen / Aït Bou Daoud			Transhumance résiduelle de 6 à 10 tentes
Tilemsine	Ihansalen / Aït Bou Iknifen du Nord		1/5	1/6	Ne l'utilisent pas
		Aït Bou Iknifen du Sud / Aït Aïssa ou Brahim / Ignaouen / Aït Bou Daoud	1/5	1/6	
D. Daou n'Izran	Aït Bou Iknifen du Nord	Aït Bou Iknifen du Sud			Surtout occupé par ceux de la zone de Boukhadel / Les anciennes dates sont tombées en désuétude
Almou n'Talmest	Aït Bou Iknifen du Nord		15/4	6/7	Tolérance pour le gros bétail pendant la mise en défens
		Aït Bou Iknifen du Sud	15/4	6/7	
Agoudal n'Ilamchane	Ihansalen		1/4	mi-sept.	En réalité seuls les Aït Taghia + 7 familles de Zaouia
Jbel Tamerroucht		Ilamchane	1/4	vers le 1/8	Utilisé au passage
Tadrarat et Bou Ighlalne	Ihansalen	Ilamchane			Uniquement Aït Taghia
		Aït Atta Msemrir			Passage pour tous Aït Atta du Sud

254

ethniques de ces régions ont lutté âprement pour la conquête de leurs territoires, leurs querelles étant plus ou moins arbitrées par les différents marabouts, nombreux dans la région, qui jouaient le rôle d'intermédiaires. La situation actuelle, qui fut à grand peine clarifiée par les autorités coloniales tant le temps d'occupation fut court dans ces contrées (1932–56), est le résultat de ces rapports de force.

Les différents groupes ethniques disposent donc de territoires aux limites précises; chaque berger les connaît et, contrairement à d'autres régions (Moyen Atlas), les contestations sur les frontières ne sont pas très vives. Bien sûr il y a des revendications, mais les collectifs n'étant en général pas délimités ou homologués, on aurait tort d'accorder trop d'importance à des luttes purement juridiques sans véritables conséquences sur le terrain. Les conflits, quand il y en a, sont souvent d'une autre nature, comme nous le verrons plus loin.

Il faut aussi considérer que ces territoires ne sont pas toujours à usage exclusif et qu'il faut distinguer:

o les territoires pastoraux intertribaux, par exemple entre Ihansalen et Aït Atta à Zaouia Ahansal (Haut Atlas central), ou entre Aït Haddidou et Aït Ameur sur le plateau des lacs à Imilchil (Haut Atlas oriental);
o les territoires tribaux, comme l'Almou n'Talmest de l'exemple détaillé plus haut qui est aux seuls Ait Bou Iknifen;
o les territoires de fraction qui sont les plus courants, par exemple chez les Aït Yahya où les parcours sont partagés entre Aït Fdouli, Aït Moussa et Aït Ameur ou Hami (Haut Atlas oriental);
o les collectifs de village.

Ces derniers se développent dans un certain nombre de secteurs trop excentrés pour être utilisés par l'ensemble des villages de la fraction et seuls les plus proches finissent par établir dessus un droit d'usage exclusif que vient renforcer l'installation de bergeries et des cultures. Ce peut-être aussi un *modus vivendi* entre semi-nomades et sédentaires, comme sur les parcours collectifs des villages Ait Zekri (versant sud). On y voit ainsi les grands transhumants (Rahali) occuper les parcours saisonniers, tout en ménageant des périmètres ayant chacun un village pour centre au sein duquel pâturent en priorité les éleveurs sédentaires installés dans leurs abris privés, *azib,* et auxquels viennent éventuellement se mêler des éleveurs voisins installés sous la tente ou dans des grottes non attribuées, et qui pâturent 'un peu plus loin'. Les petits éleveurs des villages ont donc un espace collectif bien à eux.

Sauf cas particuliers, les limites ne sont pas des barrières infranchissables et les bergers les transgressent couramment en pâturant chez les voisins à condition de n'y pas dormir, éventuellement de n'y pas boire. Enfin, sur le parcours ils évitent avec soin les 'aires de respect', *itissaa,* qui balisent les environs immédiats d'une tente, d'une bergerie, d'une parcelle de céréale ou d'un point d'eau. Chacun connaît les limites de ces domaines momentanément privés et respecte les règles de la bienséance.

Les ayants droit et l'accès aux ressources On sait que l'appartenance à une communauté – être ayant droit Aït Atta ou Aït Ameur – est soit héréditaire et automatique en référence à une ascendance, soit le résultat d'un processus d'intégration et d'assimilation plus ou moins long selon l'importance numérique et le poids social du groupe d'immigrés et de ceux qui acceptent de les accueillir. Peu de problèmes se posent à ce sujet dans cette partie du Haut Atlas plus isolée et plus hermétique aux 'étrangers' qu'ailleurs.

Plus au Nord, vers le Moyen Atlas, ce problème de reconnaissance de populations étrangères venues en l'occurrence il y a plus de 40 ans de la Moulouya lors des grandes sécheresses, se louant comme bergers puis se fixant sur place, est devenu considérable et provoque de graves conflits. Devant la multiplication des troupeaux 'étrangers', les autochtones adoptent une définition très restrictive de la notion d'ayants droit sans pour autant pouvoir régler le contentieux. Refoulées sans ménagements, certaines familles séjournent toute l'année en montagne sous des abris précaires faits de bâches de plastique et installés dans des endroits peu disputés aux limites entre deux fractions.

Les règles d'usage et la gestion des ressources

Multiplicité des règles

L'institution coutumière ne se contente pas de garantir des territoires et d'identifier des ayants droit, mais multiplie les règles et les pratiques particulières. La coutume reconnaît ainsi, selon les cas, le droit de couper ou non de l'herbe à l'ouverture des *Agdal*, celui d'y mettre des vaches ou des moutons, celui de permettre ou de proscrire la construction d'un abri en dur ou *azib,* celui de cultiver, celui de pâturer, avec ou sans l'installation de la tente, avec ou sans le droit de faire *ifrilil*.

Faire *ifrilil* c'est pour un berger partir seul avec le troupeau quelques jours avec un matériel minimum – une petite tente ou un simple *burnous,* un sac de farine, du thé et du sucre – et faire son pain soi-même. C'est une pratique courante pendant les beaux jours que seuls les jeunes bergers, auxquels on ne demande pas leur avis, supportent. C'est donc un moyen commode de transgresser discrètement les règles, tout en affichant clairement la non-revendication sur le parcours.

L'institution de l'*Agdal*

Le Haut Atlas est certainement la région du Maroc où cette institution est la plus vivante. On en connaît le principe, mettre en défens au printemps ou en début d'été une zone bien délimitée du parcours dans sa partie la plus productive. Cette interdiction de pâturer pendant la période la plus sensible pour les plantes qui puisent à cette époque dans leurs réserves puis entrent en floraison, est tout à fait judicieuse puisqu'elle permet de renforcer la vigueur de la végétation et d'assurer un report sur pied de la biomasse disponible en fin de saison.

L'organisation se plie toujours au même schéma traditionnel, fermeture et ouverture à des dates convenues, arrêtées par la coutume mais pouvant souffrir quelques modifications à la demande de l'une ou l'autre des parties selon l'état des ressources, surveillance exercée par des gardiens.

Ils sont payés par la communauté des éleveurs, ou le cas échéant par les éleveurs de la fraction la plus éloignée qui craint le plus les délits. Leur rôle se limite à informer la *Jemaa*[3] pendant les deux ou trois mois que dure leur mandat sur l'identité des délinquants. S'ils sont de la tribu, ils seront sanctionnés comme le veut la coutume, autrefois sacrifice d'un mouton, à présent paiement d'une amende ou, plus rarement, préparation d'un repas collectif (*izmaz*) offert à une dizaine de personnes désignées par la *Jemaa* et que le délinquant doit offrir le soir du jour du prochain *souk*[4]. Si ces délinquants sont 'étrangers', ils seront refoulés sans ménagement, voire dénoncés aux autorités locales.

Au même titre que les territoires pastoraux, on distingue des *Agdal* intertribaux, de tribu, de fraction ou de quelques villages seulement. Les droits d'accès sont basés (exemple de Zaouia Ahansal) sur des droits historiques et sur l'appartenance ethnique, qui pèsent bien sûr d'un grand poids. Mais ils peuvent dépendre aussi de privilèges particuliers comme chez les Rhiraya du Haut Atlas de Marrakech où le fameux *Agdal* d'Oukaïmeden qui est approprié collectivement par deux tribus (Rhiraya et Ourika), est mis sous le patronage d'un saint local, Sidi Fares, qu'on fête par une cérémonie rituelle le premier vendredi après l'ouverture (Mahdi, 1993). Deux groupes se démarquent alors, les serviteurs du Saint dont les campements sont situés dans la meilleure partie de la prairie, et les non-serviteurs installés vers le torrent et qui doivent quitter l'*Agdal* après 15 jours seulement d'utilisation. Contrairement à ce que nous dirons plus loin à propos des abris sur collectif, l'accès à cet *Agdal* n'est offert qu'aux éleveurs qui possèdent un abri (*azib*) car la tente est inconnue dans cette partie de l'Atlas.

Les modalités de l'exploitation des *Agdals*

Pendant la fermeture, les troupeaux quittent le secteur concerné y compris les sédentaires qui séjournent le restant de l'année dans les *azib* dont la construction a été permise ou tolérée. A l'ouverture, les *Agdal* les plus riches sont fauchés par les familles de transhumants (et même de non transhumants) qui font du foin (par exemple sur le grand *Agdal* d'Islane, quatre à cinq *achlif*[5], soit environ 0,5t par famille en année normale). Certains *Agdal* ne sont ouverts dans un premier temps qu'aux vaches et aux mulets, puis un mois après aux petits ruminants (Almou n'Igri, Tidemt chez les Ait Fdouli), d'autres ne sont permis que dans la journée, le berger ne devant pas 'faire *ifrilil*' (éleveurs de Tirdhouine à l'*Agdal* n'Inouzane, contrairement aux éleveurs Aït Brahim de Tilmi qui ont le droit d'y planter leurs tentes et d'y cultiver). Beaucoup d'éleveurs ne viennent sur l'*Agdal* que pour profiter de l'herbe des premiers jours et le quittent après trois à cinq semaines. Ne restent sur place que les gros troupeaux, les semi-nomades, les voisins immédiats, etc.

Ainsi, pour la seule région du Parc du Haut Atlas Oriental qui intéresse une trentaine de villages, on ne compte pas moins d'onze *Agdal* (Tableau 10.2). Les

---

[3] *Jemaa*: assemblée villageoise des chefs de familles.
[4] *Souk*: marché hebdomadaire.
[5] *Achlif*: charge portée par un mulet.

Tableau 10.2. Les différents *Agdal* du Parc du Haut Atlas Oriental

Utilisateurs (fractions et villages)	Agdal	Période de fermeture	Coupe	Azib
Aït Ameur, Aït Brahim de Tilmi/ Imilchil/Outerbat, quelques éleveurs Ouderhour	Islane E	24/3–14/6	Coupe d'herbe	Quelques *azib* et peu de cultures
Taghighactht, Aït Yazza d'Imilchil/ Tarribant	Islane W	24/3–14/6	Coupe d'herbe	Beaucoup d'*azib* et de cultures
Anefgou, Arheddou, Tirrhist	Ioualghizen	14/5–14/7	Coupe d'herbe[1]	Beaucoup d'*azib* et de cultures[1]
Anefgou	Amalou n'Fazzaz	10/6–10/8	Pas de coupe	Pas d'*azib*
Aït Fdouli	Almou n'Igri Tidemt	28/3–23/6[2] 28/3–23/7[3]	Pas de coupe	*Azib* et culture
Aït Brahim de Tilmi, Tirdhouine	Assameur et Amalou n'Inouzane	24/3–10/7	Coupe pour Aït Haddidou	Pas d'*azib* Quelques cultures[4]
Tirrhist, Tirdhouine	Anzad	14/6–14/7	Pas de coupe	Pas d'*azib*
Arheddou	Taoudalt	14/6–19/8	Pas de coupe	Pas d'*azib*
Aït Brahim d'Outerbate	Ouiyalzane – Iffer	15/4–5/7	–	–
Michlifen, Aït Hattab, Aït Ali ou Assou	Akdar	5/4–5/7	Pas de coupe	*Azib* et cultures sauf Michlifen

1. Coupe de foin et *azib* interdits pour le village d'Arheddou.
2. 23/6 pour les vaches, 23/7 pour les moutons.
3. 23/7 pour les vaches, 23/8 pour les moutons.
4. Cultures interdites aux Aït Ameur qui n'ont le droit que de pâturer de jour (*ifrilil* interdit).

dates indiquées sont celles fixées par la coutume et susceptibles d'être repoussées de une à deux semaines en cas d'année à fort enneigement. Exceptionnellement, les années très sèches, l'*Agdal* peut-être abandonné.

Les systèmes d'*Agdals* combinés

Chez les Aït Ameur et les Aït Fdouli du Haut Atlas oriental, il existe un système, tout à fait exceptionnel au Maroc, d'*Agdals* combinés en rotation à l'image des plans de pâturage conçus par les pastoralistes. Par exemple, à Anefgou les troupeaux pâturent autour de Ioualghizen jusqu'à la fermeture de l'*Agdal* (14 mars) et vont dans les forêts de Tarhiouine ou de Tallount ou vers le Fazzaz jusqu'à l'ouverture d'Islane (14 juin), puis reviennent le 14 juillet à l'ouverture d'Ioualghizen, pour finir par pâturer Amalou n'Fazzaz fermé du 10 juin au 10 août. Dès l'automne, les troupeaux s'éparpillent vers les différents sites d'*azibs* (Carte 10.3 'systèmes d'*Agdal* combinés').

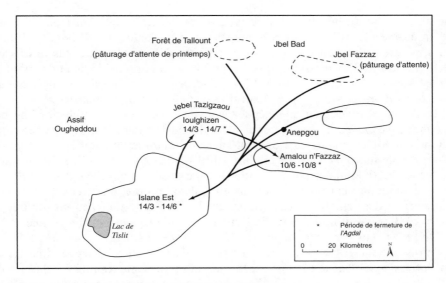

Carte 10.3 *Système d'*Agdals *combinés (Maroc)*

Les *Agdal* spécialisés pour bovins et mulets et le gardiennage collectif ('*tiwili*')
L'organisation pastorale coutumière s'exprime également au travers de formules parfaitement rodées de gardiennage collectif sur des *Agdal* spécialisés qui permettent de s'affranchir du gardiennage des vaches et des mulets dans les périodes de gros travaux agricoles.

   Par exemple, à Aït Ali ou Ikkou (Haut Atlas oriental), à partir du mois de mai, le troupeau du village composé de 100 vaches et de 50 mulets est mené chaque jour sur des prairies collectives (*ilmouten*) spécialement mises en défens. Un '*moqqadem*' spécialisé s'occupe de l'organisation du '*tiwili*'. Six bergers sont désignés chaque jour selon un tour de rôle (*nouba*) de 22 jours (132 foyers sont donc impliqués). Chaque foyer dépêche une personne – fille ou garçon – quel que soit le nombre d'animaux exploités. Au troupeau des vaches s'ajoutent deux taureaux collectifs qui le soir sont gardés séparément dans deux étables du village selon un double tour de rôle où sont impliqués les 132 foyers d'éleveurs. Lorsque le '*tiwili*' prend fin après l'*Agdoud* (*moussem d'Imilchil* en mi septembre), l'un des taureaux poursuit sa carrière sur le même mode, l'autre est alors vendu aux enchères au plus offrant des éleveurs du village.

Les institutions   Il est certain que, quoi qu'il en soit des conditions d'accès aux ressources collectives, leur utilisation se fait individuellement et la participation à leur gestion ne se traduit pas par l'existence d'institutions fortes. L'organisme qui gère le collectif est en principe la *Jemaa*, sans existence légale, et à propos de laquelle il est utile de rappeler que le mot désigne un ensemble de personnes liées par des intérêts communs, et que par conséquent la *Jemaa* n'est pas toujours la même sur un espace donné. Il y a donc une *Jemaa* de tribu, de fraction, de village,

259

de quartier ou de lignage selon le type de problème traité. Ainsi quand il est déclaré que 'la tribu a décidé des dates d'ouverture de l'*Agdal*', il faut comprendre qu'il y ait eu simplement réunion des éleveurs les plus intéressés, le plus souvent à la mosquée après la prière du vendredi. Même chose pour le tirage au sort des *azib* ou l'accord pour l'accueil d'un troupeau étranger qui ne concerne qu'un groupe très restreint d'usagers directement concernés.

La *Jemaa* peut désigner un délégué, *amghar n'tuga* (c'est à dire chef de l'herbe) ou un simple *moqqadem* (vague équivalent du garde-champêtre) chargé de veiller au bon déroulement de la transhumance (installation des tentes, installation d'une 'tente-mosquée-lieu de réunion', utilisation des *azib* collectifs, entraide et recherche d'animaux perdus). Elle désigne aussi les gardiens des *Agdal* qui sont rétribués par la collectivité et qui surveillent les mises en défens. Elle veille à l'exécution des sanctions qui punissent les auteurs d'infractions (autrefois sacrifice d'un mouton ou repas collectif, maintenant paiement d'une amende).

De façon plus officielle, la *Jemaa* de chacun des lignages ou de chacune des fractions de la tribu peut-être amenée à désigner un 'délégué aux terres collectives' agréé par le *Caïd*. C'est le *Naïb* qui représente les intérêts du groupe au sein de la '*Jemaa* des terres collectives'. Celle-ci donne en particulier son avis sur le partage des terres et l'installation des abris.

Les *Jemaa* ne prennent leurs décisions que par consensus et le débat dure tant qu'il n'y a pas unanimité. Une telle procédure ne permet pas en général de faire face aux conflits graves qui autrefois se réglaient par la force et qu'on porte maintenant devant les tribunaux. C'est notamment le cas pour tous les conflits qui ont pour origine la construction d'abris et les mises en culture.

Les stratégies individuelles pour le contrôle des ressources

Ces organisations coutumières visent donc à gérer équitablement les ressources collectives, et nombreuses sont les déclarations qui proclament que les droits sur le parcours sont les mêmes pour tous. Pourtant, ces vertueuses professions de foi ne résistent pas à l'analyse car, au-delà de ce principe formel, se développent de vigoureuses stratégies individuelles, mais aussi de lignages ou de villages, qui introduisent de fortes inégalités.

Pour un individu, le seul vrai stratagème pour asseoir sa maîtrise sur une portion de parcours collectif, c'est la prise de possession d'un *azib* qui sert de prélude à un contrôle définitif par le défrichement, la mise en culture . . . ou le creusement d'un puits. Il est donc important pour un éleveur de conforter sa place dans le territoire par l'installation d'*azib* situés dans des milieux différents et complémentaires.

Normalement, l'accord pour une installation nouvelle devrait se faire à l'échelon de la tribu (la *Jemaa* des terres collectives) et sous couvert du *Caïd*. Mais dans les faits, il y a reconnaissance effective d'aires d'influence plus restreintes sur lesquelles des groupes de taille variable ont leur avis à donner: le lignage, le village, la fraction et plus rarement la tribu. L'espace est donc beaucoup plus segmenté que ne le laissent croire les déclarations car la liberté théorique de circulation d'un troupeau et les autorisations de construire un *azib* sont en permanence entravées par un contrôle strict du parcours à ces différents niveaux.

De plus, n'obtient pas un *azib* qui veut. Quand les intéressés déclarent que 'c'est la tribu qui a décidé', il faut plutôt imaginer un processus complexe où jouent à la fois le poids politique du demandeur, l'accord de quelques voisins influents, voire l'intervention de la *Jemaa* des terres collectives ou du *Caïd* lui-même. La décision finale est souvent couronnée par un repas offert à un certain nombre de chefs de famille de la tribu ou du village. Obtenir un accord relève donc d'un processus subtil et non mesurable puisque entrent en jeu le degré d'influence et le poids politico-économique d'une personne ou d'un lignage (Gregg & Geist, 1987).

On distingue plusieurs types d'*azib*. Ceux de premier type sont des ronds de pierre où l'on installe la tente ou de simples grottes; ils sont collectifs et occupés chaque saison par le premier arrivé, ou tirés au sort. Ceux de deuxième type sont aménagés de petits murs et d'un toit; ils sont appropriés et peuvent être prêtés à condition d'y laisser le fumier (Aït Zekri).

Les *azib* sont en principe installés en dehors des *Agdal*, mais cette règle souffre beaucoup d'exceptions et provoque un certain nombre de conflits. A Islane par exemple, la partie Ouest comporte de nombreux *azib* entourés de cultures appartenant aux Aït Yazza, suite à un partage qui date de l'époque du Protectorat. Plus à l'Est, les gens de Tilmi et d'Ali ou Daoud ont commencé à labourer mais en ont été empêchés alors qu'en 1975, quatre ou cinq familles des Aït Ali ou Ikkou se sont installées au bord du lac d'Isli, ont construit des *azib* et mis en culture une dizaine de parcelles. Malgré les menaces, ils résistent farouchement ('tuez-nous, mais nous resterons!') et étendent chaque année leur emprise. Chez les Aït Zekri, les places des *azib* à l'ouverture de l'*Agdal* sont tirées au sort ('*ilan*'), ce qui présente le double avantage de ne favoriser personne et surtout de prévenir toute tentative d'appropriation; des accords amiables permettent ensuite de réajuster les choses si besoin est.

On tire ainsi de l'analyse de ces organisations traditionnelles un sentiment double, l'un de cohérence et d'équilibre que peut suggérer cette 'production communautaire du droit' (Tozy, 1990) au service d'une gestion solidaire, souple et étroitement adaptée à un milieu complexe, l'autre plus tumultueux à l'image des conflits et des pratiques individualistes que les éleveurs développent pour s'approprier l'espace. Quel bilan établir sur la capacité de ces organisations à bien gérer les ressources? Quels enseignements tirer de leur étude, quels principes retenir pour une meilleure gestion de la mobilité? Nous nous efforcerons pour conclure de répondre à ces questions.

Bilan et perspectives

On a vu que partout au Maghreb, les formes traditionnelles d'organisation collective pour l'exploitation des ressources pastorales ont périclité sous la pression d'un essor démographique irrésistible et d'une forte montée de l'individualisme. Confrontés à un cadre juridique de moins en moins adapté, à l'effet pervers de certaines décisions économiques telles que les subventions sur les aliments concentrés et aux interventions massives de l'État dans le processus de transformation de l'espace rural, les éleveurs sur parcours sont engagés dans une course aux effectifs et une compétition sur l'espace telles, qu'on les accuse de ne plus savoir gérer leurs ressources quand elles sont communes. C'est pourquoi de très fortes pressions s'exercent pour privatiser les collectifs et sédentariser les troupeaux comme c'est maintenant le cas en Tunisie.

Résistance des organisations traditionnelles 'reliques'

A travers ces quelques exemples tirés de la montagne marocaine, nous avons souhaité démontrer l'intérêt de situations qui témoignent d'une certaine résistance, puisqu'on rencontre encore des modes de gestion tantôt officiels réglant la mobilité des hommes sur de vastes espaces (Zaouia Ahansal), tantôt paisibles et discrets à l'échelle plus modeste de la fraction ou du village. Comment expliquer dans un contexte si peu favorable la force de résistance de ces organisations traditionnelles du Haut Atlas? On peut identifier plusieurs raisons.

L'isolement Il a certainement protégé ces systèmes d'au moins deux facteurs très déstabilisateurs, le contrôle étatique étroit et l'arrivée de capitaux urbains. Les projets étatiques d'aménagement des parcours qui ont vu le jour ces trente dernières années ont fait preuve en général d'une grande maladresse et ont montré leur incapacité à prendre en compte la gestion coutumière là où elle résistait encore. Leur mise en oeuvre a le plus souvent accéléré la désorganisation en confortant la classe des grands éleveurs. Le Haut Atlas de ce point de vue est resté à l'écart du développement, et s'est trouvé également protégé des capitaux urbains qui viennent habituellement s'investir en têtes de moutons sur les parcours par le biais des associations, minant de l'intérieur le système par l'accroissement incontrôlé des effectifs.

Les qualités intrinsèques de ces organisations La première des qualités est la simplicité qui fait reposer l'organisation sur quelques principes: des territoires délimités et des ayants droit identifiés, des restrictions (et non des interdictions) sur les droits de construire des abris, de mettre en culture, de prendre des animaux en association, des droits d'abreuvement, l'instauration d'*Agdal* qui suppose une instance de décision, un système de gardiennage et des sanctions. Sur un plan purement technique, l'*Agdal* ou mise en défens saisonnière, est la formule la plus simple qu'un pastoraliste puisse proposer . . . mais aussi la seule admissible en l'absence de tout contrôle des effectifs animaux.

On soulignera aussi la souplesse et la tolérance à propos des modes d'utilisation. Les limites sont bien identifiées mais leur franchissement par un troupeau 'étranger' est toléré à l'échelle de la journée. Même attitude *a priori* bienveillante pour l'abreuvement, 'on ne refuse pas l'eau à un berger'. Sur les *Agdal*, les dates de fermeture et d'ouverture sont discutées chaque année; *Agdal* est abandonné les années de sécheresse, l'ouverture retardée si la neige est tardive.

Autre qualité, qui n'est pas antinomique de la simplicité, la richesse des détails d'une gestion quotidienne qui s'adapte à toutes les situations: le *tiwili*, les autorisations de campement (tente, *guitoun*, faire *ifrilil*), le passage sur *Agdal* des vaches avant les moutons, la fauche d'herbe sur *Agdal*, le paiement des sanctions par un repas collectif, le prêt d'*azib* contre du fumier, etc.

Les institutions qui interviennent dans la gestion sont d'une grande légèreté. On relève une grande diversité de formules selon l'emprise sur le territoire: grande ou petite *Jemaa*, avec ou sans délégué, avec ou sans recours au *Caïd*. Chacun peut faire entendre son point de vue puisque la *Jemaa* ne prend ses décisions que par consensus. Mais on en sait les limites.

Il faut enfin faire état de la solidarité entre éleveurs et plus encore entre bergers qui semblera en apparence contradictoire avec ce que nous dirons plus loin de la concurrence. Sans doute est-ce surprenant, mais les mêmes qui s'empoignent sur le parcours pour un *azib* indûment construit, s'entraideront pour la tonte et célébreront ensemble le *maarouf*[6].

On peut même affirmer qu'il y a une certaine acceptation des inégalités tant qu'elles restent dans des limites supportables par la communauté, car le principe de gestion collective intègre, on l'a vu, la possibilité d'inégalité de l'usage. De plus, chacun a bien conscience que la survie de la communauté dans ces milieux marginaux repose sur l'émergence de personnes assez puissantes pour défendre les intérêts du groupe auprès des autorités. On ne conteste donc pas (trop) aux notables les abus auxquels ils se livrent. C'est à ce prix que la distorsion entre logiques individuelles et logique collective reste supportable (Chiche, 1992).

Fragilité et déséquilibres Pourtant, à des degrés divers, ces organisations coutumières du Haut Atlas sont menacées et les territoires qu'elles contrôlent risquent de se réduire comme peau de chagrin, jusqu'à disparaître comme ils ont disparu ailleurs des espaces maghrébins.

Il est bien clair que l'affirmation selon laquelle les droits sur le collectif sont les mêmes pour tous est totalement erronée. Aucune limitation d'effectif n'est appliquée, les prises d'animaux en association et les pratiques d'achats spéculatifs d'animaux engraissés rapidement sur les *Agdal* les plus accessibles se font sans réel contrôle au seul profit des gros éleveurs.

C'est donc un système fort peu égalitaire puisque chacun met sur le parcours tous les animaux qu'il peut et tente par tous les moyens (citernes transportées, campements d'altitude, annexion de parcours) de récupérer le maximum de ressources. Aucun esprit coopératif au sens moderne du terme, car l'ayant droit revendique pour lui un droit qu'il partage de bon gré ou mal gré avec d'autres. Dans ces conditions, 'le principe de gestion n'est pas la mise en valeur en commun des ressources mais le contrôle de la concurrence pour leur usage individuel' (Chiche, 1992).

Mais le plus grave, c'est la multiplication des *azib* qui prépare la privatisation d'une partie des parcours ou leur contrôle par des groupes restreints. Bien que la

---

[6] *Maarouf*: sacrifice rituel suivi d'un repas collectif.

menace se fasse pressante, l'institution coutumière se révèle incapable de la maîtriser. La règle du consensus, qui par certains aspects est réellement démocratique, est battue en brèche dans les conflits les plus graves qui sont alors portés devant les autorités administratives. Souhaitant avant tout calmer les esprits, ces dernières évitent de trancher, si bien que la plupart des conflits actuels qui portent sur la construction d'abris et la mise en culture font l'objet de procès multiples qui ne règlent rien. Certaines personnes font même état du procès-verbal qui leur a été adressé nommément pour prouver la légitimité de leur présence sur le site contesté, le paiement de l'amende prenant à leurs yeux valeur de titre foncier!

Que retenir de ces modes de gestion traditionnels pour un développement futur? Comment s'en inspirer pour de nouveaux projets? N'est-ce pas utopique de vouloir en tirer un enseignement applicable à d'autres contextes?

Suite à la succession d'échecs que les projets de développement sur parcours ont connus depuis près de trente ans, nombreux sont les opérateurs qui maintenant reconnaissent qu'il faut plutôt promouvoir une gestion souple, flexible et participative des ressources naturelles à l'image des organisations traditionnelles. Mais on aura compris, au travers des exemples donnés, que derrière ces concepts de 'souplesse ', de 'flexibilité' et de 'participation' se cachent des modes de gestion et d'organisation dont on peut attendre le pire ou le meilleur selon la manière dont on les applique. Il faut donc se garder des éloges excessifs adressés aux modèles traditionnels et n'en retenir que le meilleur.

Un territoire et des usagers

C'est la règle essentielle de toute organisation qui fonctionne correctement, délimiter précisément le territoire ouvert au parcours et identifier les usagers et autres ayants droit en respectant ou en s'inspirant des droits traditionnels, et en appliquant des règles d'inclusion et d'exclusion précises. La difficulté réside dans la finesse de l'analyse 'socio-foncière' qui doit s'appliquer à enregistrer les modes d'occupation réels de l'espace pastoral, tout en anticipant quelque peu sur la dynamique sociale. Cas par cas, il faudra donc identifier l'échelle pertinente, la tribu, la fraction, le lignage (majeur ou mineur), le village ou le hameau.

Des restrictions sur les privatisations occultes

Il faut d'abord apurer les prises de contrôle sur le parcours (*azib* et mises en culture) afin d'en bloquer l'extension future car c'est la cause principale de la mort des organisations coutumières et des échecs des nouveaux projets. Il faut cependant reconnaître aux usagers, dont le nombre croît chaque année, le droit légitime de construire et de cultiver là où c'est possible et reconnaître, dans bien des situations, les faits établis. On peut espérer dans un tel cas des interventions plus déterminées de la part des autorités administratives.

Le contrôle de l'accès aux ressources

Pour limiter le surpâturage et par souci d'équité, il paraîtrait normal de limiter autoritairement les effectifs des plus gros troupeaux et de contrôler les associations sur le bétail. L'expérience des organisations traditionnelles montre que ce n'est pas possible et qu'il est plus judicieux d'adopter des méthodes moins directes. Les mises en défens, c'est à dire l'*Agdal* appliqué selon des modalités un peu plus variées, permettent d'agir sur le temps de pâturage et de diminuer la charge. La gestion de *Agdal* doit respecter la souplesse et l'inventivité des formules coutumières (coupe d'herbe, différences de traitement des bovins et des ovins, modes de campement, organisation du gardiennage, paiement des sanctions). Le paiement, au bénéfice du groupe d'usagers, d'une redevance liée à la taille du troupeau ou l'octroi de baux de pâturage concédés par l'État, sont des formules envisageables à terme pour contrôler les effectifs.

Des institutions plus fortes

Les institutions actuelles (la *Jemaa*, les *Naïb* et autres délégués), ne sont pas toujours officielles et leur statut mérite d'être reconnu, voire renforcé afin de poursuivre l'effort de décentralisation et d'accroître le rôle des organisations d'éleveurs. Sur le plan de leur fonctionnement, la règle du consensus qui pourrait passer pour un modèle de démocratie, ne permet pas dans la réalité de faire face aux problèmes nouveaux et aux pressions de plus en plus fortes qu'exercent certains notables, ou plus récemment les élus[7]. Qu'est-il possible d'imaginer? Peut-on remplacer cette règle, vieille comme la société elle même, par des procédures plus modernes? Les récentes tentatives pour installer des coopératives pastorales inspirées du droit moderne sur les steppes de l'Oriental marocain invitent à la plus grande prudence, car tous les pouvoirs au sein des conseils d'administration sont rapidement tombés dans les mains des plus gros éleveurs qui les ont manipulés à leur convenance et à leur seul profit.

Ainsi, au-delà des quelques règles déjà connues (un territoire, des usagers, la reconnaissance de la mise en valeur, la maîtrise de la charge de pâturage), on peut conclure que l'adoption de pratiques traditionnelles au sein d'organisations modernes qui seraient à construire *ex nihilo*, paraît bien utopique et semée d'embûches. Ce qu'il faudrait reprendre des pratiques traditionnelles est en effet le moins facilement transmissible: la flexibilité et l'art du compromis dans l'exécution et la mise en oeuvre. Par contre, là où ces organisations résistent, il faut savoir les conforter et reconnaître leurs qualités qui résident dans ce subtil mélange entre l'esprit de concurrence et la solidarité entre usagers, la souplesse des règles et un laxisme apparent, le clientélisme sans scrupule des notables et l'égalitarisme proclamé, l'amabilité des arrangements locaux et la violence des revendications formelles, ou entre la simplicité de façade et la complexité des détails.

'Der Teufel steckt im Detail' (*'Le diable se cache dans les détails'*).
Il faut donc apprivoiser le diable.

---

[7] Les élus au conseil communal jouent à présent un rôle nouveau qui vient contrebalancer le poids des agents nommés par l'État (*moqqadem, cheikh*).

11

Toward a synthesis of guidelines for legitimizing transhumance

Vers une synthèse des orientations permettant de légitimer la transhumance

MARYAM NIAMIR-FULLER

ABSTRACT

A synthesis of the adaptive mechanisms that have evolved over thousands of years within pastoral societies, coupled with 'modern' notions of law and order, is useful in showing the range of options available for managing mobility in arid Africa. Those customary mechanisms that have survived the harsh tests of recent constraints are particularly important, as they provide the basis upon which new forms of institutions can be evolved. This chapter distils the fundamental principles necessary for a legitimization of transhumance, such as transience and flexibility, reciprocity, negotiation, priority of use, and cost-effectiveness. It also provides a set of practical and immediately applicable guidelines for the design of pastoral development in arid Africa, in the key areas of: property rights and legal reform, strengthening common-property management-regimes, key-site management, developing and enforcing rules and regulations for natural-resource management, mobile services, improved communication, and conflict-management.

RÉSUMÉ

Une synthèse des mécanismes d'adaptation élaborés en plusieurs milliers d'années au sein des sociétés pastorales, avec des notions 'modernes' de loi et d'ordre, est utile pour dégager la gamme d'options qui s'offrent pour gérer la mobilité des troupeaux dans les zones arides d'Afrique. Les mécanismes coutumiers qui ont survécu à la dure période récente sont particulièrement importants car ils peuvent servir de base pour mettre en place de nouvelles institutions. Ce chapitre présente les principes fondamentaux nécessaires pour légitimer la transhumance, par exemple le mouvement et la souplesse, la réciprocité, la négociation, la priorité d'utilisation et la rentabilité. Il fournit également une série d'orientations pratiques immédiatement applicables pour préparer le développement pastoral dans les zones arides d'Afrique: droits de propriété et réforme juridique, renforcement des régimes de gestion de propriété commune, gestion des sites clés, élaboration et application de règlements pour la gestion des ressources naturelles, services mobiles, amélioration des communications et règlement des conflits.

THE PREVIOUS CHAPTERS have, we hope, provided a realistic interpretation of pastoral evolution and the various ways in which pastoral development has been implemented. Several of the case studies have provided practical recommendations on how to improve pastoral livelihoods. The central theme has been the mobility paradigm – but to what extent is it applicable? In which situation is it a good idea to advocate a resurgence of livestock mobility? Based on the mobility paradigm, we propose that increasing or reinstating mobility is an appropriate tool in situations where there are combinations of the following factors:

o a non-equilibrium or multiple-equilibrium ecosystem
o aridity and low primary-productivity leading to a dependence on common-pool resources
o land degradation due to over-concentration of animals
o economic marginality.

The case studies in this book can be distilled into two different strategic models of pastoral production:

o physical mobility, where livestock are taken to the resource, but in which mobility is being curtailed (see, for example, cases in Chapters 2, 3, 4, 5, 6 and 7)
o virtual mobility, where the resource is taken to the livestock, leading to land degradation as a result of both sedentarization and importation of feed (Chapter 10).

In both cases, the general principles and set of practical solutions suggested in this final chapter can be applied towards improved and sustainable production systems.

Elite pastoralists consider, somewhat nostalgically, that transhumance is a distinct way of life, a culture that breeds autonomy, inner strength, and hardiness. But it is not necessarily an activity that withstands the pull of modern amenities such as schools and markets. The trend towards spontaneous settlement is partly due to a conscious choice by the younger generation to take advantage of the benefits of industrialization. Even if the people settle, however, it does not mean that the animals should. Managing the mobility of animals can be seen as separate from the mobility of people. What are some of the design principles and tools that can be used to strengthen livestock mobility, without inadvertently denying people the fruits of the modern world?

A synthesis of the adaptive mechanisms that have evolved over thousands of years within pastoral societies, coupled with 'modern' notions of law and order, is useful in showing the range of options available for managing livestock mobility in arid Africa. Those mechanisms that have survived the harsh tests of recent constraints are particularly important, as they provide the basis upon which new forms of institutions can be evolved. The following sections describe two sets of mechanisms: fundamental principles that could guide the design of pastoral development, and a set of practical, immediately applicable, solutions.

The definitions and explanations of concepts and terms that follow have been provided in Chapter 2. Before concluding, this chapter reviews the implications for further research, and suggests a few areas in which more work is necessary on methodologies, data collection, modelling and analysis.

Some underlying design principles

The underlying principles for managing mobility are location-specific, as they are very much tied in with each group's cultural and social history. However, a few can be distilled into a general set of design principles. They should be seen within the wider context of contemporary trends for creating or strengthening social capital, civil society and political institutions at the intermediate level between the state and the local customary institutions. The same design principles can also be applied for strengthening local customary institutions, and reforming state-level policies and laws.

> KEY WORDS: Transience and flexibility; reciprocity; priority of use; participation; decentralization; social sensitivity; managing risk; negotiation; confrontation; cost-effectiveness

The fundamental design principles have been distilled from the evolution of both traditional pastoral societies, and lessons learnt from development assistance. In the long run, this evolution has been a progressive and iterative process, and should continue to provide new insights and tools.

Transience and flexibility

The term 'transient' in the West has a negative connotation with 'homeless people'. However, here (in the context of 'transience institution') it is used in a positive sense to denote a mechanism for temporary, opportunistic and adaptive use of variable resources. Transient institutions can be informal and *ad hoc* – such as scouting and tracking, and a community organizing itself to build a well – or they can be formal – such as yearly negotiations for alliance building, and development projects. Transient institutions normally disband spontaneously after their mandate has been achieved. But in some cases, they can be transformed into more permanent structures when and if the need arises.

The variable and unpredictable ecosystem requires flexible response mechanisms for long-term sustainability, be they transient institutions, flexible rules and regulations, or flexible property boundaries. The scientific paradigm, and efficiency-led administrations, are ill-equipped to deal with the concepts of uncertainty and flexibility. Despite the rhetoric in the last two decades, flexibility is still not fully practised. Donors continue to have rigid budgeting and accounting requirements, governments continue to create rigid administrative structures, and scientists continue to develop static models. Participatory planning, heralded as a flexible tool, is not possible when one set of stakeholders (donors and governments) continue to maintain fixed rules of the game.

Part of the problem stems from the fear that there is a fine line between flexibility and chaos. Structural rigidity is imposed whenever there is a perceived danger of chaos. The strength of pastoral societies lies in the fact that even formal structures maintain flexibility in how they function. Where both formal and informal transient institutions are functioning well, it is important to give them due recognition in a judicial or administrative sense. But care must be taken not to unnecessarily or prematurely 'formalize the informal' (Vedeld, 1993b). Putting customary institutions into a straitjacket of formal, bureaucratic rules and laws, meets the needs of governments more than those of customary decision-makers.

Reciprocity

As discussed in Chapter 2, reciprocity has been shown to be the backbone of interdependence and collective action in pastoral societies.

> The flexibility of reciprocal arrangements in traditional systems, often defy the narrow conceptual categories within which we want them to fit (Potkanski, 1994).

Is traditional reciprocity of no use in the context of modern, individualistic behaviour, ruled by market forces? Exchange of favours, political alliances, gift-giving, and other forms of reciprocity still survive as a necessary part of collective action even in our modern lifestyles. Reciprocity is fundamentally the basis of market exchanges, which is codified into contracts and monetary transactions. In the context of managing mobility, therefore, transactions can be either negotiated through customary institutions of reciprocity and political alliances, or they can be based on a system of reciprocated fees and permits. Both reciprocation, and the threat of denial of reciprocation, are powerful tools for ensuring respect of collective rules and regulations.

Priority of use

'Priority of use' is an important concept that can be used to mediate between the requirements of resource inclusivity and exclusivity. The community to which a common resource is vested has the primary right and responsibility for using and managing the resource-flow from the 'home base' (or 'terroir d'attache', the term used in West Africa). Each institution within the hierarchy of customary institutions can have its own home base. The home base of the entire tribe geographically encompasses the home bases of the lineage, clan or sub-tribe. Secondary users are defined as those who regularly seek access to the resources of the primary user. Tertiary users are those who need to use the resource-flow infrequently (e.g. during droughts). Outsiders who have secondary or tertiary rights, can use the resource-flow from someone else's home base, but only according to the conditions imposed by the primary right-holder. Governance by a community is applicable to its own home base, while access rights are negotiated to someone else's home base. In recent years, the two sides of the coin, right to use and responsibility to manage, have become separated. The concept of priority of use can be translated into a legal mechanism to reinstate effective governance, not only in pastoral situations, but anywhere where there is multiple use of natural resources.

Current administrative and legal procedures are ill-placed to consider the problems of administrative scale, as evoked by Zeidane (Chapter 3, this volume), and of 'inclusivity'. The former refers to the proposition that the larger the administrative scale, the more difficult it is to manage the resources within it. The latter refers to the fact that the issue of inclusive rights to resources is contradictory to present laws that are based on exclusionary principles. However, these problems are diminished once the concept of 'priority of use' is considered. Small and fairly homogeneous pastoral units can easily be identified, and given primary responsibility for their home area. Still, each group will also have the right to negotiate access, as a secondary or tertiary user, to the resources of another group's home area. In other words, the social and administrative boundaries of a community need not coincide with its ecological ones.

Participation

The definition of sustainability is primarily egocentric and ethnocentric, depending on the land-user and stakeholder's values, needs and expectations. Stakeholders[1] may have different objectives for pastoral development, but those of the pastoralist should be allowed to be paramount[2]. In its ideal sense, participatory planning provides the context within which the objectives and decisions of all relevant stakeholders are expressed and compromises achieved between varying interests. The first generation of 'participatory projects' used the method of direct questions to elicit people's objectives for development. This method was not very successful for various reasons.

o Many pastoralists still view outside facilitators as pro-government, and therefore likely to repeat the same government slogans that have already been thrown at them.
o Far from being homogeneous, pastoral societies are very heterogeneous; individuals may profess common goals in public, but may have ambitions that are incompatible with the welfare of the society.
o Opinions of customary leaders may or may not reflect the common perceptions, because of perturbations within the socio-cultural and political structures.

---

[1] A stakeholder is defined as (i) one who has or needs access to or control of a resource, (ii) one who is affected by the use of a resource by others, (iii) one who wishes to influence the decision of others with regard to the use of a resource for scientific, ethical or other reasons (FAO/UNEP, 1997).
[2] Despite the fact that in the 1990s some pastoralists are effectively participating both at the project level and in lobbying efforts, for the majority the term 'development' has come to mean that which has been promised to them by governments, officials and projects (Sandford, 1983), i.e. synonymous with settlement, destocking, and crop farming: a prospect that many pastoralists would resist if they had a real choice.

True participatory planning requires an implicit recognition and acceptance of pastoralists as partners in the development process. Each stakeholder is well aware of his/her own objectives and will base his decisions on those, whether or not he is able to verbalize the objectives as a list.

As part of this participatory planning process, the term 'co-learning' has been used in the sense of informed discussion of development options, and inviting both 'insiders' and 'outsiders' to benefit from participatory appraisal and planning sessions (Cousins, 1995). Where a co-learning approach is used, the process of planning and implementing natural resource-management options becomes more viable and successful. Another term more familiar to those involved with Integrated Assessment and Ecological Economics is that of 'scoping', which refers to the process by which the objectives, needs, interests and ideas of stakeholders are identified and put up for public debate.

Participatory planning among mobile pastoralists cannot be expected to be as rapid and easy as with sedentary populations. For example, experience involving the Kenyan Forestry Department and the Turkana in the late 1980s showed that the process of planning and building up the appropriate institution took six years, while another four years were spent in investment costs (Barrow, 1991). Similar experiences have been reported elsewhere, e.g. among the Fulani in Mali (Marty, 1993). There are several problems of participatory planning specific to mobile pastoralists, including: political marginalization, imposition of foreign concepts, extraction of ideas rather than real participatory analysis, and the limited patience and financial resources of government or NGO representatives to follow the transhumance cycle (Waters-Bayer & Bayer, 1995).

Rule by consensus, whereby a decision is only taken if there is full agreement by all concerned, is practised by many pastoralists and other traditional societies (with a few notable exceptions in extremely hierarchical societies). Rule by consensus may appear to be a thing of the past given the requirements and precepts of 'representational democracy'. In the latter, decisions are based on majority rule, and those in opposition have to accept the decision. Today, full consensus is difficult in situations where the customary system is breaking down. In addition, gathering all mobile and dispersed peoples on a regular basis involves both logistical and opportunity costs. Bourbouze (Chapter 10, this volume) suggests that one can, instead, talk of 'consensus of those concerned'. Those pastoralists who have the most at stake in any given endeavour, must be able to take the lead in decision-making. Others who may have an indirect interest should then be involved at relevant points in the process.

Decentralization

In discussing decentralization, it is useful to make a distinction between 'devolution', where power is given to new institutions, and 'de-concentration' where power is given to existing peripheral institutions (Carney, 1995). These terms are normally applied to the decentralization of government structures, but in the spirit of participatory development and co-management, they can also apply to the transfer of power to customary pastoral institutions. Thus, de-concentration is applied to cases where viable pastoral institutions continue to

271

exist. De-concentration would then imply a transfer of power, with relatively less need for capacity building than in devolution, where a new pastoral institution needs to be created. The concept of subsidiarity is applicable to both devolution and de-concentration. This concept specifies that 'tasks should be carried out as near as possible to the level of actual users of resources or beneficiaries as is compatible with efficiency and accountability' (Swift, 1993).

Decentralization in Africa has been accompanied by 'democratization', i.e. the modification of both customary and post-colonial structures towards a western-style representational democracy. As mentioned above, it has been difficult to apply democratic processes, such as electioneering and voting, in mobile pastoral areas, because of their dispersion and mobility, but also because of the difficulties of defining the constituency to begin with – since ethnic boundaries are not synchronized with administrative boundaries. So far, representational democracy has largely involved the élite and urban-based pastoralists.

In some cases, democracy has proven to be a useful tool for change, such as among the Karamajong of Uganda (Niamir-Fuller, Chapter 7, this volume). This is not so in other cases, particularly where pastoralists are a social and political minority. For example, the 'national debates', established as the necessary process for developing participatory plans for the Convention to Combat Desertification, have in large part failed to identify appropriate representatives of mobile pastoralists. As a result, this important group of stakeholders has been left out of the planning process (UNSO, 1995). Full public participation will be weak so long as mobile pastoralists see no immediate gain in participating in local- and national-level representational politics. Representational democracy can only be successful when local customary institutions are re-created or strengthened, empowered with local governance, and allowed to define their own communal administrative boundaries, constituencies and objectives.

Both decentralization and participation are viable concepts that require close attention being paid to the social dimension of natural-resource management. The definition of the 'community' is one of the more important, and yet, more difficult tasks. Internal heterogeneity makes it difficult to define membership boundaries. In addition, in non-egalitarian societies, it is difficult to ensure full representation by all socio-economic groups. In cases where customary institutions no longer persist, the community behind those institutions is difficult to identify. Needless to say, the identification of the 'community', to which the responsibility for designing a common-property regime would be given, should be made by the pastoralists themselves.

The pitfalls of most projects have been in rigidly defining the boundaries of a particular community, and inadequate participation of surrounding communities. Isolated projects that focus on only one or two communities create more harm than good since they increase the chances of conflict in the long run. It would be helpful if there could be a concerted, nationwide definition of social units that does not rigidly classify people into ethnic categories, but rather into an agreed-upon set of socio-geographical communities. A nested hierarchy of socio-geographical units, reflecting the nested nature of communal property, would ensure that there are a series of institutional structures, from local to regional or federal, to accommodate the needs of mobility. Exclusive and inclusive land tenure can then be assigned accordingly. This is a long process of administrative reorganization that only a few countries up to now, notably Senegal, have attempted. In Morocco, such an attempt in pastoral areas has opened up old wounds and rivalries. An appropriate, neutral, forum to allow negotiation of not only existing rights, but also what are perceived to be future entitlements[3], may reduce the inherent conflicts in this process.

Social sensitivity

Men have always seen themselves as being the main engines in pastoral systems. However, women not only have a major complementary role, but in many cases, actually have the dominant role. In most transhumance systems in Africa, married women do not herd the animals, although young girls often do. Rather, women are responsible for small stock, calves, poultry, and sick animals kept at the homestead. They are also responsible for checking the grazing animals as they return from pasture, for watering them, for collecting or producing supplemental feed for them, and for milking and processing milk. In addition, women's role as the 'nurturers and healers' often gives them the primary responsibility for traditional healing of animals, although men control the use of modern veterinary care (Niamir, 1994c). In some cases, such as in areas with high male migration, or with women-headed households, or where women control the use of natural resources, women may be in a better position to manage common-pool resources.

Given the breakdown of customary institutions, and the increasing fragmentation of local authority, the process of decentralization and building local-level natural resource-management capacity cannot be expected to be smooth. Internal divisions and factions, and the emergence of new stakeholders such as women and the younger generation, widen the potential power base. Efforts to create new decentralized common-property management regimes (CPRs) therefore, are often marked with internal conflict (Cousins, 1993). Including younger members and women can sometimes lead to 'silent' conflicts within the family structure (Faure, 1995).

Customary institutions are not necessarily egalitarian organizations (Ouedraogo & Rochette, 1996; Ribot, 1996). They sometimes reflect only the interests of the élite, or are used as vehicles for promoting inter-tribal enmity and competition, and have been known to protect members who violate State or ethnic laws (Catley, 1996). Strengthening local decision-making processes, in the form of selective capacity building, creation of new structures, and/or imposition of new rules and regulations, is appropriate only when it ensures sensitivity to social dynamics and full participation of all relevant stakeholders.

There is an inherent contradiction, however, in asserting on the one hand that sustainable development should be based on participatory processes and existing structures and dynamics, and on the other, pushing from the outside for issues of

---

[3] We use the term 'entitlements' according to its simple definition in a standard dictionary as a title, right or claim to something, rather than as it has been used recently in the context of a theory of entitlements and endowments (Leach *et al.*, 1997).

273

social sensitivity and equity. How far can the western world-view of human rights be imposed before the process is labelled, once again, 'top-down'? Development workers, too, should be seen as stakeholders in the process of 'negotiating' pastoral development.

Managing uncertainty and risk

Uncertainty in arid lands is a fact of life (Scoones, 1994). Uncertainty in the natural environment is manifested in the patchiness and variability of natural resources. Only through daily monitoring, frequent re-negotiation of access rights, and flexible institutions for managing the natural resources can such uncertainty be reduced. Risk management includes both risk-buffering and risk minimizing. Risk minimizing refers to preventive actions, i.e. ways in which economic risk can be reduced, such as alternative-income generation, land security, and increasing livestock productivity[4].

Risk buffering refers to reducing the impact of risks, and allowing a rapid come-back. These could include communal insurance schemes, collective guarantees for bank loans, and local savings banks. Experiences with such techniques in agropastoral and sedentary situations have been mixed, but generally positive (see e.g. *Haramata* Newsletters, IIED). Innovative ways should be found to apply them to mobile situations.

Traditional risk-management techniques among pastoralists remain viable if they function within the immediate circle of the household or group of neighbouring households. With the strengthening of CPRs, it is to be expected that there will be an increasing venue or basis upon which to reinstate or create risk-management techniques that are managed at higher institutional levels. Several experiences with 'drought contingency planning' have concentrated on creating risk-buffering mechanisms in Africa against the effects of droughts, such as restocking schemes (e.g. OXFAM's 'Habbanae' programme in Niger) and credit schemes (e.g. UNSO's programme in Dori, Burkina Faso). Their success is correlated with the social and institutional strength of the local CPR, with minimal external stress (e.g. crop encroachment, range shortage), and with supporting external institutions, such as banks.

Negotiation and confrontation

Granting common-property rights to local institutions and organizations is a national political process of empowerment that pulls against resistance from the centre. There are two alternatives for pushing the process through: confrontation (lobbying, advocacy, litigation) and negotiation (compromise).

The choice of strategy will depend on the particular set of circumstances

---

[4] The case of Machakos, Kenya provides an insight as to how the reduction of risk can help solve environmental problems. The large-scale land degradation and famine predicted in the 1930s for Machakos District never occurred despite a three-fold increase in livestock numbers, a five-fold increase in population, and a 35 per cent increase in cropland. Risks were minimized through alternative-income generation (dairy and horticulture), land-tenure security, and greater access to markets, which in turn generated enough local capital for investment in soil and water conservation (English *et al.*, 1992).

274

surrounding the process of change. In cases where there is unequal power, or where the State is reluctant to relinquish power, lobbying and advocacy for pastoral rights would be a necessary precursor to change. Litigation is an important tool where feasible. In cases where the process of decentralization and participatory development is well advanced, establishing the procedures for negotiation would be necessary.

As discussed in Chapter 2, there has recently been greater attention given to the role of negotiation in CPR and conflict resolution. The generic objectives of any process of negotiation are to:

o obtain access rights to resources
o establish or reinforce one's status or priority in the process
o maintain peace and good relations.

Negotiation and compromise between individuals in a community, between groups, or between pastoralists and the State, require a strong bargaining position, access to information flow and an effective judiciary, things that are rarely accessible to marginal and minority groups in developing countries. Procedures for negotiation, whether for resource-allocation or for conflict-resolution, cannot be divorced from customary rules. They are therefore very site-specific. However, they should include the following:

o a formal or informal institutional structure within which negotiating parties can meet peaceably
o appropriate representation of each party
  – in the case of negotiation between groups, representation by leaders allows the expression of collective needs
  – in the case of negotiation by individuals, representation by 'advocates' minimizes the emotional contact between parties
o means through which parties can strengthen their representation (capacity-building, prior collective-consultations)
o public discourse and openness as appropriate to each situation, acting as a tool for informal enforcement of the accord
o fact finding, or the investigation of key issues, either by a third party in the case of a conflict, or by the respective parties in the case of natural-resource management agreements and accords
o facilitation, or assistance by a third-party mediator in running the meeting, particularly useful in case of conflicts
o record keeping, whether oral or written, both for establishing precedence and for legitimizing subsequent enforcement of the accord.

Cost-effectiveness

There are costs associated with any change, whether they are planned or unforeseen. Costs affect the individual, the community and the nation. Costs are calculated not only in economic terms, but also in terms of losing social capital (set of values and mores), and environmental resilience. Costs may be distributed differently depending on the heterogeneity of the population (the poor or women

275

may bear more of the cost). The community may bear more of the cost than the nation. Cost–benefit analysis of expected changes, based on both quantitative and qualitative values, is a useful tool for helping local decision-makers make appropriate choices among paradigms and tools. Local communities should not be expected to bear undue costs if the benefits accrue primarily to the nation.

Some practical solutions for managing mobility

Today, there is a growing interest in pursuing development paths that confer a greater role on civil society. In this spirit, there is a greater interest in the decentralization of political authority and the strengthening of indigenous common-property institutions. Given the history of pastoral development, these new emphases are positive developments. Using the principles distilled from pastoral systems, we can now focus on the practical aspects of change. The solutions discussed below subsume one or more of the principles discussed above. They are a general list of activities that could be applied depending on local needs.

Property rights and legal reform
In most official tenure laws, there is very little recognition of common ownership of land, or of 'joint usage rights'. Indirect recognition is given only by way of recognizing 'customary ownership' in some colonial and post-colonial statutes. Finally, there is no jurisprudence developed in writing on these issues (Vedeld, 1994).

Historically, land-tenure reform in pastoral areas has consisted of either nationalization or privatization of the commons (land-titling or group ranches). These efforts have not only been resisted by pastoral peoples but have proven unsuccessful by either economic and ecological criteria, since they have led to a weakening of common-property institutions and have not escaped the strong environmental requirements of livestock mobility.

Careful attention needs to be paid to the distinction between sovereignty (e.g. all common lands are vested in the State), ownership (local communities own the resource) and usufruct (neighbouring groups can use the resource). Table 11.1 categorizes the types of nested rights typical to pastoral systems. The customary pastoral territory is the land area claimed by the tribe or other higher social unit as its home base, as distinct from other claims. The boundaries of these territories are relatively fixed from year to year and identifiable by landscape features. Buffer zones and overlapping areas refer to resources systems in between customary pastoral territories. Overlapping areas are zones over which neighbouring tribes have dual rights, and which are subject to negotiation and cooperative management, or conflict. Buffer zones are contested areas where permanent claims are not recognized, and the area is rarely used except in times of need (drought, epidemics, etc.).

Each clan, sub-tribe, or fraction has an annual grazing area, covering its seasonal movements, that usually includes and extends outside the home base.

276

Table 11.1. Typology of customary property rights and management regimes

Nested property right	Type of boundary	Management regime
Overlapping territories	Relatively fixed	Tribal council
Buffer zones	Relatively fixed	Tribal council
Customary pastoral territory (home base)	Relatively fixed	Tribal council
Annual grazing area	Extremely fluid	Clan, fraction, sub-tribe (mixture of primary and secondary rights)
Range reserves	Fixed	Clan, fraction, sub-tribe, tribe
Key sites	Fixed	Clan, fraction, sub-tribe
Cropland, special resources	Relatively fixed	Camp, village
Cropland, special resources	Fixed	Individual

The geographical boundary of this area is extremely fluid from year to year because of variability in rainfall. Depending on the customary political system, secondary access rights are established either through dictum from the higher level institution, or through yearly negotiation at more local levels. Most pastoral groups establish well-defined range reserves within their annual grazing area in order to provide a 'savings bank' of forage. Most reserves are communal, such as drought reserves (e.g. Odell, 1982) and sacred sites (Schlee, 1987), but some are also private, such as the immediate surroundings of Maasai camps (Ole Kuney & Lendiy, 1994). These reserves perform the dual functions of risk-reduction and maintaining ecosystem resilience.

Key sites are special areas of relatively high value and either limited or extensive geographical scale, such as water points and the forage immediately around it, communal salt licks, special high-value pastures, etc. By being special patches, the boundaries of key sites are relatively fixed, although the importance of the patch will change seasonally and inter-annually. Each annual grazing area will try to be self-sufficient in a good array of key sites, but in times of need access by other clans/fractions is defined the same way as access to annual grazing areas, i.e. through negotiation. Crop-land, man-made water points and other special resources (e.g. special trees for apiculture) can be owned and managed either by an individual household or by a group of households (neighbourhood, camp, or village). The boundaries of these resources are fixed.

Land reform in its generic sense is much needed in order to increase the security of pastoral claims to land. Still, the institutional requirements of live-stock production on arid lands, most notably the need for livestock mobility, need to be seriously considered. These requirements can be met through legal recognition of nested property rights, fluid boundaries, inclusivity, and the concept of priority of use.

Agricultural encroachment and weakened customary institutions governing pastoral usufruct are widespread throughout the continent. The growing problem of land expropriation is particularly acute in north and east Africa, where there have been calls for a complete moratorium on land titling until the rights of all can be surveyed and recorded – a sort of 'freeze' on the current situation (e.g. see proceedings of PANET[5] workshops in Tanzania, 1991–1995). Some pastoral advocates have called for the creation of 'Pastoral Reserves' (ideas ranging from the northern Nigerian model, to the North American Indian model). Others have called for the development of a 'Pastoral Code or Law', similar to the forestry laws of many nations.

Land-tenure reform takes time to develop. The government of Niger, for example, has been working since 1986 to elaborate a 'Rural Code' guaranteeing land security for all rural peoples, and is still not ready to pass it as law. Therefore, it has been suggested that while 'substantive' law is being gradually reformed, greater attention be paid also to 'procedural law'. Substantive law refers to rules of right which the courts are called on to apply, while procedural law applies guidelines within which land conflicts or land needs could be addressed:

> Instead of legislatively dictating detailed property rights to pastoral and agricultural resources, the procedural law could specify the framework within which the concerned parties could legitimately put forward their claims to a certain resource. . . Over time, a jurisprudence would develop and competence in the processing governments and organizations [would] be built. (Vedeld, 1994)

In considering reforms, policy makers should consider carefully the roles that more informal institutions have played in providing controllable but flexible resource-access in arid rangelands. A significant hazard is the imposition of overly rigid substantive regulations that specifically allocate resource ownership to different groups. Through greater specification, such regulations may reduce the channels through which pastoralists can gain access to important resources. Increased security is necessary but that security should not be attained solely through prescriptive rules that actually increase the exclusivity of pastoral resources. Resource holders should retain authority to grant temporary use-rights to secondary and tertiary users. Without such flexibility, increased tenure-security may be associated with increased economic insecurity.

Procedural law is more consistent with such requirements. Strengthening procedural law would include developing administrative and judicial institutions at the local level, establishing through local dialogue and participation the principles and guidelines for judging on claims, and creating the means and procedures for enforcing the rulings. Appropriate legal terminology should be developed to recognize temporary rights of usage under procedural laws. The 'splitting' of tenurial rights into primary, secondary or tertiary could be legalized and applied specifically in transhumance areas, but also in other areas with multiple land-use systems. A clear distinction should also be made and legally

---

[5] PANET (Pastoral Network) is an NGO based in Dar es Salaam whose meetings were sponsored by different donors, including UNSO, FAO and Danida.

codified between rights to use (usufruct) and stewardship rights and duties, thus allowing the flexibility inherent in the duality of owning the resource system, and controlling access to the resource-flow. This 'learning by doing' approach based in a participatory process implies that tenure reforms would be situation-specific, and that new tenure regimes would evolve through local competition and power struggles, thereby acquiring legitimacy.

Strengthening management regimes

A commitment to devolution, de-concentration and participatory processes, while important, may prove insufficient to maintain and reinvigorate the management of the pastoral commons. Strengthening CPRs rests on effective leadership, as well as appropriate structure and function of institutions. The process by which traditional indigenous knowledge is transformed into new action is of primary importance for strengthening management regimes. The inherent capacity of leaders, the degree of social cohesion and discipline, a collective image or ideal for the future, and the ability to innovate and experiment, are the main ingredients for a successful transformation. In many cases, however, indigenous knowledge, economic logic, and political leadership of pastoral communities have changed so drastically as to make reinvigoration of 'customary' institutions extremely difficult.

In these cases where community cohesion has eroded, the imposition of new, more formalized common-property institutions may be the only recourse. Creation of local institutions is not always a straightforward process. In many cases it can be so alien a structure that vestiges of the customary institutions continue to function alongside it. In other cases, it can be constrained by local factional politics. For example, in Leliefontein Reserve in South Africa, the various committees set up within the village to plan and implement projects have devoted a great deal of time to politicking and setting up rival committees, all of which have tended to create an 'attitude of passivity towards controversial issues among the majority of residents' (Cousins, 1995).

Where possible, policy makers and developers may have more success by looking to build upon newly formed transient institutions. For example, for the past seven years, the Messeriya of northern Sudan have established agreements with the Sudanese People's Liberation Army (SPLA), the main rebel group in southern Sudan, to use pastures in the southern 'Sudd'. At the boundary between north (Messeriya land) and south (Dinka land), the Messeriya are requested to deposit their guns, in exchange for a receipt and permit, and then to enter to graze for a specified period of time. Several Messeriya men, either neutral or loyal to the SPLA, are stationed at this 'border' to liaise with the tribesmen, and to ensure the orderly transaction of guns-for-receipts (Jamma'a, personal communication). This institution, clearly a transient one that may disband when the civil war terminates, is responding to a present need (the need for peaceful access to pastures), but has the potential for being used in the future as a nucleus of an institution for peaceful cooperation and development. Transient institutions can act as 'catalytic' structures that pave the way for more appropriate and less transient institutions.

279

As discussed earlier, viable transhumance systems must function within a nested hierarchy of institutions, linked through legitimate and streamlined procedures for negotiated access, with appropriate sets of functions based on the concept of subsidiarity. Decentralization and devolution cannot mean the atomization of pastoral and agricultural usufruct among individuals or small communities. Higher-order institutions are needed both to facilitate and co-ordinate broader-scaled movements (within and between countries) and to manage conflict.

Co-management arrangements form the basis of the newest experiments in CPR in Africa. The state has an important role in initiating and fostering co-management. It must assist with the internal workings of the local institution to confer legitimacy for local-level decision-making, and with law enforcement and conflict-management wherever necessary. Through substantive and procedural laws at both the national and local levels, the state must ensure that the boundaries of management regimes, however fluid, will be protected against expropriation and violation. Management of livestock mobility requires multiple institutions working at multiple spatial scales, authorities, and functions.

Key-site management

Key sites within the annual grazing area of pastoral groups are the 'attractors' in the system. The degree of concentration around a key site defines the amount of pressure on the surrounding land (e.g. Dougill and Cox, 1995). Pastoral mobility is defined by the capacity and variability of these key sites. Depending on their characteristics, key sites are an essential safety net in periods of environmental stress (Moorehead, Guèye and Toulmin, 1996). Key sites can have multiple purposes, thus strengthening the justification for investment in the key site. For example, creation of strategic water points can also be combined with establishment of seasonal service-outposts.

Range improvement, ordinarily quite expensive because of the heavy need for labour, capital or external inputs, is only feasible if carried out in specific key sites. Pastoral groups traditionally employed such techniques as bush fires, over-browsing, and strategic water-point development to improve the capacity of specific key sites in their rangelands (see e.g. Jacobs, 1980; McDermott and Ngor, 1983; Reisman, 1984). Modern rangeland improvement techniques[6] on key sites can be within the reach of pastoralists only if they:

o are based on indigenous knowledge and practices
o respond to the production system needs
o require low external inputs
o set modest and progressive objectives
o focus on key sites selected by the community

---

[6] Range-improvement is used in its wide sense to denote any investments in rangeland resources that are intended to increase productivity and sustain environmental health. This can include water-point development, pasture modification, landscape management, or maximization of heterogeneity in the ecosystem.

o involve secondary and tertiary users in the process (either in the implementation stage or by paying user fees)

o are used where the community has, if necessary, access to credit and other financial resources.

Key-site management involves not just improvement of the resource, but also maintaining the control over its use. Allocation of the responsibility of key sites to well-identified communities is a primary prerequisite. Clarifying and securing tenure of key sites would have a greater impact on local-level natural-resource management than would security of the overall customary territory, because access to the land area is defined by access to key sites. In addition, it is easier and cheaper to survey and cadastre key sites than entire territories.

Classical land-use planning, wherein only the boundary of the entire territory is surveyed, marked and titled, is not suitable for pastoral management, since boundaries are fluid, and productivity is defined by patches and key sites, rather than the whole territory. However, land-use planning is extremely useful as a safeguard against illegal expropriation. Therefore, as a practical measure and to avoid undue cost, it is proposed that land titling be carried out in two complementary ways. The first is to survey and title all key sites, as identified by pastoralists in their 'home base', and the second is to mark the boundary of the home base, especially where it is subject to agricultural and other encroachment, or in areas subject to dispute between neighbouring socio-geographical units.

Local capacity-building

Capacity building may be needed for strengthening the negotiating power of socio-geographical communities in conflicts and for securing access rights to resources. It may also be needed for establishing and maintaining CPRs, for increasing the level of social discipline, for key-site management, for reducing and buffering against risk and uncertainty, and improving the capacity of individual members to carry out intended functions. Traditional institutions have demonstrated their ability to survive and adapt to eons of change (both environmental and anthropic change). Their inherent resilience is a trump card that should not be thrown out by the zeal of 'capacity-building' programmes. There is a need to confirm the important role of local notables, and at the same time to prepare appropriate procedural guidelines adapted to each situation for implementation of CPRs and conflict-management mechanisms.

Training of individual members may be necessary, depending on the situation, in such aspects as accountability and accounting, planning, procedural rules, and negotiation. Formal training is an important component of capacity building, but it is not the only one. Learning by doing, or the practical side of institutional strengthening, is just as important. The act of simply creating institutional structures (e.g. local associations, village governments) is not a sufficient activity for capacity building. Institutions need time to work out internal conflicts, difficulties and inertia. Progressive handling of functions, e.g. starting with the establishment of a cooperative and ending over several years with yearly development planning meetings, is necessary to build the internal capacity of institutions to function within the modern context.

It may be necessary, depending on the context, to focus on the 'professionalization' of herders and scouts. This is intended not only as a capacity-building measure, but also as a means through which the exodus of the young from the pastoral profession can be reduced. Professionalization could be carried out through training courses, awarding of diplomas, certificates and badges of merit, local and regional competitions, as well as through provision of material inputs (horses, radios, etc.).

Rules and regulations for natural-resource management

The combination of informal and formal institutions for natural-resource management should be able to cover: scouting and tracking resources, resource-management strategies, monitoring access and utilization, negotiation process, conflict-management and enforcing procedures. Three important types of rules are needed for successful common-property management (Ostrom, E., 1993; Agrawal, 1994):

o membership rules that define who can use the resources, their rights and responsibilities
o appropriation rules that specify how much of what type of resource can be extracted, or the condition of the resource after extraction
o procedural rules that empower monitoring, sanctioning, arbitration and negotiation.

Given the requirement for inclusivity, membership rights and responsibilities should be graduated according to priority of use. Appropriation rules can assign quotas, fees, and other means of limiting use-levels. The specific rules would be developed based on frequent monitoring and tracking of the ecological *and* political environment before and after use. Procedural rules would be developed to determine the means and the processes of tracking, monitoring and evaluation. All these rules and regulations should be based on indigenous knowledge systems and be both formal and informal, in order to maintain flexibility in the system. Informal rules and regulations can be communicated through various 'awareness' techniques, such as informal education and theatrical shows, and formal ones through public edicts, signed accords and contracts.

The development and enforcement of rules and regulations should rely on a few important mechanisms:

o a decision-making body with the requisite social and moral authority
o the means to communicate the decisions to the entire community (information exchange, participation by users in decision-making)
o guards and enforcers of law to monitor and enforce the regulations
o means to punish transgressors
o popular involvement and participation in contributing inputs to the process.

Developing the rules requires full and consensual participation by members of the community (hosts) and other temporary users (guests). Provisions must be made in case of droughts, and for accommodating guests and immigrants. Flexibility has to be maintained in how natural resource-management rules are developed, and applied. The rules must be subject to continual revision and re-negotiation, using the same participatory process, depending on the exigencies of the physical and social environment. Scale is also a very important consideration. Rules that function well at one level, may not be applicable at other levels. For example, social ostracism works very well at the local level, but loses its value at the regional level. Thus the design of rules and regulations must remain appropriate to the context.

Rules that restrict ('don't do this') are preferable to those that impose ('do this'), because it is easier to enforce the former. The institution engaged in collective choice could design rules that are too lax or too restrictive. If too lax, then the cost of monitoring 'free-rider' problems becomes too high, and the resource may be over-exploited. If too restrictive, the amount of violations may outstrip the capacity of the institution to monitor and impose sanctions. Only through participatory planning, and subsequent trial and error would a newly created institution be able to strike an appropriate balance.

Finally, rules must have a buffer capacity (Ostrom, E., 1993). In other words, there must be some tolerance for a certain level of infringement, to allow flexibility and dynamism, as Bourbouze points out in his case study (Chapter 10) In addition, sanctions and punishments must be graduated not only to fit the crime, but also to fit the frequency of offence. First-timers would get relatively lower sanctions than repeating violators.

Law enforcement

The previous sections have described how appropriate rules and laws can be developed. In this section we recognize that creating a rule is not enough; it needs to be enforced. The breakdown of traditional patterns of rangeland use is partly due to the breakdown of social discipline, and customary institutions for popular enforcement of rules and regulations. 'Rebuilding' the social capital and a sense of community is not an easy task, and may take a generation or more. Several solutions need to be pursued simultaneously. On the one hand, both formal and informal customary law-enforcement mechanisms, such as social ostracism, sanctions, incentives, fees, etc., need to be strengthened through social-awareness programmes and public discourse. On the other hand, formal contracts, codes and laws can be passed to legitimize local-level decisions.

These formal documents are important tools for legitimizing new paradigms and concepts, and for guaranteeing the 'rules of the game' between neighbouring groups. However, given the requirement for flexibility in the management of pastoral resources, it is important that these instruments be used only where appropriate, and be adjusted as the need arises. One way is through constant monitoring, which must be designed in such a way as to provide timely information and feedback on how well the rules are functioning, and whether there are any anomalies, such as: free-riders, rent seeking, asymmetric access to information, and corruption. In addition, there must be a mechanism by which 'undue loss', as a result of obedience to the rules, is identified among members of the community or among neighbouring communities, and appropriate compensation given.

In remote and mobile pastoral systems, the customary judicial system is often the only recourse for adjudication of conflicts. Can the customary and modern judicial systems be reconciled and effectively work together? The studies by Zeidane and Bourbouze in this volume (Chapters 3 and 10) indicate yes. It is necessary to identify an appropriate division of responsibilities that shows the respective jurisprudence of each system, and to provide the mechanisms by which cases can be referred from one system to the other.

Improved communication

Without adequate access to information, many of the adaptive mechanisms used by pastoralists would be impossible. Tracking resource variability, negotiating resource access, resolving conflicts, and more, requires the free flow of information. Information in pastoral areas is disseminated through five types of sources:

o indigenous experts (those who do better than others)
o indigenous professionals (specialists)
o innovators (the same as experts but who experiment and take risks)
o formal intermediaries (those who are formally designated to communicate information but who rarely generate information themselves, such as mediators), and
o informal intermediaries (all other members of the group).

There are at least five types of indigenous channels for communicating information to others: folk media (actors, story tellers), indigenous organizations and formal gatherings, deliberate instruction (passing information to younger generation or newcomers), oral or written records, and unstructured talk (in informal settings such as markets, road, wells) (Mundy and Compton, 1995). The so-called 'rural radio' is very strong in pastoral areas, precisely because of mobility.

Individualistic behaviour, internal divisions, and State hegemony are able to restrict access to communication channels and information. Both state governments and in some cases customary authorities do not always have a strong capacity to impart information equitably. In strengthening CPRs it is important to make provisions for improved communication, between members, between groups, and between co-managers of resources. Negotiation between far-flung mobile groups for access to resources requires communication channels that may function only once a year. Democratic representation requires easy access by the constituency to its leaders and representatives. Communication channels therefore have to be embedded within an appropriate institutional structure.

Communication in pastoral areas has to be a mixture of both formal and informal. Popular communication needs to be encouraged, such as the itinerant 'criers', folk artists, gatherings such as markets or wells, and scouting parties, or it could be formalized within newspapers, magazines, and radio broadcasts, and eventually satellites and electronic mail. Recent work by several NGOs in West Africa is remarkable, exemplified by their efforts in translating the 'news' into local languages, using Arabic or Latin script[7].

Mechanisms for conflict resolution

As discussed previously, conflict management involves a large portion of conflict-prevention. In the case of natural-resource management among pastoralists, this implies the existence of institutional frameworks that can effectively regulate and allocate resource-flows to multiple users. Such institutions are subsumed under the title of 'common property regimes' and have already been discussed. In addition, managed mobility systems should help strengthen existing economic and environmental symbiosis between mobile and sedentary land-users. One example is the 'host–client' relationship as proposed by Turner (Chapter 5, this volume).

Once conflicts do occur, then dispute-settlement and conflict-resolution processes become more important. A paradigm shift on conflict-resolution would be helpful: i.e. a shift from seeing conflicts as only negative, to recognizing their positive impacts (Cousins, 1996; Ouedraogo & Rochette, 1996). Conflicts can be a learning experience for the community, by drawing their attention to underlying causes of conflict such as inequality, social disintegration, etc., and to developing appropriate responses to the underlying problems. The very process of conflict-resolution is important in building up a 'mass of mediators' available in the community to help on future cases.

Inequality limits the usefulness of negotiation, mediation and other joint problem-solving mechanisms (Cousins, 1993). Inequality may pre-empt the very process of discussion; weaker parties may not even arrive at the bargaining table. Very often, mobile pastoralists have been weak minorities. Several techniques can be employed to equalize power, such as modifying the procedures used to resolve the conflict, legal interdictions designed to create an even playing field, mobilizing and organizing of local institutions, forming alliances with external organizations for lobbying and capacity building. In almost all cases, the intervention of a neutral third party is indispensable.

Third-party mediators can be customary leaders and institutions, or modern NGOs, but the latter should not be allowed to substitute for the former. Local-government structures must also be strengthened so as to provide a supporting role, especially in cases of regional conflict between tribes, or in cases of preventing conflict arising from local political machinations. Institutional strengthening toward equitable conflict-resolution is only possible when the political will exists to create an even playing field.

Although most of the work on conflict-management will have to be done at the local level, there are a few 'up-stream' issues that need to be tackled in order to support successful conflict-management at the local level. According to the

---

[7] See for example, *Jewdi men*, published by APESS in Burkina Faso, and the work of ARED in Senegal.

285

conclusions of the Conflict Management Workshop held in Bobo-Dioulasso (Burkina Faso) in June 1996[8], the following conditions must be met:

o revision of land-tenure laws in order to integrate specific needs of mobile pastoralists, and to assure their tenure security
o imposition of a balance between agriculture and livestock in terms of policies, development projects, and in the conflict-management process
o enhanced dialogue and collaboration between different stakeholders (e.g. pastoralists, NGOs, governments, donors).

Mobile services

The delivery of services to mobile pastoralists is one of the aspects of development that has been greatly neglected. In the past, delivery systems were constructed on the basis of settled, concentrated communities, thus justifying the per capita costs. Both the content of the services (particularly true in the case of farming products and education) and their structure were designed for sedentary life. Typically in most local administrations, agricultural services were better stocked than livestock services. Schools taught pastoral children the value of a settled life – school practicals were on small plots of crops or stall-fed animals, not on extensive livestock production. By being concentrated in the capitals of wards, cantons, and provinces, these services rarely reached remote areas. Mobile pastoralists benefited on an *ad hoc* basis, i.e. whenever they passed by the service centres on their transhumance routes.

In a few countries boarding schools were set up for pastoral children, however these had a high dropout rate. Not only was the curriculum inappropriate to herders, but much needed labour was taken away from the household (Swift & Lane, 1989). Only recently have experiments with mobile and so-called 'long distance' services been initiated, some quite successfully (Ezeomah, 1985). Mobile services include itinerant service givers, and community workers.

The concept of community workers, whereby members of the community are selected and given training and equipment to cover basic services, has been implemented in many countries among pastoralists and agropastoralists. Problems emerging from these experiences are related to inadequate community participation in planning, and inadequate monitoring and supervision of the workers by professionals. However, they do appear to fill a perceived gap in service delivery (Swift, 1988).

Long-distance methods have been applied more particularly to providing education over the radio. Mobile units, seasonal outposts, radios and satellites are making interesting contributions. Increasing decentralization and empowerment of local communities, as well as privatization of such services as health and education, should help in widening the reach of delivery services. The State will continue to have a role in standardization, quality control, and in ensuring coverage of the less 'profitable' (i.e. remote) sectors and areas.

---

[8] 'Atelier sur la Gestion des Conflits liés aux ressources pastorales', 1996, PRASET/GTZ.

Implications for research

There are ample grounds for arguing that pastoralism is one of the most intensely researched fields. However, given the current paradigm change, many of the old methodologies and analyses need to be reviewed. Well-targeted research would then have the objectives of developing appropriate methodologies, collecting data on factors and variables not considered important in the past, analysing the information through the lens of the new paradigm, establishing appropriate monitoring-mechanisms, and contributing to the development of analytical models. However, if research ignores local-level participation, then it could run the risk of producing tools and recommendations not relevant to pastoral development. Local resources can contribute to designing a research protocol, to collecting the information (at a lower price), and to helping researchers analyse the information.

Some of the main areas in which further research is needed are:

Ecological issues
o ecological monitoring of long-term variability and uncertainty in primary productivity
o appropriate methodologies for studying land degradation
o understanding the indigenous knowledge and systems behind tracking
o indigenous indicators for environmental monitoring, especially for monitoring variability
o integrated assessment (ecological economics) of the value of extensive range-lands in all its aspects, production, biodiversity, climate change, etc. (e.g. what are the relative costs and benefits when rangelands are ploughed up?)

Institutional and land tenure issues
o documenting and analysing informal institutions (flexibility, transience, rules, etc.)
o analysing the nested hierarchy of indigenous institutions (relations, efficiencies, leadership issues, etc.)
o 'mapping' multiple-use patterns of extensive common lands (priority of use, precedence, choreography of movements, etc.)
o documenting the costs and benefits of institutional change
o documenting conflict-management mechanisms, both formal and informal.

The objectives of research should not be to extract information, as is too often the case, but to allow a better understanding of pastoral systems by outsiders, and to develop appropriate activities and policies in a participatory process of analysis, diagnosis, and problem-solving.

Conclusions

In recent years there has been a growing understanding of how complex the management of natural resources in communal areas of arid and semi-arid lands in Africa can be. The role of transhumance and animal mobility is increasingly being considered as ecological, economically and institutionally important for sustainable development in these areas. Given the unpredictable nature of the ecosystem, whatever is done to legitimize transhumance must be transient in nature to allow flexibility and modification. This should not be seen as a sign of weakness, but as way of building resilience into the system. There cannot be any guarantee even that transhumance itself will remain intact; other forms of production and livelihood may replace it in the long run. In the short run, however, it is a viable technique, adapted to the ecosystem, and capable of being modified to meet 'modern' needs.

The call to recreate the 'local community' and common property regimes is not confined to pastoralists, nor to Africa. We are witnessing a growing trend in both developing and developed countries toward devolution of authority, de-centralization of administration, and local-level empowerment, due to a revival of the 'small is beautiful' philosophy, to perceived dangers of globalization, and to perceived benefits of localization.

Public administration in the United States in the twentieth century is based on the 'rational' model of bureaucratic planning, 'in which efficiency, not represen-tation, is the goal' (Moote & McClaran, 1997). Very few legal outlets exist for citizens to voice their opinions and objections to public-sponsored activities. There is an increasingly vocal call for the re-emergence of local communities, popular participation, and changing the distribution of power. Proponents are not advocating a return to some earlier age '. . . which we may romanticize but perhaps never existed' (Viederman, 1996: 6), but a re-creation or renovation of an age-old idea.

Increased public participation in governance was stimulated in the 1960s in the United States, and has gathered momentum since. From a long-term point of view, efficiency and accountability are not incompatible or mutually exclusive. It may take more time to reach popular consensus, but in the end the decision will be accepted and followed by all, resulting in a more efficient process. In addition, both efficiency and accountability require information about the preferences of citizens, and institutions that effectively and equitably aggregate these pre-ferences will be both efficient and accountable (Ostrom et al., 1993). There are quite a few examples in the United States where control over natural resources has successfully been devolved to local governance, the oyster beds of Long Island being one good case in point[9].

At a gathering of 50 'model communities' during the fiftieth anniversary of the United Nations, participation was diverse – from the richest cooperative in Spain, to the association of street dwellers of Bombay – but they shared a positive vision of community development, running counter to the seeming trend of globalization. The group called for strong community bonds, local control over common-pool resources, consensus-based decision-making, environmental integrity, and respect for human dignity (IISD, 1996).

---

[9] Property Rights and Performance of Natural Resource Systems Workshop, The Beijer Inter-national Institute of Ecological Economics, September 1993.

In Europe we are also witnessing a return to common-property management among herders and livestock owners of Spain, Italy and France. Local-level natural-resource management is seen as a necessity, as is a revival of trans-humance between summer (high) and winter (low) mountain (Besombes, 1996; Buffière, 1996; Raffin, 1996). The association 'Pastoralisme du Monde' is active in lobbying for changes in land tenure laws in Europe to accommodate transhumance.

In former communist countries, where the traditional pastoral system was presumed to have been destroyed by the collectives and communes, there has been a spontaneous revival of parts of the traditional systems. Tribal and clan hierarchies are too weak to revive spontaneously, but neighbourhood/village communities based on kin networks are becoming important institutions for co-managing mobility of livestock. In Buryati (Russian Republic) mobility managed by the extended family has increased (Humphrey & Sneath, 1996). In Mongolia, local authorities have recognized this trend, and are relying on it for administrative activities, such as services, provision of inputs, famine relief, etc. (Mearns, 1993a). However, poverty is a constraint to the revival of mobility in Central Asia, much as it is for post-drought African pastoralists. Privatized small-scale pastoral units do not have enough family labour and transport to affect the long-distance transhumance needed for efficient production, and their small herds are not economically viable enough to pay for hired shepherds (Kerven et al., 1998).

Mexico can be said to be at the forefront of this movement, as it officially legalized a hybrid form of common property called *ejidos* (hybrid between Aztec and Spanish systems) as early as 1915 with its First Agrarian Reform Act. Today, *ejidos* cover about one third of Mexico's rangelands (Wilson & Thompson, 1993). However, for various political, demographic and economic reasons, the power of local authorities to regulate and manage the *ejidos* has drastically reduced in the last few decades, and the problem of free-riders has re-emerged. Yet some herders (the exception rather than the rule) are creating *ad hoc* coalitions of groups of families, who stake legal claims to remote and less-used pastures, and are re-establishing small, fragmented institutions and common property regimes. This would indicate that common-property regimes may have an optimal scale in each ecological context, one that Mexican herders are finding through trial and error.

Many examples of co-management can be found in Africa, outside the pastoral context, such as the CAMPFIRE project in Zimbabwe which promotes co-management of wildlife resources between local communities and local government (Niesten, 1996), the Mount Elgon National Park in Uganda, Ngorongoro Crater in Tanzania (Danida), forest management in Cameroon (Thomas, 1998), and coastal-zone management in Tanzania (Nurse, 1998). Many have teething problems that have not been completely resolved as yet. Part of the problem lies in the overall institutional weakness: local communities are not well organized, and local governments are hampered by lack of resources and biased attitudes.

Many pastoral advocacy groups in Africa have also emerged in the past two

289

decades to champion the needs of pastoralism, for example, the Miyetti Allah Association of northern Nigeria advocating for the Fulani (Ezeomah & Egbe, 1988), APESS (Association pour la promotion de l'Elevage au Sahel et en Savane) in Burkina Faso, and AREN in Niger, just to name a few.

This Chapter has presented an analytical framework and several key principles and guidelines which could assist in the legitimization and regulation of transhumance in Africa. These principles and guidelines are helpful in defining the design elements for promoting sustainable development, for establishing the relative roles of co-managers, and for highlighting the processes of resilience, adaptation and evolution in institutional design. Previous attempts at creating 'Pastoral Codes' have largely been ineffective mainly because they have been created in an institutional and legal vacuum. Simultaneous strengthening of several important parameters, within a flexible and participatory framework, is necessary. The choice of a specific solution will depend on the needs of the community, and the particular strategic model it wishes, and is able, to adopt.

In some cases the solutions are intimately intertwined, making it difficult to separate one from the other. Both institutional reform and land-tenure reform, for example, are like the proverbial chicken and egg. Appropriate communal institutional frameworks are seen as indispensable for successful natural-resource management (e.g. Shanmugaratnam et al., 1992), as are tenure regimes (Runge, 1981; Feeny et al., 1990; Moorhead, 1993). Conferring a set of property rights on an arbitrarily defined social unit or 'community' does not begin to solve the problem, and may even exacerbate it (Niamir, 1994a; Sivaramakrishnan, 1996). There has been more concern with establishing the structure of the institutions, than with considering their role and functions in relation to management needs. As a result, most institutions have been static and unable to adapt to emerging needs (see e.g. Swift, 1988; Bonfiglioli & Watson, 1992; Bourbouze, Chapter 10, this volume). Creating or identifying an existing 'community' without giving it the authority over the resource, will result in an empty shell. Both the 'composition axiom' (i.e. allocating control of resources to well-defined communities) and the 'authority axiom' (i.e. action of a group for a unified purpose) need to be strengthened, preferably simultaneously (Larson & Bromley, 1990).

Perhaps the most important first step, however, is to convince development practitioners and government officials that what the pastoralists knew all along – namely that mobility is an ecological and economic adaptation to arid lands – is a viable concept upon which to base pastoral development.

References

Abaab A., Bedrani, S., Bourbouze, A. and Chiche, J. (1995) 'Les politiques agricoles et la dynamique des systèmes pastoraux au Maghreb', pp. 139–168 in *Les agricultures maghrébines à l'aube de l'an 2000*, Options Méditer-ranéennes, Série B, No. 14.

Adams, M. (1995) 'Land reform: new seeds on old ground?', *ODI Natural Resource Perspectives*, 6: 1.

Adams, M. (1996) 'When is ecosystem change land degradation?', *Pastoral Development Network*, 39e.

Adegboye, R.O. et al. (1978) *A Socio-economic Study of Fulani Nomads in Kwara State*, Federal Livestock Department, Kaduna, Ibadan, Nigeria.

Adepetu, A.A. et al. (1988). 'Demographic aspects', in *Education of Nomadic Families*, Research Report vol. 5, UNDP/UNESCO/University of Jos project, Nigeria.

Agrawal, A. (1994). 'Rules, rule making, and rule breaking: examining the fit between rule systems and resource use', pp. 267–282 in *Rules, Games and Common-Pool Resources*, eds E. Ostrom, R. Gardner and J. Walker, University of Michigan Press, Ann Arbor.

Ahmed, A.G.M. (n.d.) 'Nomadic competition in the Funj area', *Sudan Notes and Records*, Khartoum.

Allen, T.F.H. and Hoekstra, T.W. (1992) *Toward a unified ecology*. New York: Columbia University Press.

Anderson, J., Cauthier, M., Thomas, G. and Wondelleck, J. (1996) 'Addressing natural resource conflicts through community forestry: setting the stage', Paper for the Electronic Conference on 'Addressing natural resource conflicts through community forestry', FAO, Rome.

Anim, F.D.K. and Lyne, M.C. (1994) 'Econometric analysis of private access to communal grazing lands in South Africa: A case study of Ciskei'. *Agricultural Systems*, 46: 461–471.

An-Na'im, A.A. (1987) 'National Unity and the Diversity of Identities'. pp. 71–77 in *The Search For Peace and Unity in the Sudan*, eds F. Deng and Professor Gifford, Wilson Center Press, Washington D.C.

APRU (Animal Production Research Unit) (1978–90) *Annual Reports*, Ministry of Agriculture, Gabarone.

Armstrong, C.L. (1940) 'Dry Season Cattle Camps of Eastern Jikany Nuer Sections', quoted by Evans-Pritchard, pp. 57–58 in *The Nuer: A Description of Modes of Livelihood and Political Institutions of a Nilotic People*, Clarendon Press, Oxford.

Artz, N., O'Rourke, J.T., Gilles, J.L. Aro, R.S. and Narjisse, H. (1985) 'The development implications of heterogeneity in a Moroccan communal grazing system', Utah University, Range Sciences Department.

Ayoo, S.J (1995) *The Ik Research*, OXFAM, Kampala.

Azarya, V. (1988) 'Jihad and Dyula States in West Africa', pp. 109–133 in *The Early State in African Perspective: Culture, Power and Division of Labour*, Studies in Human Society Vol. 3, eds S.N. Eisenstadt, M. Abitbol and N. Chazan, E.J. Brill, Leiden.

Ba, A.S. (1982) *L'art vétérinaire des pasteurs Saheliens*, ENDA serie Études et Recherches, No. 73–82, Dakar.

Baker, P.R. (1967) *Environmental Influences on Cattle Marketing in Karamoja*, Occasional Paper No. 5, Department of Geography, Makerere University College, Kampala.

Baker, P.R. (1974) *Perception of Pastoralism*, Discussion Paper No.3, School of Development Studies, University of East Anglia, Norwich.

Barrow, E. (1991) 'Evaluating the effectiveness of participatory agroforestry extension programmes in a pastoral system, based on existing traditional values', *Agroforestry Systems*, 14: 1–21.

Barrows, R. and Roth, M. (1990) 'Land tenure and investment in African agriculture: Theory and evidence' *The Journal of Modern African Studies*, 28: 265–297.

Barry, I. (1993) 'Le royaume de Bandiagara (1864–1893), Le Pouvoir, le Commerce et le Coran dans le Soudan Nigérien au XIXe siècle', PhD thesis, EHESS, Paris.

Bartels, G.B., Norton, B.E. and Perrier, G.K. (1990) 'The applicability of the carrying capacity concept in Africa', *Pastoral Development Network* 30e.

Barton, T. and Wamai, G. (1994) *Equity and Vulnerability: A Situation Analysis of Women, Adolescents and Children in Uganda, 1994*, Uganda National Council for Children, Kampala.

Bassett, T.J. (1986) 'Fulani herd movements', *Geographical Review*, 76: 233–248.

Bassett, T.J. (1994) 'Hired herders and herd management in Fulani pastoralism (Northern Côte d'Ivoire)', *Cahiers d'Études africaines*, 34: 147–173.

Baxter, P.T.W. and Hogg, R. eds (1990) *Property, Poverty, and People: Changing Rights in Property and Problems in Pastoral Development*, University of Manchester, Manchester, UK.

Bayer, W. and Waters-Bayer, A. (1995). *Planning with Pastoralists: PRA and more. A review of Methods Focused on Africa*. GTZ Working Paper, GTZ, Eschborn.

Bazaara, N. (1994) 'Land reforms and agrarian structures in Uganda: retrospect and prospect', *Nomadic Peoples*, 34/35: 37–54.

Beauvilain, A. (1977) *Les Peul du Dallol Bosso, Etudes Nigeriennes 42*, Institut de Recherche en Sciences Humaines, Niamey, Niger.

Behnke, R.H. (1994) 'Natural resource management in pastoral Africa', *Development Policy Review*, 12: 5–27.

Behnke, R.H. (1997) 'Carrying capacity and rangeland degradation in semi-arid Africa: clearing away conceptual rubble', paper presented at the Rangeland Desertification Workshop, 16–19 September 1997, Reykjavik, Iceland.

Behnke, R.H. and Abel, N. (1996) 'Revisited: the overstocking controversy in semi-arid Africa', *World Animal Review*, 87: 4–27.

Behnke, R.H. and Scoones, I. (1993) 'Rethinking range ecology: implications for rangeland management in Africa', pp. 1–30 in *Range Ecology at Disequilibrium*, eds R. Behnke, I. Scoones and C. Kerven, Overseas Development Institute, London.

Behnke, R.H, Scoones, I. and Kerven, C., eds (1993) *Range Ecology at Disequilibrium. New Models of Natural Variability and Pastoral Adaptation in African Savannas*, Overseas Development Institute, London.

Behnke, R.H., Koruhama, K. and Kaurimuje, J. (1998) *Range and Livestock Management in the Etanga Development Area, Kunene Region: Final Report*, Northern Regions Livestock Development Project: Windhoek.

Beidi, B.H (1993) *Les Peuls du Dallol Bosso*, Paris: Editions Sépia, Paris.

Bellot, J.-M (1980) 'Kel Tamasheq du Gourma Nigerien et peul du Torodi: Sociétés Agropastorales en Mutation', Thèse de troizième cycle, Université de Bordeaux III, Bordeaux.

Belshaw, D., Avery, S., Hogg, R. and Obin, R. (1996) *Report of the Evaluation Mission*, Integrated Development in Karamoja, Uganda project, Overseas Development Group, Norwich and UNCDF, New York.

Bencherifa A. and Johnson, D.L. (1990). 'Adaptation and intensification in the Pastoral Systems of Morocco', pp. 394–416 in *The World of Pastoralism*, eds J. Galaty and D.L. Johnson, Guilford Press, New York.

Benoit, M. (1979) *Le chemin des Peuls du Boobola: Contribution à l'Écologie du Pastoralisme en Afrique des Savanes*, Travaux et documents de l'ORSTOM 101, ORSTOM, Paris.

Berkes, F. (1995) 'Community-based management and co-management as tools for empowerment', pp. 138–146 in *Empowerment: Towards Sustainable Development*, eds. N. Singh and V. Titi, Zed Books, London.

Berkes, F. (1997) 'New and not-so new directions in the use of commons: co-management', *Common Property Resource Digest*, No. 42.

Berry, S. (1989) 'Social institutions and access to resources', *Africa*, 59: 41–54.

Berry, S. (1993) *No Condition is Permanent. The Social Dynamics of Agrarian Change in Sub-Saharan Africa*, University of Wisconsin Press, Madison.

Besombes, M. (1996) 'Une coopérative de transhumance en Cantal (France). Comment une coopérative agricole peut contribuer à régler de façon durable le problème du foncier en faisant participer les éleveurs à la gestion de l'espace pastoral', pp. 121–128 in *Proceedings of seminar Pastoralisme et foncier: impact du régime foncier sur la gestion de l'espace pastoral et la conduite des troupeaux en régions arides et semi-arides*, CIHEAM/IRA, Gabès 17–19 October 1996.

Bollig, M. (n.d.) 'Resource Management and Pastoral Production in the Epupa Project Area (the Kunene Drainage System from Swartbooisdrift to Otjinungwa)', unpublished manuscript, University of Cologne.

Bollig, M. (1996) 'Power and trade in precolonial and early colonial times in northern Kaokoland, c. 1860–1950' in P. Hayes, J. Silvester, M. Wallace, W. Hartmann and B. Fuller, *Trees Never Meet: Mobility and Containment in Namibia, 1915–1946*, Longman: Windhoek.

Bollig, M. (1997) 'Contested places: graves and graveyards in Himba culture', *Anthropos*, 92: 35–50.

Bollig, M. and Mbunguha, T.J. (1997) *'When War Came the Cattle Slept . . .': Himba Oral Traditions*, Rudiger Koppe Verlag, Koln.

Bonfiglioli, A.M. (1982) *Ngaynaka: Herding according to the Wodaabe*, USAID/NRL, Niamey, Niger.

Bonfiglioli, A.M. (1985) 'Évolution de la propriété animale chez les Wodaabe au Niger', *Journal des Africanistes*, 55: 29–37.

Bonfiglioli, A.M. (1988) *Dudal: Histoire de Famille et Histoire de Troupeau chez un Groupe de Wodaabe du Niger*, Cambridge University Press, Cambridge, UK.

Bonfiglioli, A.M. (1990) 'Pastoralisme, agro-pastoralisme et retour: itinéraires sahéliens', *Cahiers des Sciences Humaines*, 26: 255–266.

Bonfiglioli, A.M. and Watson, C. (1992) *Pastoralists at a Crossroads*, UNICEF/UNSO Project for Nomadic Pastoralists in Africa (NOPA), New York.

Borrini-Feyerabend, G. (1996) *Collaborative Management of Protected Areas: Tailoring the Approach to the Context*, IUCN Gland, Switzerland.

Botkin, D. (1990) *Discordant Harmonies: A New Ecology for the Twenty-First Century*, Oxford University Press, Oxford, UK.

Botterweg, R. (1994) *Dryland areas in sub-saharan Africa*, UNSO, New York.

Boukhobza, M. (1976) 'Nomadisme et crise de la société pastorale en Algérie. Quelques points de repère historiques', pp. 207–21 in *L'élevage en Méditerranée occidentale*, CNRS, Paris.

Bourbouze, A. (1991) 'Les aspects socio-économiques et législatifs relatifs à l'exploitation des parcours des pays en voie de développement d'Afrique et d'Asie', Rapport Général, pp. 1186–1188 in *Colloque F15*, Congrès International des Terres de Parcours, Montpellier.

Bourn, D. and Wint, W. (1994) 'Livestock, land-use and agricultural intensification in sub-Saharan Africa', *Pastoral Development Network*, 37a.

Bovin, M., and Manger, L. (1990) *Adaptive Strategies in African Arid Lands*, Scandinavian Institute of African Studies, Uppsala.

Bradbury, M., Fisher, S. and Lane, C. (1995) *Working with Pastoralist NGOs and Land Conflict in Tanzania*, IIED, London.

Bredon, R.M. (1963) 'The chemical composition and nutritive value of grasses from semi-arid areas of Karamoja as related to ecology and types of soil', *East African Agricultural and Forestry Journal*, 29: 134–142.

Breman, H. and de Wit, C.T. (1983) 'Rangeland production and exploitation in the Sahel', *Science*, 221: 1341–1347.

Bremaud, O. and Pagot, J. (1962) 'Grazing lands, nomadism and transhumance in the Sahel' in *The problems of the Arid Zone*, Part II, UNESCO, Paris.

Bromley, D.W. (1989) 'Property relations and economic development: the other land reform', *World Development*, 1: 867–877.

Brown, W.A. (1969) 'The Caliphate of Hamdullahi ca 1818–1864: A Study of African History and Tradition', PhD thesis, University of Wisconsin, Madison.

Bruce, J. (1993) 'Do indigenous tenure systems constrain agricultural development?', pp. 35–56 in *Land in African Agrarian Systems*, eds T.J. Bassett and D.E. Crummey, University of Wisconsin Press, Madison.

Buffière, D. (1996) 'Propriété foncière et festion de l'espace collectif haut-pyrénéen', pp. 109–111 in *Proceedings of seminar Pastoralisme et foncier: impact du régime foncier sur la gestion de l'espace pastoral et la conduite des troupeaux en régions arides et semi-arides*, CIHEAM/IRA, Gabès 17–19 October 1996.

Burke, A. (1999) 'Vegetation Resources of NOLIDEP Pilot Communities: Etanga and Orongoto, Kunene Region', *EnviroScience*, in press.

Burnham, P. (1979) 'Spatial mobility and political centralization in pastoral societies', pp. 349–360 in *Pastoral Production and Society*, ed. L'Équipe Ecologie et Anthropologie des Societes Pastorales, Cambridge University Press, Cambridge.

Burton, J. and Dukes, F. (1990) *Conflict: practices in management, settlement and resolution*, St Martin's Press, New York.

Carney, D. (1995) *Management and Supply in Agriculture and Natural Resources: is Decentralization the Answer?*, Natural Resource Perspectives, No. 4, Overseas Development Institute, London.

Casimir, M. and Rao, A. eds (1992) *Mobility and Territoriality: Social and Spatial Boundaries Among Foragers, Fishers, Pastoralists and Peripatetics*, Berg, Oxford.

Catley, A. (1996) 'Pastoralists, paravets and privatization: experiences in the Sanaag region of Somaliland', *Pastoral Development Network*, 39d.

Chambers, R. (1992) *Rural Appraisal: Rapid, Relaxed and Participatory*, IDS Discussion Paper 311, Institute of Development Studies, Sussex.

Channock, M. (1991) 'Paradigms, policies and property: a review of the customary law of land tenure' in *Law in Colonial Africa*, eds K. Mann and R. Roberts, James Currey, London.

Chiche, J. (1992) 'Pratiques d'utilisation des terres collectives au Maroc', pp. 41–56 in *Terres collectives en Méditerranée*, eds A. Bourbouze and R. Rubino FAO/Réseau Parcours, Ars Grafica, Rome.

CIPEA (1983) *Recherche d'une Solution aux Problèmes de l'Élevage dans le Delta intérieur du Niger au Mali*, Direction Nationale de l'Élevage, Bamako, Mali.

Cissé, S. (1982) 'Les leyde du delta central du Niger: Tenure traditionnelle ou exemple d'un amenagement de territoire classique', pp. 178–189 in *Enjeux fonciers en Afrique Noire*, eds E.L. Bris, E. le Roy and F. Leimdorfer, Karthala, Paris.

Cissoko, S.M. (1968) 'Famines et épidémies à Tombouctou et dans la boucle du Niger du XVIe au XVIIIe siècle', *Bulletin de l'IFAN, Serie B*, 3: 806–21.

Cisterino, M. (1979) '*Karamoja: the human zoo*', PhD thesis, University of Swansea, UK.

Colson, E. (1971) 'The impact of the colonial period on the definition of land rights', in *Profiles of Change. African Society and Colonial Rule*, ed. V. Turner, Cambridge University Press, Cambridge, UK.

Commons, J.R. (1959) *Legal Foundations of Capitalism*, University of Wisconsin Press, Madison. (First published in 1924, Macmillan, New York).

Cossins, N.J. (1985) 'The productivity and potential of pastoral systems', *ILCA Bulletin* 21: 10–15.

Cousins, B. (1987) *A Survey of Current Grazing Schemes in the Communal Lands of Zimbabwe*, Centre for Applied Social Sciences, University of Zimbabwe, Harare.

Cousins, B. (1992) *Managing Communal Rangeland in Zimbabwe: Experiences and Lessons*, Commonwealth Secretariat, London.

Cousins, B. (1993) *Inappropriate Technology, Key Resources and Unstable Institutions: A Case Study of Mutakwa Grazing Scheme*, Overseas Development Institute, London.

Cousins, B. (1995) 'Range management and land reform policy in post-apartheid South Africa', *Proceedings of 5th International Rangeland Congress*, Salt Lake City, Utah.

Cousins, B. (1996) 'Conflict management for multiple resource users in pastoralist and agro-pastoralist contexts', *Proceedings of 3rd International Technical Consultations on Pastoral Development*, EU/UNSO, New York.

Cox, J., Kerven, C., Werner, W. and Behnke, R. (1998) *The Privatization of Rangeland Resources in Namibia: Enclosures in Eastern Oshikoto*, Overseas Development Institute, London.

Crandall, D.P. (1991) 'The strength of the OvaHimba patrilineage', *Cimbebasia*, 13: 45–51.

Crandall, D.P. (1991/2) 'The importance of Maize among the OvaHimba', *Cimbebasia*, 43: 7–19.

Dahl, G. (1979) 'Ecology and equality: the Boran case', pp. 261–282 in *Pastoral*

Production and Society, ed. L'Équipe Ecologie et Anthropologie des Societes Pastorales. Cambridge University Press, Cambridge, UK.

Dahl, G., and Hjort, A. (1976) *Having herds: Pastoral Herd Growth and Household Economy*, Department of Anthropology, Stockholm University, Stockholm.

Dalli, A.L. and Ezeomah, C. (1988) 'Socio/cultural aspects' in *Education of Nomadic Families*, Research Report Vol. 2, UNDP/UNESCO/University of Jos project, Nigeria.

Danckwerts, J. (1974) *A Socioeconomic Study of Veld Management in the Tribal Areas of Victoria Province*, The Tribal Areas of Rhodesia Research Foundation, Salisbury.

de Angelis, D. and Waterhouse, J. (1987) 'Equilibrium and non-equilibrium concepts in ecological models', *Ecological Monographs*, 57: 1–21.

de Haan, C. (1994) 'An overview of the World Bank's involvement in pastoral development', *Pastoral Development Network*, 36b.

de Leeuw, P.N. and Tothill, J.C. (1993) 'The concept of rangeland carrying capacity in Sub-Saharan Africa – myth or reality', pp. 77–88 in *Range Ecology at Disequilibrium*, eds R.H. Behnke, I. Scoones and C. Kerven, Overseas Development Institute, London.

de Leeuw, P.N., Diarra L. and Hiernaux, P. (1993) 'An analysis of feed demand and supply for pastoral livestock: the Gourma region of Mali', pp. 136–152 in *Range Ecology at Disequilibrium: New Models of Natural Variability and Pastoral Adaptation in African Savannas*, eds R.H. Behnke, I. Scoones and C. Kerven, Overseas Development Institute, London.

de Ridder, N. and Wagenar, K.T. (1984) 'A comparison between the productivity of traditional livestock systems and ranching in eastern Botswana', *ILCA Newsletter*, 3: 5–6.

de Wispelaere, G. (1980) 'Les photographies aériennes témoins de la dégradation du couvert ligneux dans un ecosystème Sahélien Sénégalais. Influence de la proximité d'un forage', *Cahiers de l'ORSTOM*, XVIII (3–4): 155–166.

Demsetz, H. (1967) 'Toward a theory of property rights', *American Economic Review*, 57: 347–359.

Despois, J. and Raynal, R. (1967) *Géographie de l'Afrique du Nord Ouest*, Payot, Paris.

Diallo, A. (1978) *Transhumance: Comportement, Nutrition et Productivité d'un Troupeau Zébus de Diafarabé*, PhD thesis, Centre Pédagogique Supérieure, Bamako, Mali.

Digard, J.-P., Landais, E. and Lhoste, P. (1993) 'La crise des sociétés pastorales. Un regard pluridisciplinaire', *Revue Élevage et Medecine Veterinaire des Pays Tropicaux*, 46: 683–692.

Dodd, J.L. (1994) 'Desertification and degradation in Sub-Saharan Africa. The role of livestock', *Bioscience*, 44: 28–34.

Dougill, A. and Cox, J. (1995) 'Land degradation and grazing in the Kalahari: new analysis and alternative perspectives', *Pastoral Development Network*, 38c.

Dowling, J.H. (1975) 'Property relations and productive strategies in pastoral societies', *American Ethnologist* 2: 419–425.

Duany, W. (1992) 'The Nuer Concept of Covenant and Covenantal Way of Life', *Publius: The Journal of Federalism*, 22: 67–89.

Duany, W. (1994) Working paper, Jikany–Lou Nuer Reconciliation Conference,

Akobo, South Sudan, July–October 1994, Workshop in Political Theory and Policy Analysis, Indiana University, Bloomington.

Duany, W. and Duany, J. (1995) 'Genesis of the Crisis in the Sudan', Working Paper, Workshop in Political Theory and Policy Analysis, Indiana University, Bloomington.

Dupire, M. (1970) *Organisation Sociale des Peul*, Recherches en Sciences Humaines 32, Librairie Plon, Paris.

Dupire, M. (1972) *Les Facteurs Humains de l'Economie Pastorale*, Etudes Nigeriennes No. 6, Centre Nigerien de Recherches en Sciences Humaines, Niamey, Niger.

Dyson-Hudson, R. (1966) *Karimajong Politics*, Clarendon Press, Oxford.

Dyson-Hudson, R. and Smith, E.A. (1978) 'Human territoriality: an ecological reassessment', *American Anthropologist*, 80: 21–41.

Ekechi, F.K. (1989) *Tradition and Transformation in Eastern Nigeria*, Kent State University Press, Kent, Ohio.

El-Arifi, S.A. (1979) 'Some aspects of local government and environmental management in the Sudan', pp. 36–39 in *Proceedings of the Khartoum Workshop on Arid Lands Management*, ed. J.A. Mabbutt, University of Khartoum, Sudan.

Elazar, D.J. (1980) 'Political Theory of Covenant', *Publius*, 10: 3–30.

Ellis, J.E. and Swift, D.M. (1988) 'Stability of African pastoral ecosystems: alternate paradigms and implications for development', *Journal of Range Management*, 41: 450–459.

Engberg-Pedersen, L. (1995) *Creating Local Democratic Politics from Above: The 'Gestion de Terroirs' Approach in Burkina Faso*, IIED Issues, Paper No. 54, IIED, London.

English, J., Tiffen, M. and Mortimore, M. (1992) *Land Resource Management in Machakos District, Kenya 1930–1990*, World Bank Environment Paper No. 5, Washington, DC.

Es Sadi, A. (1900) *Tarikh es Soudan*, trans. O. Houdas, Maisonneuve/Leroux, Paris.

Euroconsult (1983) *Dairy Development Study*, European Union, Brussels.

Evans-Pritchard, E.E. (1940) *The Nuer: A Description of Modes of Livelihood and Political Institutions of a Nilotic People*, Clarendon Press, Oxford.

Ezeomah, C. (1985) 'Land tenure constraints associated with some recent experiments to bring formal education to nomadic Fulani in Nigeria', *Pastoral Development Network*, 20d.

Ezeomah, C. and Egbe, E.N. (1988) 'Language and communication aspects', in *Education of Nomadic Families*, Research Report Vol. 4, UNDP/UNESCO/University of Jos project, Nigeria.

FAO/UNEP (1997) *Negotiating a Sustainable Future for Land*, FAO, Rome.

Farah, M.I. (1993) *From Ethnic Response to Clan Identity: A Study of State Penetration Among the Somali Nomadic Pastoral Society of Northern Kenya*, Almqvist & Wiksell, Stockholm.

Faure, A. (1995) *Private Land Ownership in Rural Burkina Faso*, IIED Issues, Paper No. 59, IIED London.

Fay, C. (1994) 'Le Maasina', pp. 363–382 in *La Pêche dans le Delta Central du Niger: Approche Pluridisciplinaire d'un Système de Production Halieutique*, ed. J. Quensiere, ORSTOM–Karthala, Paris.

Fay, C. (1995) 'La démocratie au Mali, ou le pouvoir en pâture', *Cahiers d'Études africaines*, 137, XXXV–1, 19–53.

Feder, G. and Noronha, R. (1987) 'Land rights systems and agricultural development in sub-Saharan Africa', *Research Observer*, 2, 143–169.

Feeny, D., Berkes, F., McCay, B.J. and Acheson, J.A. (1990) 'The tragedy of the commons: 22 years later', *Human Ecology*, 18: 1–19.

Finch, V.A. and King, J.M. (1979) 'Adaptation to undernutrition and water deprivation in the African zebu: changes in energy requirements', in *Proceedings of Research Coordination Meeting on Water Requirements of Tropical Herbivores Based on Measurements with Titrated Water*, IAEA Nairobi and Vienna.

Forbes, R.G. and Trollope, W.S.W. (1991) 'Veld management in the communal areas of Ciskei', *Journal of the Grassland Society of Southern Africa*, 8: 147–152.

Fratkin, E. (1994) 'Pastoral land tenure in Kenya: Maasai, Samburu, Boran and Rendille experiences, 1950–1990', *Nomadic Peoples*, 34/35: 55–68.

Froude, M. (1974) 'Veld management in the Victoria Province Tribal areas', *Rhodesian Agriculture Journal*, 71, 29–33.

Funtowitcz, S.O. and J.R. Ravetz (1993) 'Science for the post-normal age', *Futures* 25(7): 739–755.

Galaty, G.D. (1988) 'Scale, politics and cooperation in organization for East African Development', pp. 282–308 in *Who shares? Cooperatives and Rural Development*, eds D.W. Attwood and B.S. Baviskar, Oxford University Press, Oxford, UK.

Galaty, G., Aronson, D. Salzman, P.C. (1981) *The Future of Pastoral Peoples, Proceedings of Conference* 4–8 August 1980, Kenya. IDRC, Ottawa, Canada.

Gallais, J. (1958) 'La vie saisonnière au sud du lac Débo (territoire du Soudan)', *Les Cahiers d'Outre-Mer*, 42, pp. 117–141.

Gallais, J. (1967) 'Le Delta intérieur du Niger, Étude de géographie régionale, Dakar, Sénégal, *IFAN Mémoires de l'Institut Fondamental d'Afrique Noire*, No. 78.

Gallais, J. (1975). *Paysans et Pasteurs du Gourma. La Condition Sahélienne.* CNRS, Paris.

Gallais, J. (1984) *Hommes du Sahel. Espaces-temps et pouvoirs. Le Delta intérieur du Niger, 1960–1980*, Flammarion, Paris.

Gallais, J. and Boudet, G. (1980) *Projet de Code Pastoral Concernant plus Specialement la Region du Delta Central du Niger au Mali*, Fonds d'Aide et de Coopération de la République Française, Paris.

Gammon, M. (1978) 'A review of experiments comparing systems of grazing management on natural pasture', *Proceedings of the Grassland Society of Southern Africa*, 13: 75–82.

Gavian, S. (1993) *Land Tenure and Soil Fertility Management in Niger*, PhD thesis, Stanford University.

Giddens, A. (1984) *The Constitution of Society*, Polity, Cambridge.

Gilles, J.L. (1988) 'Slippery grazing rights: using indigenous knowledge for pastoral development', pp. 1159–1166 in *Arid Lands Today and Tomorrow*, eds E.E. Whitehead, C.F. Hutchinson, B.N. Timmerman and R.G. Varady, Westview Press, Boulder, Colorado.

Gilles, J.L. and Jamtgaard, K. (1982) 'Overgrazing in pastoral areas: the commons reconsidered', *Nomadic Peoples*, 10: 1–10.

Gorse, J.E. and Steeds, D.R. (1987) *Desertification in the Sahelian and Sudanian Zones of West Africa*, Technical Paper No. 61, World Bank, Washington, D.C.

Gray, R. (1961). *A History of the Southern Sudan, 1839–1889*, Oxford University Press, London.

Grayzel, J.A. (1990) 'Markets and migration: a Fulbe pastoral system in Mali', pp. 35–68 in *The World of Pastoralism. Herding Systems in Comparative Perspective*, eds J.G. Galaty and D.L. Johnson, Guilford Press, New York.

Gregg, G.S. and Geist, A. (1987) *The Socio-economic Organization of the Ait Imeghrane*, Working Paper for ORMVA of Ouarzazate Province, Morocco.

Gritzner, J. (1988) *The West African Sahel. Human Agency and Environmental Change*, Geography Research Paper No. 226, University of Chicago.

Gubert, R., ed. (1988) *La Sfida Dello Sviluppo: in Una Societá Pastorale dell'Africa Orientale, Karamoja – Uganda*, Jaca Books, Milan.

Gulliver, P.H. (1970) *The Family Herds: A Study of Two Pastoral Tribes in East Africa, the Jie and Turkana*, Negro University Press, Westport.

Gulliver, P.H. (1975) 'Nomadic movements: causes and implications', pp. 369–386 in *Pastoralism in Tropical Africa*, ed. T. Monod, Oxford University Press, Oxford, UK.

Hanna, S. (1995) 'Efficiencies of user participation in natural resource management', pp. 59–67 in *Property Rights and the Environment. Social and Ecological Issues*, eds. S. Hanna and M. Munasinghe, Beijer International Institute of Ecological Economics/World Bank, Washington, D.C.

Hanna, S., Folke, C. and Mähler, C. (1995) 'Property rights and environmental resources', pp. 15–29 in *Property Rights and the Environment: Social and Ecological Issues*, eds. S. Hanna and M. Munasinghe, Beijer International Institute of Ecological Economics/World Bank, Washington D.C.

Hardin, G. (1968) 'The tragedy of the commons', *Science*, 162: 1243–1248.

Hardin, G. (1991) 'The tragedy of the unmanaged commons: population and the disguises of providence', pp. 162–185 in *Commons Without Tragedy*, ed. R.V. Andelson, Barnes & Noble, Savage, Maryland.

Haywood, M. (1980) *Changes in Land Use and Vegetation in the ILCA/Mali Sudano–Sahelian Project Zone*, ILCA Working Document No. 3, International Livestock Centre for Africa, Addis Ababa, Ethiopia.

Heady, H.F. and Child, R.D. (1994) *Rangeland Ecology and Management*, Westview Press, Boulder.

Hellden, U. (1991) 'Desertification – time for an assessment,' *Ambio*, 20: 372–383.

Hiernaux, P. (1993) *The Crisis of Sahelian Pastoralism: Ecological or Economic?*, International Livestock Centre for Africa, Addis-Ababa, Ethiopia.

Hiernaux, P. (1996) 'The crisis of Sahelian pastoralism: ecological or economic?', *Pastoral Development Network*, 39b.

Hiernaux, P. and Turner, M.D. (1996) 'The effect of the timing and frequency of clipping on nutrient uptake and production of Sahelian annual rangelands', *Journal of Applied Ecology*, 33: 387–399.

Hobsbawm, E. and Ranger, T. (1983) *The Invention of Tradition*, Cambridge University Press, Cambridge, UK.

Holleman, J. (1969) *Chief, Council and Commissioner: Some Problems of Government in Rhodesia*, Koninkelijke van Gorcum & Co., Netherlands.

Homewood, K. and Rodgers, W.A. (1987) 'Pastoralism, conservation and the overgrazing controversy', pp. 111–128 in *Conservation in Africa*, eds. D. Anderson and R. Grove, Cambridge University Press, Cambridge, UK.

Howell, P.P. (1954a) *A Manual of Nuer Law: Being an Account of Customary Law, Its Evolution and Development in the Courts Established by the Sudan*

Government, Oxford University Press, London (published for the International African Institute).

Howell, P.P. (1954b) 'Property Rights', pp. 178–189 in *A Manual of Nuer Law*, ed. P.P. Howell, Oxford University Press, London.

Humphreys, C. and Sneath, D. (1996) 'Pastoralism and institutional change in inner Asia: comparative perspectives from the MECCIA research project', *Pastoral Development Network*, 39b.

Huntley, B.J. and Walker, BJ. eds (1979) *Ecology of Tropical Savannas*, Ecological Studies 42, Springer-Verlag, Berlin.

Hutchinson, S. (1992) 'The cattle of money and the cattle of girls among the Nuer, 1930–83', *American Ethnologist*, 19: 294–316.

IISD (1996) *Developing Ideas*, Issue 1: 2.

Imperato, P.J. (1986) *Historical Dictionary of Mali*, Scarecrow Press, London.

Institute de Recherche en Sciences Humaines (1977) *Étude de Say: Rapport Final*, Ministère de l'Education Nationale and Ministère du Plan, Niamey, Niger.

IUCN (1989) *Rainfall in the Sahel*, IIED Issues, Paper No. 10, IIED, London.

Jacobs, A.H. (1980) 'Pastoral Maasai and tropical rural development', pp. 275–300 in *Agricultural Development in Africa: Issues of Public Policy*, eds R.H. Bates and M.F. Lofchie, Praeger, New York.

Jal, G.G. (1987) *The History of the Jikany Nuer Before 1920'*, PhD thesis, School of Oriental and African Studies, University of London.

Johnson, D. (1980) *History and Prophecy among the Nuer of Southern Sudan* (Parts 1 and 2), PhD thesis, University of California at Los Angeles.

Johnson, D. (1994) *Nuer Prophets: A History of Prophecy from the Upper Nile in the Nineteenth and Twentieth Centuries*, Clarendon Press, Oxford.

Joof, A.E., Diallo, A. Amali, E. and Ezeomah, C. (1988) 'Economic Aspects', in *Education of Nomadic Families*, Research Report Vol. 3, UNDP/UNESCO/University of Jos project, Nigeria.

Kerven, C., Lunch, C. and Wright, I. (1998) *First Fieldwork Report on Impacts of Privatization on Range and Livestock Management in Semi-Arid Central Asia*, Overseas Development Institute, London.

King, J.M. (1983) *Livestock Water Needs in Pastoral Africa in Relation to Climate and Forage*, ILCA Research Report No. 7, International Livestock Centre for Africa, Addis Ababa, Ethiopia.

KPIU (1996) *Karamoja Project Implementation Unit: Inception Report*, KDA, Ministry of State for Karamoja, Kampala.

KTA (1996) *Proceedings of Workshop on Land Tenure, Natural Resources and Wildlife in Karamoja*, Ministry of State for Karamoja, Kampala.

Laine, F. (1987) *Un programme Vétérinaires sans Frontières, La Regénération de Bourgoutières dans le Cercle de Tombouctou (Bilan après deux ans)*, Thesis in Veterinary. University of Claude Bernard de Lyon I, Lyon.

Lamphear, J. (1976) *The Traditional History of the Jie of Uganda*, Clarendon Press, Oxford.

Lamprey, H.F. (1983) 'Pastoralism yesterday and today: the overgrazing problem', pp. 643–666 in *Tropical Savannas: Ecosystems of the World*, ed. F. Bourliere, Elsevier, Amsterdam.

Landais, E. and Lhoste , P. (1990) 'L'association agriculture-élevage en afrique intertropicale: un mythe techniciste confronté aux réalités du terrain', *Cahiers des Sciences Humaines*, 26: 217–235.

Lane, C.R. (1991) 'Alienation of Barabaig pastureland', PhD thesis, University of Sussex.

Lane, C. and Moorehead, R. (1993) 'New directions in African Range Management, Natural Resource Tenure and Policy', in *Proceedings of a Workshop on New Directions in African Range Management and Policy*, ODI/IIED/ Commonwealth Secretariat, Woburn, London.

Larson, B.A. and Bromley, D.W. (1990) 'Property rights, externalities, and resource degradation', *Journal of Development Economics*, 33: 235–262.

Lawry, S. (1990) 'Tenure policy towards common property natural resources in sub-Saharan Africa', *Natural Resources Journal*, 30, 403–422.

Laya, D., ed. (1984) *La Voie Peul: Solidarité et Bienséances Sahéliennes*, Nubia, Paris.

Le Houérou, H.N. (1989) *The Grazing Land Ecosystems of the African Sahel*, Ecological Studies 75, Springer-Verlag, Berlin/New York.

Le Houérou, H.N. (1995) *Bioclimatologie et biogéographie des steppes arides du nord de l'Afrique*, Options méditerranéennes, Série B, No. 10.

Leach, M., Mearns, R. and Scoones, I. (1997) *Environmental Entitlements: A Framework for Understanding the Institutional Dynamics of Environmental Change*, Institute of Development Studies, Brighton, UK.

Lercollais, A. and Faye, A. (1994) 'Des troupeaux sans pâturages en pays Sereer au Sénégal', pp. 165–196 in *A la Croisée des Parcours: Pasteurs, Éleveurs, Cultivateurs*, eds. C. Blanc-Pamard and J. Boutrais, ORSTOM, Paris.

Lewis, B.H. (1940) 'Sketch showing movements in dry season of the Seraf communities', pp. 61–62 in *The Nuer: A Description of Modes of Livelihood and Political Institutions of a Nilotic People*, ed. E. Evans-Pritchard, Clarendon Press, Oxford.

Lewis, J.v.D. (1981) 'Land use and Fulbe social organization the view from Macina', in *Image and Reality in African Interethnic Relations: The Fulbe and their Neighbors*, eds. M. Zamora, V. Sutlive and N. Altshuler, Department of Anthropology, College of William and Mary, Williamsburg.

Little, P.D. (1985a) 'Social differentiation and pastoral sedentarization in northern Kenya', *Africa*, 55: 243–261.

Little, P.D. (1985b) 'Absentee herd owners and part-time pastoralists: the political economy of resource use in northern Kenya', *Human Ecology*, 13: 131–151.

Little, P.D., Horowits, M.M. and Nyerges, A., eds (1987) *Lands at Risk in the Third World*, Westview Press, Boulder, Colorado.

Long, N. and Long, A., eds (1992) *Battlefields of Knowledge. The Interlocking Theory and Practice in Social Research and Development*, Routledge, London.

Long, N. and van der Ploeg, J. (1989) 'Demythologizing planned development: an actor perspective', *Sociologia Ruralis*, XXIX: 227–49.

Lugard, F.D. (1930) *The Dual Mandate in British Tropical Africa*, William Blackwood & Sons, Edinburgh/London.

Mace, R. (1993) 'Transitions between cultivation and pastoralism in sub-Saharan Africa', *Current Anthropology*, 34: 363–382.

Mahdi, M. (1993) *L'Organisation Pastorale chez les RHERAYA du Haut Atlas. Production Pastorale, Droit et Rituel*, PhD thesis, University of Hassan II, Casablanca.

Malan, J.S. (1973) 'Double-descent among the Himba of South West Africa', *Cimbebasia*, 2: 81–112.

Malan, J.S. (1974) 'The Herero-speaking peoples of Kaokoland', *Cimbebasia*, 2: 113–129.

Malan, J.S. and Owen-Smith, G.L. (1974) 'The ethnobotany of Kaokoland', *Cimbebasia*, 2: 131–178.

Mamdani, M., Kasoma, P.M.B. and Latende, A.B. (1992) *Karamoja: Ecology and History*, Centre for Basic Research Working Paper No. 22, CBR publications, Kampala.

Marty, A. (1993) 'La gestion des terroirs et les éleveurs: un outil d'exclusion ou de négociation?', *Revue Tiers Monde*, 34: 327–344.

McDermott, J. and Ngor, M.D. (1983) *Grazing Management Strategies among the Tuic, Nyarraweng and Ghol Dinka of Kongor Rural Council: Prospects for Development*, Kongor Integrated Rural Development Project, FAO, Rome.

McIntire, J., Bourzat, D. and Pingali, P. (1992) *Crop–Livestock Interactions in Sub-Saharan Africa*, World Bank, Washington, D.C.

Mearns, R. (1993a) *Pastoral Institutions, Land Tenure and Land Policy Reform in Post-Socialist Mongolia*, PALD Research Report 3, Mongolian Institute of Agricultural Economics/Institute for Development Studies, London.

Mearns, R. (1993b) 'Territoriality and land tenure among Mongolian pastoralists: variation, continuity and change', *Nomadic Peoples*, 33: 73–103.

Mearns, R. (1996) 'Community collective action and common grazing: the case of post-socialist Mongolia', *Journal of Development Studies*, 32, 283–325.

Mendes, L. (1988) 'Private and communal land tenure in morocco's western High Atlas mountains: complements, not ideological opposites', *Pastoral Development Network*, 26a.

Migot-Adholla, E.S., Hazell, P., Blarel, B. and Place, F. (1991) 'Indigenous land rights systems in sub-Saharan Africa: a constraint on productivity?', *World Bank Economic Review*, 5: 155–175.

Monod, T., ed. (1975) *Pastoralism in Tropical Africa*, Oxford University Press, London.

Moorehead, R. (1991) *Structural Chaos: Community and State Management of Common Property in Mali*, PhD thesis, Institute for Development Studies, University of Sussex.

Moorehead, R. (1993) 'Policy options for pastoral resource tenure in non-equilibrium environments', pp. 17–25, in *Proceedings of Workshop on Pastoral Natural Resource Management and Pastoral Policy in Africa, 1993*, ed. M. Niamir. UNSO, New York.

Moorehead, R. and Lane, C. (1995) 'Nouvelles orientations en matière de politique et de tenure foncières des ressources pastorales', pp. 421–453 in *Terre, Terroir, Territoire, Les Tensions Foncières*, eds C. Blanc-Pamard and L. Cambrézy, ORSTOM, Paris.

Moorehead, R., Guèye, B. and Toulmin, C. (1996) *Making Local Level Planning Work*, Drylands Programme, IIED, London.

Moote, M.A. and McClaran, M. (1997) 'Viewpoint: implications of participatory democracy for public land planning', *Journal of Range Management*, 50: 473–481.

Mosse, D. (1994) 'Authority, gender and knowledge: theoretical reflections on the practice of participatory rural appraisal', *Development and Change*, 25: 497–526.

Moulin, C.-H. (1993) *Performances Animales et Pratiques d'Elevage en Afrique Sahélienne*, PhD thesis, Institut National Agronomique, Paris-Gignon.

Mukamuri, B. (1988) 'Rural environmental conservation in South-central Zimbabwe. An attempt to describe Karanga thought patterns, perceptions and environmental control', paper presented at *African Studies Association Conference, September 1988, Cambridge.*

Mundy, P.A. and Compton, J.L. (1995) 'Indigenous communication and indigenous knowledge', pp. 112–123, in *The Cultural Dimension of Development: Indigenous Knowledge Aystems,* eds D.M. Warren, L.J. Slikkerveer and D. Brokensha, Intermediate Technology Publications, London.

Murombedzi, J. (1992) 'The need for appropriate level common property resource management institutions in communal tenure regimes', pp. 39–58 in *Institutional Dynamics of Communal Grazing Regimes in Southern Africa,* ed. B. Cousins, Centre for Applied Social Sciences, University of Zimbabwe, Harare.

Mutizwa-Mangiza, N. (1991) 'Decentralisation in Zimbabwe: problems of planning at the district level', pp. 418–443 in *Rural Development and Planning in Zimbabwe,* eds. N. Mutizwa-Mangiza and A. Helmsing, Avebury, Aldershot, UK.

Ndagala, D.K. (1982) '"Operation Imparnati": the sedentarization of the pastoral Maasai in Tanzania', *Nomadic Peoples,* 10: 28–39

Ndagala, D.K. (1991) *Pastoralism and rural development: the Ilparakuyu experience,* Reliance Publishing House, New Delhi.

Neumann, R.A. (1995) 'Local challenges to global agendas: conservation, economic liberalization and the pastoralists' rights movement in Tanzania', *Antipode,* 27: 363–406.

Niamir, M. (1987) *Grazing Intensity and Ecological Change in Eastern Senegal: Implications for the Monitoring of Sahelian Rangelands,* PhD dissertation, University of Arizona, Tucson.

Niamir, M. (1990) *Herder's Decision-Making in Natural Resource Management in Arid and Semi-Arid Africa,* FAO Community Forestry Note 4, FAO, Rome.

Niamir, M., ed. (1994a) *Proceedings of Workshop on Pastoral Natural Resource Management and Pastoral Policy in Africa,* Arusha, Tanzania, UNSO, New York.

Niamir, M., ed. (1994b) *Proceedings of 1st International Technical Consultations on Pastoral Development,* Paris, UNSO, New York.

Niamir, M. (1994c) *Women Livestock Managers in the Third World: A Focus on Technical Issues Related to Gender Roles in Livestock Production,* IFAD Technical Issues in Rural Poverty Alleviation, Staff Working Paper 18, IFAD, Rome.

Niamir, M. (1997) 'The resilience of pastoral herding in Sahelian Africa', in *Linking social and ecological systems: institutional learning for resilience,* eds F. Berkes & C. Folke, Cambridge University Press, Cambridge, UK.

Niamir, M., Lugando, S. and Kundy, T. (1994) *Barabaig Displacement from Hanang District to Usangu Plains: Changes in Natural Resource Management and Pastoral Production in Tanzania,* FTPP Working Paper, SUAS/FAO/IIED, Rome.

Niesten, E. (1996) *The Role of Markets in Wildlife Management: A Description of the CAMPFIRE Program in Zimbabwe,* Food Research Institute, Sandford University.

Novelli, B. (1988) *Aspects of Karimajong Ethnosociology,* Museum Combonianum No. 44, Comboni Missionaries, Kampala.

Nurse, M. (1998) 'Defining institutions for participatory natural resource management: a case study of the Participatory Management of Mangroves from Tanga, Tanzania', paper presented at the Workshop on Participatory Natural Resource Management, Manchester College, Oxford, 6–7 April 1998.

Oakerson, R.J. (1986) 'A model for the analysis of common property problems' pp. 13–30 in *Proceedings of the Conference on Common Property Resource Management*, National Academy Press, Washington.

Oakerson, R.J. (1992) 'Analyzing the commons: A framework', pp. 41–59 in *Making the Commons Work*, ed. D.W. Bromley, ICS Press, San Francisco.

Ocan, C.E. (1992) *Pastoral Crisis in Northeast Uganda: The Changing Significance of Cattle Raids*, Working Paper No. 21, Centre for basic Research, Kampala.

O'Connor, T.G. (1985a) *A Synthesis of Field Experiments concerning the Grass Layer in the Savanna Regions of Southern Africa*, South African National Scientific Programmes Report No. 114, CSIR, Pretoria.

O'Connor, T. (1985b) *A Synthesis of Field Experiments concerning the Grass Layer in the Semi-Arid Regions of Southern Africa*, South African Natural Science Programme Report No. 144, CSIR, Pretoria.

O'Connor, T.G. and Roux, P.W. (1995) 'Vegetation changes (1949–1971) in a semi-arid, grassy dwarf shrubland in the Karoo, South Africa: influence of rainfall variability and grazing by sheep', *Journal of Applied Ecology*, 32: 612–26.

Odell, M.J. (1982) 'Local institutions and management of communal resources: lessons from Africa and Asia', *Pastoral Development Network*, 14e.

Okoh, A.E.J., Uza, D.V., Ajayi, J. and Ezeomah, C. (1988) 'Animal health and husbandry aspects', in *Education of Nomadic Families*, Research Report Vol. 8, UNDP/UNESCO/University of Jos project, Nigeria.

Ole Kuney, R. and Lendiy, J. (1994) 'Pastoral institutions among the Maasai of Tanzania', paper presented at the 6th PANET Workshop, FAO, Dar es Salaam.

Ostberg, W. (1987) *Ramblings on Soil Conservation: An Essay from Kenya*, SIDA, Vernamo.

Ostrom, E. (1989) 'Microconstitutional change in a multiconstitutional political system', *Rationality and Society*, 1: 11–50.

Ostrom, E. (1990) *Governing the Commons: The Evolution of Institutions for Collective Action*, Cambridge University Press, New York.

Ostrom, E. (1993) 'The evolution of norms, rules and rights', paper presented at the Property Rights and the Performance of Natural Resource Systems Workshop, September 1993, Beijer International Institute of Ecological Economics, Stockholm.

Ostrom, E., Walker, J. and Gardner, R. (1992) 'Covenants with and without a sword: self-governance is possible', *American Political Science*, 86: 404–417.

Ostrom, E., Schroeder, L. and Wynne, S. (1993) *Institutional incentives and sustainable development: infrastructure policies and perspectives*, Westview Press, Boulder, Colorado.

Ostrom, E., Gardner, R. and Walker, J. (1994) *Rules, Games and Common Pool Resources*, University of Michigan Press, Ann Arbor.

Ostrom, V. (1980) 'Artisanship and Artifact', *Public Administration Review*, 40: 309–317.

304

Ostrom, V. (1987) *The Political Theory of a Compound Republic: Designing the American Experiment*, 2nd edn, ICS Press, San Francisco.

Ostrom, V. (1991) *The Meaning of American Federalism: Constituting a Self-Governing Society*, ICS Press, San Francisco.

Ouedraogo, H. and Rochette, R. (1996) *Atelier de Dakar sur la Gestion des Conflits Liés aux Ressources Pastorales*, PRASET/GTZ, Ouagadougou.

Oxby, C. (1982) 'Group ranches in Africa', *Pastoral Development Network*, 13d.

Oxby, C. (1990) 'The "living milk" runs dry: the decline of a form of joint ownership and matrilineal inheritance among the Twareg (Niger)', pp. 222–228 in *Property, Poverty and People: Changing Rights in Property and Problems of Pastoral Development*, eds P.T.W. Baxter and R. Hogg, University of Manchester Press, Manchester, UK.

Pagiola, S. (1993) *Soil Conservation and the Sustainability of Agricultural Production*, PhD thesis, Stanford University.

Painter, T., Sumberg, J. and Price, T. (1994) 'Your terroir and my "action space": Implications of differentiation, mobility and diversification for the Approche Terroir in Sahelian West Africa', *Africa*, 64: 447–463.

PARTICIP (1992) *Evaluation of Karamoja Development Project, Phase II*, European Development Fund Project No. 5100.33.42.46, PARTICIP GmbH, Wehingen.

Payne, W.J.A. (1965) 'Specific problems of semi-arid environments', *Qualitas Plantarum et Materiae Vegetabiles*, 12: 269–294.

Pazzaglia, A. (1982) *The Karimajong: Some Aspects*, EMI press, Bologna, Italy.

Peluso, N.L. (1993) 'Coercing conservation? The politics of state resource control', *Global Environmental Change*, 3: 199–217.

Pendzich, C. (1994) 'Conflict management and forest disputes – a path out of the woods?', *Forest, Trees and People Newsletter*, No. 20, FAO, Rome.

Penning de Vries, F.W.T. and Djitièye, M.A., eds (1982) *La Productivité des Paturages Saheliens: une Étude des Sols, des Végétations et de l'exploitation de cette Ressource Naturelle*, Report No. 918, Wageningen Center for Agricultural Publications and documentation, Wageningen, The Netherlands.

Perevolotsky, A. (1995) 'Conservation, reclamation and grazing in the Northern Negev: contradictory or complementary concepts?', *Pastoral Development Network*, 38a: 1–22.

Peters, P. (1987) 'Embedded systems and rooted models: the grazing lands of Botswana and the commons debate', pp. 171–194 in *The Question of the Commons. The Culture and Ecology of Communal Resources*, eds B. McCay and J. Acheson, University of Arizona Press, Tuscon.

Peters, P. (1994) *Dividing the Commons: Politics, Policy and Culture in Botswana*, University Press of Virginia, Charlottesville.

Picardi, A.C. and Siefert, W.W. (1976) 'A tragedy of the commons in the Sahel', *Technology Review*, May: 42–51.

Place, F. and Hazell, P. (1993) 'Productivity effects of indigenous land tenure systems in sub-saharan Africa', *American Journal of Agricultural Economics*, 75: 10–19.

Platteau, J-P. (1995) *The Evolutionary Theory of Land Rights as applied to sub-Saharan Africa: A Critical Assessment*, Cahiers de la Faculté des sciences economiques et sociales No. 145, Facultes Universitaires Notre Dame de la Paix, Namur, Belgium.

Potkanski, T. (1994) *Property Concepts, Herding Patterns and Management of*

Natural Resources among Ngorongoro and Salei Maasai of Tanzania, Pastoral Land Tenure Series No. 6, IIED, London.

Quezel, P. and Barbero, M. (1982) 'Definition and characterization of Mediterranean type ecosystems', *Ecologia Mediterranea*. 7: 15–29.

Raffin, Y. (1996) 'Alpages et transhumance en Isère. Spécificités dauphinoises, évolution des effectifs, aspects législatifs et structures foncières', pp. 117–121 in *Proceedings of seminar Pastoralisme et foncier: impact du régime foncier sur la gestion de l'espace pastoral et la conduite des troupeaux en régions arides et semi-arides*, CIHEAM/IRA, Gabès, 17–19 October 1996.

Randhir, T.O. and Lee, J.G. (1996) 'Managing local commons in developing economies: an institutional approach', *Ecological Economics*, 16: 1–12.

Ranger, T. (1993) 'The communal areas of Zimbabwe', pp. 354–385 in *Land in African Agrarian Systems*, eds T.J. Bassett and D.E. Crummey, University of Wisconsin Press, Madison.

Rattray, J.M. and Byrne, F.O. (1963) *Report to the Government of Uganda on a Reconnaissance Survey of the Karamoja District*, Project No. UGA/TE/AN, FAO, Rome.

Raymond, L. (1997) 'Viewpoint: are grazing rights on public land a form of private property?', *Journal of Range Management*, 50: 431–438.

Ribot, J. (1996) 'Participation without representation: Chiefs, councils and forestry law in the West African Sahel', *Cultural Survival Quarterly*, Fall: 40–44.

Riddell, J.C. (1982) *Land Tenure Issues in West African Livestock and Range Development Projects*, Research Paper No. 77, Land Tenure Center, Madison, Wisconsin.

Riesman, P. (1984) 'The Fulani in a development context: the relevance of cultural traditions for coping with change and crisis', pp. 171–191 in *Life before the drought*, ed. E.P. Scott, Allen & Unwin, Boston.

Riker, W.H. (1964) *Federalism: Origin, Operation, Significance*, Little Brown, Boston.

Ring, M.M. (1990) 'Dinka stock trading and shifts in rights in cattle', pp. 192–205 in *Property, Poverty and People: Changing Rights in Property and Problems of Pastoral Development*, eds P.T.W. Baxter and R. Hogg, University of Manchester Press, Manchester, UK.

Roe, E.M. (1994) 'New frameworks for an old tragedy of the commons and an aging common property resource management', *Agriculture and Human Values*, 11: 29–36.

Roeder, P. (1996) 'Livestock disease scenarios of mobile vs. sedentary pastoral systems', in *Proceedings of 3rd International Technical Consultations on Pastoral Development*, Brussels. UNSO, New York.

Rose, L.L. (1992) *The Politics of Harmony: Land Dispute Strategies in Swaziland*, Cambridge University Press, Cambridge, UK.

Ross, M. (1995) 'Interests and identities in natural resource conflicts involving indigenous peoples', *Cultural Survival Quarterly*, 19, No. 3.

Rothman, J. (1995) 'Pre-negotiation in water disputes', *Cultural Survival Quarterly*, 19, No. 3.

Roux, P.J. Le (1971) *The Common Names and a Few Uses of the Better Known Indigennous Plants of South West Africa*, Department of Forestry, Government Printer, Pretoria.

Rugege, S. (1995) 'Conflict resolution in African Customary Law', *Africa Notes*, October.

Runge, C.F. (1981) 'Common property externalities: isolation, assurance and resource depletion in a traditional grazing context', *American Journal of Agricultural Economics*, November: 595–606.

Runge, C.F. (1986) 'Common property and collective action in economic development', *World Development*, 14: 623–635.

Rupesinghe, K. (1995) 'Multi-track diplomacy and the sustainable route to conflict resolution', *Cultural Survival Quarterly*, 19, No. 3.

Sabiiti, E.N. (1990) 'The place and role of fire in pasture and rangeland management in Uganda', in *First Rangeland Improvement for Beef Production Workshop*, Mkarere University, Uganda.

Salzman, P.C. (1994) 'Afterword: reflections on the pastoral land crisis', *Nomadic Peoples*, 34/35: 159–163.

Sanankoua, B. (1990) *Un Empire Peul au XIXe Siècle, La Diina du Maasina*, Karthala-ACCT, Paris.

Sanderson, G.N. and Sanderson, L.M. (1981) *Education, Religion, and Politics in Southern Sudan*, Ithaca Press, London/Khartoum.

Sandford, S. (1982) 'Pastoral strategies and desertification: opportunism and conservatism in dry lands', pp. 61–80 in *Desertification and Development: Dryland Ecology in Social Perspective*, eds B. Spooner and H. Mann, Academic Press, London.

Sandford, S. (1983) *Management of Pastoral Development in the Third World*, John Wiley, Chichester.

Sawyer, A. (1992) *The Emergence of Autocracy in Liberia: Tragedy and Challenge*, ICS Press, San Francisco.

Schlee, G. (1987) 'Holy grounds', paper presented at the *Workshop on Changing Rights in Property and Problems of Pastoral Development in the Sahel*, Manchester University, Manchester, UK.

Schmitz, J. (1986) 'L'État géomètre: les leydi des Peul du Fuuta Tooro (Sénégal) et du Maasina (Mali)', *Cahiers d'Etudes Africaines*, 26: 349–394.

Schoffleers, J. (1979) *Guardians of the Land*, Mambo Press, Gweru.

Scoones, I. (1989) 'Economic and ecological carrying capacity: implications for livestock development in Zimbabwe's communal areas', *Pastoral Development Network*, 27b..

Scoones, I. (1990) *Livestock Populations and the Household Economy: A Case Study from Southern Zimbabwe*, PhD thesis, University of London.

Scoones, I. (1991) *Wetlands in Drylands: The Agroecology of Savanna Systems in Africa*, IIED, London.

Scoones, I. (1992) 'Coping with drought: responses of herders and livetock in contrasting savanna environments in southern Zimbabwe', *Human Ecology*, 20: 293–314.

Scoones, I. (1993) 'Exploiting heterogeneity: habitat use by cattle in dryland Zimbabwe', *Journal of Arid Environments*, 29: 221–237.

Scoones, I., ed (1994) *Living with Uncertainty – New Directions in Pastoral Development in Africa*, Intermediate Technology Publications, London.

Scoones, I. (1995) 'Exploiting heterogeneity: Habitat use by cattle in the communal areas of Zimbabwe', *Journal of Arid Environments*, 29: 221–237.

Scoones, I. and Wilson, K. (1989) 'Households, lineage groups and ecological dynamics: issues for livestock development in Zimbabwe's communal lands' in *People, Land and Livestock*, ed. B. Cousins, Centre for Applied Social Sciences, University of Zimbabwe, Harare.

Scoones, I. and Thompson, J., eds (1994) *Beyond Farmer First: Rural People's Knowledge, Agricultural Research and Extension Practice*, Intermediate Technology Publications, London.

Scoones, I. et al. 1996. *Hazards and Opportunities. Farming Livelihoods in Dryland Africa. Lessons from Zimbabwe*, Zed Press, London.

Seligman, N.G. and Perevolotsky, A. (1994) 'Has intensive grazing by domestic livestock degraded the Old World Mediterranean rangelands?', in *Plant–Animal Interactions in Mediterranean-Type Ecosystems*, eds M. Arianoutsou and R.H. Groves, Kluwer, Dordrect.

Sevier, C.E. (1975) *The Anglo-Egyptian Condominium in the Southern Sudan 1918–1939*, Phd thesis, Princeton University, Princeton, New Jersey.

Shanmugaratnam, N., Vedeld, T., Mossige, A. and Bovin, M. (1992) *Resource Management and Pastoral Institution Building in the West African Sahel*, Discussion Paper No. 175, World Bank, Washington D.C.

Shipton, P. (1994) 'Land and culture in tropical Africa: Soils, symbols and the metaphysics of the mundane', *Annual Review of Anthropology*, 23: 347–377.

Simpson, J.R. and Sullivan, G.M. (1984) 'Planning for institutional change in utilization of Sub-Saharan Africa's common property range resources', *African Studies Review*, 27: 61–78.

Sinclair, A.R.E. and Frywell, J.M. (1985) 'The Sahel of Africa: Ecology of a disaster', *Canadian Journal of Zoology*, 63: 987–994.

Sithole, B. (1995) *Access to and Use of Dambo Resources in Mutoko and Chiduku Communal Lands in Zimbabwe*, Stockholm Environment Institute, Stockholm.

Sivaramakrishnan, K. (1996) 'Co-management for forests: are we overly preoccupied with property rights?', *Common Property Resource Digest*, 37: 6–8.

Smith, A.B. (1992) *Pastoralism in Africa: Origins and Development Ecology*, Ohio University Press, Athens.

Sobania, N.W. (1990) 'Social relationships as an aspect of property rights: Northern Kenya in the pre-colonial and colonial periods', pp. 1–19 in *Property, Poverty, and People: Changing Rights in Property and Problems in Pastoral Development*, eds P.T.W. Baxter and R. Hogg, University of Manchester, Manchester, UK.

Spencer, P. (1965) *The Samburu: A Study of Gerontocracy in a Nomadic Tribe*, Routledge & Kegan Paul, London.

Stebbing, E.P. (1935) 'The encroaching Sahara: the threat to the West Africa colonies', *Geographical Journal*, 85: 506–524.

Steinfeld, H., de Hahn, C. and Blackburn, H. (1997) *Livestock–Environment Interactions: Issues and Options*, European Union, Brussels.

Stenning, D.J. (1960) 'Transhumance, migratory drift, migration: Patterns of pastoral Fulani nomadism', pp. 139–159 in *Cultures and Societies of Africa*, eds S. Ottenburg and P. Ottenburg, Random House, New York.

Stoddart, L.A., Smith, A.D. and Box, T.W. (1975) *Range Management*, 3rd edn, McGraw-Hill, New York.

Sutter, J.W. (1987) 'Cattle and inequality: herd size differences and pastoral production among the Fulani of northeastern Senegal', *Africa*, 57: 196–218.

Swallow, B.M. (1994) The role of mobility within the risk management strategies of pastoralists and agro-pastoralists, IIED Gatekeeper Series No. 47, International Institute for Environment and Development, London.

Swallow, B.M. and Bromley, D.W. (1991) 'Co-management or no management:

the prospects for internal governance of common property regimes through dynamic contracts', *Oxford Agrarian Studies*, 22: 3–16.

Swift, J. (1988) *Major Issues in Pastoral Development with Special Emphasis on Selected African Countries*, FAO, Rome.

Swift, J (1993) 'Dynamic ecological systems and pastoral administration', presented at a *Workshop on New Directions in African Range Management and Policy*, ODI/IIED/Commonwealth Secretariat, Woburn, London.

Swift, J. (1995) 'Dynamic ecological systems and the administration of pastoral development', pp. 153–173 in *Living with Uncertainty: New Directions in Pastoral Development in Africa*, ed. I. Scoones, Intermediate Technology Publications, London.

Swift, J. and Lane, C., eds (1989) *Pastoral Land Tenure in Africa: Report of a Workshop*, 1–3 December 1988, Arusha, Tanzania. Institute for Development Studies, Brighton, UK.

Sylla, D (1995) 'Pastoral organizations for uncertain environments', pp. 1134–152. in *Living with Uncertainty: New Directions in Pastoral Development in Africa*, ed. I. Scoones, Intermediate Technology Publications, London.

Taylor, P. (1992) 'Re-constructing socio-ecologies: system dynamics modelling of nomadic pastoralists in sub-Saharan Africa', pp. 115–158 in *The Right Tool for the Job: At Work in the Twentieth Century Life Sciences*, eds A. Clarke and J. Fujimura, Princeton University Press, Princeton, New Jersey.

Thébaud, B. (1988) *Elevage et développment au Niger*, Bureau International du Travail, Geneva.

Thébaud, B. (1998) 'Entre rentabilité économique et viabilité de l'économie familiale: le rôle de l'élevage dans la microéconomie pastorale et agropastorale', presentation at the *Fourth International Technical Consultation on Pastoral Development*, Ouagadougou, 24–27 April 1998, UNSO.

Thomas, D.S.G. (1993) 'Sandstorm in a teacup? Understanding desertification', *Geographical Journal*, 159: 318–331.

Thomas, D. (1998) 'Lessons from community forest management at the Kilum-Ijim Forest Project, Cameroon', paper presented at the Workshop on Participatory Natural Resource Management, Manchester College, Oxford, 6–7 April 1998.

Thompson, E.P. (1991) *Customs in Common*, The New Press, New York.

Thorton, R. (1980) *Space, Time and Culture Among the Iraqw of Tanzania*, Academic Press, New York.

Toulmin, C. (1983) *Economic Behavior among Livestock Keeping Peoples: A Review of the Literature on the Economics of Pastoral Production in the Semi-Arid Zones of Africa*, School of Development Studies, University of East Anglia.

Tozy, M. (1990) *Les Modes d'Appropriation, Gestion et Conservation des Ressources entre le Droit Positif et Communautaire au Maghreb*, Faculté de Droit de Casablanca.

Traoré, G. (1978) 'Evolution de la disponibilité et de la qualité de fourrage au cours de la transhumance de Diafarabé', PhD thesis, Centre Pédagogique Supérieure, Bamako, Mali.

Tubiana, M.-J. and Tubiana, J. (1977) *The Zhagawa from an Ecological Perspective*, Balkema, Rotterdam.

Tucker, C.J., Dregne, H.E. and Newcomb, W.W. (1991) 'Expansion and contraction of the Sahara Desert from 1980 to 1990', *Science*, 253: 299–301.

Turner, M.D. (n.d.) 'Social institutions and community based resource management projects in dryland Africa', *Society and Natural Resources*, (in review).

Turner, M.D (1992) *Living on the Edge: FulBe Herding Practices and the Relationship Between Economy and Ecology in the Inland Niger Delta of Mali*, PhD thesis, University of California, Berkeley.

Turner, M.D. (1993) 'Overstocking the range: A critical analysis of the environmental science of Sahelian pastoralism', *Economic Geography*, 69: 402–421.

Turner, M.D. (1995) 'The sustainability of rangeland to cropland nutrient transfer in semi-arid West Africa: ecological and social dimensions neglected in the debate', pp. 435–452 in *Proceedings of an International Conference on Livestock and Sustainable Nutrient Cycling in Mixed Farming Systems of sub-Saharan Africa*, 22–26 November 1993, eds J.M. Powell, S. Fernandez-Rivera, T.O. Williams and C. Renard. International Livestock Centre for Africa, Addis Ababa, Ethiopia.

Unruh, J.D. (1995) 'The relationship between indigenous pastoralist resource tenure and state tenure in Somalia', *GeoJournal*, 36: 19–26.

UNEP (1991) *Status of Desertification and Implementation of the United Nations Plan of Action to Combat Desertification*, UNEP/GLSS, 111/3, Nairobi.

UNSO (1995) *Proceedings of the 2nd International Technical Consultation on Pastoral Development in Africa*, Eschborn, GTZ. UNSO, New York.

Urvoy, Y. (1929) 'La Mekrou et le double-v', *Afrique Francaise*, 39: 135–140.

van Keulen, H. and Breman, H. (1990) 'Agricultural development in the West African Sahelian region: a cure against land hunger?', *Agriculture, Ecosystems and Environment*, 32: 177–197.

Vedeld, T. (1993a) *Rangeland Management and State Sponsored Paastoral Institution Building in Mali*, Drylands Networks Programme Issues Paper No. 46. IIED, London.

Vedeld, T. (1993b) 'Enabling pastoral institution building in the dryland Sahel', *Pastoral Development in Africa, Proceedings of First International Technical Consultations on Pastoral Development*, Paris. UNSO, New York.

Vedeld, T. (1994) *The State and Rangeland Management: Creation and Erosion of Pastoral Institutions in Mali*, Dryland Networks Programme Paper No. 46, IIED, London

Viederman, S. (1996) 'Uncivil society', *ISEE Bulletin*, 1: 3.

Wade, R. (1987) 'The management of common property resources: a cooperative solution', *World Bank Research Observer*, 2: 219–234.

Walker, B.H., Ludwig, D. Holling, C.S. and Peterman, R.M. (1981) 'Stability of semi-arid savanna grazing systems', *Journal of Ecology*, 69: 473–498.

Waller, R.D. (1985) 'Ecology, migration and expansion in East Africa', *African Affairs*, 84: 347–370.

Warren, D.M. and Rajasekaran, B. (1993) 'Using indigenous knowledge for sustainable dryland management: a global perspective', presented at the *International Workshop on Listening to the People: Social Aspects of Dryland Management*, UNEP, Nairobi.

Waters-Bayer, A. and Bayer, W. (1995) 'Recent advances in participatory planning for pastoral development', in *Proceedings of 2nd International Technical Consultations on Pastoral Development*, Eschborn., ed. M. Niamir. GTZ/UNSO, New York.

Western, D. and Finch, V. (1986) 'Cattle and pastoralism: survival and production in arid lands', *Human Ecology*, 14: 77–94.

Westoby, M, Walker, B. and Noy-Meir, I. (1989) 'Opportunistic management for rangelands not at equilibrium', *Journal of Range Management*,
42: 266–273.
White, C. (1984) 'Herd reconstitution: the role of credit among WoDaaBe
herders in Central Niger', *Pastoral Development Network*, 18d.
White, C. (1990) 'Changing animal ownership and access to land among the
Wodaabe (Fulani) of Central Niger', pp. 240–254 in *Property, Poverty, and
People: Changing Rights in Property and Problems in Pastoral Development*,
eds P.T.W. Baxter and R. Hogg, University of Manchester, Manchester, UK.
White, F. (1983) *The Vegetation of Africa*, UNESCO, Paris.
Wilson, J.G. (1962) *The Vegetation of Karamoja District, Northern Province,
Uganda*, Memoir of Research Division, Series II, No. 5, Government Printer.
Department of Agriculture, Entebbe.
Wilson, K. (1990) 'Ecological Dynamics and Human Welfare: A Case Study of
Population, Health and Nutrition in Zimbabwe', PhD thesis, University of
London.
Wilson, P.N. and Thompson, G.D. (1993) 'Common property and uncertainty:
compensating coalitions by Mexico's pastoral *Ejidatarios*', pp 299–318 in
Economic development and Cultural Change.
Wilson, R.T., de Leeuw, P.N. and De Haan, C. (1983) *Recherches sur les Sys-
tèmes des Zones Arides du Mali, Résultats Préliminaires*, Rapport de
Recherche No. 5, CIPEA, Addis-Abeba.
Wilson, W. (1984) *Resource Management in a Stratified Fulani Community*, PhD
thesis, Howard University.
Wilson, W. (1995) 'The Fulani model of sustainable agriculture', *Nomadic
Peoples*, 36/37: 35–51
Winrock International (1992) *Assessment of Animal Agriculture in Sub-Saharan
Africa*, Winrock International Institute for Agricultural Development,
Morrilton, Arkansas.
Winter, M. (1984) 'The Twareg', pp. 531–620 in *Pastoral Development in Central
Niger: Report of the Niger Range and Livestock Project*, ed. J.J. Swift, USAID,
Niamey.
Zimmerer, K.S. (1994) 'Human geography and the 'new ecology': The prospect
and promise of integration', *Annals of the Association of American
Geographers*, 84: 108–125.

Index

Acacia nilotica 153
Acacia raddiana 53, 56
Acacia seyal 55, 153
Afar of Djibouti 44
Agdal 237–238, 252, 256–263, 265
Agro-pastoralist 6, 30
Angola 186–187

Balanites aegyptiaca 55, 57
Banditry 149, 160–161, 164, 171–172, 178–180,
   182
Benign neglect 23
Bergeries 244, 247–248, 252, 255
Boom and bust 20, 157, 172, 225
Boscia senegalensis 53, 60
Burgu 83–87, 89–90, 92–95

Capparis decidua 53
Cattle
   camps 130–133, 139, 143–144, 188, 191,
      195–198, 201, 205, 212
   kraal 163–165, 169–170, 179, 196–197, 205
Charte de transhumance 250
Co-management 18, 38, 44–45, 223, 235, 271,
   280, 289
Coefficient of variation 151
Combretum spp. 153
Commercial ranch 36, 185, 193, 212
Common pool resources
   buffer zones 276
   corridors
      fluid boundaries 38, 40–41
   fluid boundaries 277
   key resources 40
   key sites 34, 277, 280–281
   range reserves 277
   transboundary resources 38, 41
Common property management 2, 6, 9, 42, 228,
   232, 266, 273, 282, 289
   Pastoral associations 29
   Pastoral reserves 278
   Stewardship 279
Common property regimes 31, 33, 38, 45
Conflict
   boundary disputes 128–129
   mediators 130–131, 172–173, 284–285
   negotiation 7, 36, 44, 131, 173, 273, 276–277,
      282, 285
   prevention 7, 44, 173, 175, 183, 285
   resolution 7, 44, 98, 125, 130–131, 135–137,
      146–147, 149–150, 164, 173, 183, 275, 285

Conflits 149–150, 237, 245, 251–253, 255–256,
   260–261, 264, 266
Covenants 7, 36, 129
Customary
   leadership 35, 230, 279
      management 230
   management 279
   regulations 282–283

Decentralization 18, 44, 172, 268, 271–273,
   275–276, 280, 286
Desertification 24–25
Destocking 25, 103, 229
Devolution 43, 123, 271–272, 279–280, 288
Diina Code 6
Dodoth of Uganda 156–157, 161, 163, 166, 169,
   171–172, 175
Droits d'usage
   collectifs 244–245, 247, 252, 255, 259–262
   niche individuelle 248
Drought 3, 20–24, 26, 33, 38, 42, 44, 103,
   115–116, 124–125, 128, 150–151, 153, 157,
   160–161, 170, 176–177, 190, 193–194, 197,
   199–200, 202, 211, 220, 225, 227, 230, 234,
   269, 274, 276–277, 283, 289

Echinochloa stagnina 83
Ecological economics 271, 287
Ecosystem 3
   heterogeneity 218, 225, 230, 272, 275
   non-equilibrium 153, 166, 170–172, 182, 220,
      267
   resilience 275, 277, 281, 288, 290
   variability 143, 153, 287
Eragrostis nindensis 207–209
Eragrostis porosa 208
Eragrostis superba 209

Forage production 103, 185, 200, 207
Fraxinus dimorphis 241
French colonial rule 6
Fulani
   in Burkina Faso 35
   in Nigeria 35, 37
   Maasina 22, 42, 99, 107–111

Gestion
   de l'accès aux ressources 247
   de la mobilité 261
   de terroirs 29
   des ressources naturelles 11, 16–17, 74, 239,
      266

312

inclusive vs exclusive 33, 41, 43, 169, 219, 270, 273
nested 40, 272, 276–277
open access 39, 42, 167
private 128, 140, 218–219
privatization of 228
usufruct 40–41, 276–277, 279–280

Quercus ilex 241

Ranching model 7
Range management 101, 185–187, 206, 213–214
Reciprocity 7, 18, 22, 34–35, 41, 45, 173, 175, 266, 268–269
Rhigozum virgatum 208
Risk
    and uncertainty 222, 281
    management 20, 38, 274

Safety nets 36, 38, 280
Sahel 20, 26, 33, 40, 103, 109, 117, 121, 290
Scoping 271
Sedentarization 20, 23, 25–28, 163, 267
Self-governance 7, 142, 148
Semi-nomadisme 236, 241
Senegal 36, 41, 273
Shona of Zimbabwe 223, 230
Skeleton Coast National Park 187
Social capital 18, 34–35, 109, 142, 268, 275, 283
Stateless society 6, 125
Statut foncier 236, 244, 249
    certificat de possession 245
    collectif domanial 249, 253
    privé 249
    terres arch 245

terres collectives 244, 246–247
Stipagrostis pungens 53, 57
Stipagrostis uniplumis 208–209
Stocking rate 6, 30, 36–37, 200, 211
Stocking rates 101–102
Sudan 1, 6, 30, 33, 36, 41–43, 124–125, 136–137, 139–143, 146, 148, 150, 156, 164, 166, 175, 183, 279

Tracking
    of resources 169, 282, 284
Tragedy of the commons 6, 38, 99–101, 122, 167
Transhumance
    definition 1
Transhumants 1
Tunisia 26
Turkana of Kenya 42, 156, 160–161, 166–167, 171–172, 175–177, 179, 271
Twareg of Mali 41

Uganda 7, 23, 36, 42, 150, 156, 163–164, 167, 169, 172, 178, 180, 183, 272, 289
Undergrazing 3, 31

Water
    boreholes 195, 198, 202, 215
    permanent vs temporary 26, 195, 197–200, 205, 211–212
    shallow wells 200–201
    surface 115, 200

Xérophytes Épineuses 241

Zimbabwe 1, 36, 217–218, 223, 225, 227, 229, 232, 234–235, 289